Topics in
Current Physics

20

Topics in Current Physics Founded by Helmut K. V. Lotsch

Inverse Scattering Problems in Optics

Edited by H. P. Baltes

With Contributions by
H. P. Baltes   M. Bertero   W.-M. Boerner
C. De Mol   P. J. Emmerman   M. A. Fiddy
R. Goulard   E. Jakeman   M. Nieto-Vesperinas
P. N. Pusey   G. Ross   A. Selloni   G. A. Viano

With a Foreword by R. Jost

With 49 Figures

Springer-Verlag Berlin Heidelberg New York 1980

Professor Dr. Heinrich P. Baltes

LGZ Landis & Gyr Zug AG, Zentrale Forschung und Entwicklung
CH-6301 Zug, Switzerland

ISBN-13:978-3-642-81474-7 e-ISBN-13:978-3-642-81472-3
DOI: 10.1007/978-3-642-81472-3

Library of Congress Cataloging in Publication Data. Main entry under title: Inverse scattering problems in optics (Topics in current physics ; v. 20). Bibliography: p. Includes index. 1. Optics. 2. Inverse problems (Differential equations) 3. Scattering (Physics) I. Baltes, Heinrich P. II. Series.
QC355.2.I58 535'.2 80-15101

Softcover reprint of the hardcover 1st edition 1980

2153/3130-543210

Publishers note: "Acknowledgement is made of the inadvertent omission of reference in chapter 3, *Topics in Current Physics,* volume 9, of N. N. Bojarski being the originator of the "Physical Optics Inverse Scattering Identity" (Bojarski, N. N., "Three Dimensional Electromagnetic Short Pulse Inverse Scattering", Syracuse University Research Corporation, Special Projects Laboratory Rpt. SURC N. SPL 67-3, AD 845126, Feb. 1967).
The Integral Equation for Incomplete Information (for the Physical Optics Inverse Scattering Identity), and the High Frequency Solution (to this integral equation) utilizing the directional derivative (Bojarski, N. N., "Electromagnetic Inverse Scattering Theory", Syracuse University Research Corporation, Special Projects Laboratory Rpt. SFL TR 68-70, AD 509134, December 1968). The first numerico-experimental verification of the high frequency solution was reported in "Electromagnetic Inverse Scattering" (Bojarski, N. N., Naval Air Systems Command Report, June 1972)."

Foreword

When, in the spring of 1979, H.P. Baltes presented me with the precursor of this volume, the book on "Inverse Source Problems in Optics", I expressed my gratitude in a short note, which in translation, reads:

"Dear Dr. Baltes, the mere title of your unexpected gift evokes memories of a period, which, in the terminology of your own contribution, would be described as the Stone Age of the Inverse Problem. Those were pleasant times. Walter Kohn and I lived in a cave by ourselves, drew pictures on the walls, and nobody seemed to care. Now, however, Inversion has become an Industry, which I contemplate with as much bewilderment as a surviving Tasmanian aborigine gazing at a modern oil refinery with its towers, its flares, and the confusing maze of its tubes."

The present volume makes me feel even more aboriginal – impossible for me to fathom its content. What I can point out, however, is one of the forgotten origins of the Inverse Scattering Problem of Quantum Mechanics: Werner Heisenberg's "S-Matrix Theory" of 1943. This grandiose scheme had the purpose of eliminating the notion of the Hamiltonian in favour of the scattering operator. If Successful, it would have done away once and for all with any kind of inverse problem.

To my knowledge, it was Wolfgang Pauli who induced S.T. Ma to check Heisenberg's idea in simple nonrelativistic models. Was it really possible to deduce all observable quantities of a system of two particles from its scattering data? Was Heisenberg's prescription for the determination of bound states infallible? S.T. Ma found a counterexample, where the S matrix had zeros which were uncorrelated with bound states. I learnt about Ma's work from Ch. Møller in Copenhagen and it finally led me to the inverse scattering problem. The reader may find more about the continuation of this story in the Bargmann Festschrift *Studies in Mathematical Physics*, ed. by E.H. Lieb et al. (Princeton University Press 1976).

ETH Zürich, November 1979 *Res Jost*

Contents

List of Contributors

Baltes, Heinrich P.

Zentrale Forschung und Entwicklung, Landis & Gyr Zug,
CH - 6301 Zug, Switzerland

Bertero, Mario

Istituto di Scienze Fisiche dell'Università and
Istituto Nazionale di Fisica Nucleare
I - 16132 Genova, Italy

Boerner, Wolfgang-Martin

Communications Laboratory, Department of Information Engineering,
University of Illinois at Chicago Circle,
Chicago, P.O. Box 4338, IL 60680, USA

De Mol, Christine

Département de Mathématique, Université Libre de Bruxelles and
Theoretische Natuurkunde, Vrije Universiteit Brussel,
B - 1050 Bruxelles, Belgium

Emmerman, Phillip J.

School of Engineering and Applied Sciences,
George Washington University,
Washington DC 20052, USA

Fiddy, Michael A.

Department of Physics, Queen Elizabeth College, University of London,
GB - London W8 7AH, United Kingdom

Goulard, Robert

School of Engineering and Applied Sciences,
George Washington University,
Washington DC 20052, USA

Jakeman, Eric

Royal Signals and Radar Establishment
GB - Great Malvern, Worcestershire WR14 3PS, United Kingdom

Nieto-Vesperinas, Manuel

Instituto de Optica, C.S.I.C. ES - Madrid-6, Spain, and
Departamento de Fisica, Universidad Nacional de
Educación a Distancia,
ES - Madrid-3, Spain

Pusey, Peter N.

Royal Signals and Radar Establishment,
GB - Great Malvern, Worcestershire WR14 3PS, United Kingdom

Ross, George

Department of Physics, Queen Elizabeth College, University of London
GB - London W8 7AH, United Kingdom

Selloni, Annabella

Laboratoire de Physique Théorique,
École Polytechnique Fédérale Lausanne,
CH - 1006 Lausanne, Switzerland

Viano, Giovanni Alberto

Istituto di Scienze Fisiche dell'Università and
Istituto Nazionale di Fisica Nucleare
I - 16132 Genova, Italy

1. Progress in Inverse Optical Problems

H. P. Baltes

The direct problem in optical physics is to predict the emission or propagation of radiation on the basis of a known constitution of sources or scatterers. The inverse or indirect problem is to deduce features of sources or scattering objects from the emitted or scattered radiation that has propagated to a detector. A small selection out of the large number of topics pertinent to the inverse problem in optical physics was examined in a previous volume in this series, entitled *Inverse Source Problems in Optics* [1.1], namely the phase retrieval problems, the question of uniqueness in the reconstruction of scatterers, the reconstruction of subwavelength sources, the connection between coherence and radiometric quantities, and the determination of statistical features of random phase screens from scattering data. A number of topics not covered by [1.1] are discussed in the present volume, *Inverse Scattering Problems in Optics*: deterministic and stochastic structural determinations using the theory of entire functions, the photon-counting statistics of optical scintillations with emphasis on the recently discovered K distributions, the connection between the raw data of photodetection and the properties of the received radiation field (the inverse detector problem), the ubiquitous question of the numerical instability of inverse problems, the multiangular absorption approach to combustion diagnostics, and polarization effects in inverse electromagnetic scattering.

Section 1.1 is a survey of recent review-type literature on inverse problems, followed by an overview of a few newly emerged inverse problems in optics which are not further discussed in this book. Section 1.2 updates the topics examined in the previous volume [1.1], with some emphasis on the construction of Lambertian scatterers (Sect.1.3). Finally, the material to be presented in this volume is summarized in Sect.1.4.

1.1 Inverse Problems in Optics and Elsewhere

The field of optical and other inverse problems is rapidly growing, if not exploding. Almost simultaneously with the previous volume [1.1] on inverse source problems in optics, a number of books or comprehensive reports on related topics became available. It therefore seems to be desirable to start this volume with a survey of the pertinent review-type literature.

The recent comprehensive reports by BOERNER [1.2], BLEISTEIN and COHEN [1.3], and WESTON [1.4] are mainly concerned with the optical and electromagnetic inverse scattering problems. In his state-of-the-art review [1.2], BOERNER emphasizes the role of polarization effects in electromagnetic inverse scattering and the related microwave and radar problems. BLEISTEIN and COHEN [1.3] reviewed some of their own work on the physical optics far field inverse scattering (POFFIS) problem, the determination of small inhomogeneities in a medium, and the question of nonuniqueness for acoustical waves. WESTON's review article [1.4] is focussed on the electromagnetic inverse scattering problem for nonhomogeneous media. A concise overview of inverse problems in optics is also available [1.5].

Of much broader interest are the recent lecture notes on applied inverse problems edited by SABATIER [1.6]. They include lectures on quite different inverse problems ranging from geophysics to elementary particles and aim at an update of topics or methods that are new when compared to the proceedings of the 1971 NASA meeting [1.7]. A recent monograph by CHADAN and SABATIER [1.8] is devoted to inverse problems in quantum scattering theory relevant for nuclear and particle physics. Moreover, we mention the book on the mathematics of ill-posed problems by TIKHONOV and ARSENIN [1.9].

Remote sensing of the atmosphere is a specific topic in optical inverse scattering of strong current interest, as is manifested by recent reviews by ISHIMARU [1.10, 11] and books edited by DEEPAK [1.12], HINKLEY [1.13], and FYMAT and ZUEV [1.14]. A likewise topical problem is *optical data and picture processing*, the various aspects of which were reviewed in books edited by HUANG [1.15], ROSENFELD [1.16] and CASASENT [1.17]. The related art of image processing in *electron microscopy* was examined by SAXTON [1.18] and HAWKES [1.19]. In particular, the phase problem in electron microscopy was reviewed by SAXTON [Ref.1.19, Chap.2]. The works of SALEH [1.20], and BARAKAT and BLAKE [1.21] are devoted to the *detection problem* in statistical optics. The particular question of extracting spectral parameters from photon-correlation measurements is reviewed by OLIVER [1.22].

Also a number of current Ph.D. theses are devoted, wholly or in part, to inverse problems in optics. Examples are the theses by HUISER [1.23] and VAN TOORN [1.24] on the phase problem in electron microscopy, by NIETO-VESPERINAS [1.25] on the structural determination of stochastic media, by SELLONI [1.26] on the quantum optical detection problem, and by DE MOL [1.27] on the regularization of linear ill-posed problems.

In the previous volume [1.1] a tentative list of some twenty specific inverse problems in optics was compiled. As one has to expect in such a rapidly developing field, a number of further problems have been broached since then. McWHIRTER and PIKE [1.28-30] studied the *inverse problem of laser anemometry*, i.e., to extract information about the velocity distribution from photon-correlation anemometry data. They avoided the ill-conditioned numerical inversion of the pertinent Fredholm integral equation of the first kind by a stabilized model-fitting approach. ROGER and MAYSTRE [1.31,32] investigated the electromagnetic theory of gratings from the point of view of inverse diffraction; the problem is to compute the grating profile for a given efficiency curve (*inverse diffraction grating problem*). Moreover, we mention two specific remote sensing problems, viz., the inversion of scattering data in order to recover aerosol particle [1.33-35] and fibre [1.36] size spectra, and the inversion of the Kubelka-Munk expression in order to analyze the optical response of turbid media [1.37].

1.2 Survey of Recent Results

The understanding of most inverse optical problems is in so rapid a state of development that a review of the subject will inevitably date rapidly. Moreover, it is easy enough to miss relevant contributions to the field. For example, a major omission in the previous volume [1.1] was most of the work of BATES and co-workers (see also Chap.7) on the phase problem in electron microscopy [1.38,39] and interferometry [1.40-42], the reconstruction from projections [1.43-46], deconvolution [1.47,48], and stellar interferometry viewing through randomly fluctuating media [1.49-51] (see also a recent paper by ROGERS [1.52] and references therein), holographic reconstruction [1.53-55], and the reconstruction of inhomogeneous index profiles [1.56] and the near field of totally reflecting objects [1.57]. In this section, an attempt is made to update the previous volume [1.1] by summarizing some of the post 1977 results.

1.2.1 Phase, Uniqueness, and Estimation

Numerous further contributions on the *phase problems* have appeared. VAN TOORN et al. [1.58] developed phase retrieval algorithms for semi-weak objects from noisy electron micrographs based on, e.g., two image intensity distributions taken with different defocussing of the microscope. A solution of the phase problem based on the theory of entire functions was proposed by ROSS et al. [1.59]. FIDDY and GREENAWAY [1.60] proposed a method using an additional experiment designed to produce a function having a zero free half plane. MONTGOMERY [1.61] stressed the relevance of the polarization identity to the phase estimation problem. An electronic-optical method for obtaining the phase of a two-dimensional wave front was described by PSALTIS and CASASENT [1.62]. The related question of uniqueness was discussed by ROBINSON [1.63] and DEVANEY and CHIDLAW [1.64]. FIDDY and GREENAWAY [1.65] and DAINTY [1.66] pointed out that maximum entropy techniques for object reconstruction in the absence of phase information could lead to false conclusions due to the existence of ambiguities. The phase problem for the degree of spatial coherence was reinvestigated by OHTSUKA [1.67] and FIENUP [1.68].

The questions of *uniqueness* and *nonradiating sources* are currently investigated for partially coherent radiation. HOENDERS and BALTES developed the scalar theory of nonradiating random sources [1.69] as well as the theory of nonradiating stochastic current distributions [1.70] and established necessary and sufficient conditions for nonradiating source correlations. DEVANEY [1.71] showed that uniqueness of the retrieval of the coherence properties from far-zone data can be achieved for the special class of quasihomogeneous scalar sources (i.e., sources with slowly varying intensity profile). The analogous result was conjectured by BALTES and HOENDERS [1.72] for quasihomogeneous current correlations, but the same authors showed how a nonradiating linear combination of these quasihomogeneous source correlations can be constructed. Based on previous work by STEINLE and BALTES [1.73], McGUIRE [1.74] showed that the source coherence function of Shell-type sources (arbitrary intensity profile) is also uniquely determined by the far-zone intensity pattern.

New results can be reported also on the problems of *estimation of scattering surfaces*, *potentials*, or *index profiles*. COHEN and BLEISTEIN [1.75] showed how to recover both the location and the reflection coefficient of a scatterer using only high-frequency backscattered data. GILLMAN [1.76] described the reconstruction of two-dimensional emissive intensity contours from a limited number of their one-dimensional projections. The digital reconstruction of the three-dimensional scattering potential of microscopic objects, e.g., small coated spheres, from amplitude and phase data was achieved by FERCHER et al. [1.77]. GOPINATH [1.78] solved the problem of determining the average dielectric constant and thickness of dielectric layers from measurements of transmitted power at a single frequency. ROGER et al. [1.79] showed how the index profile of an inhomogeneous dielectric medium can be recovered using a regularization method described by TIKHONOV [1.9].

1.2.2 Radiometry and Coherence

The torrent of papers (as HAWKES [1.80] calls it) on the relationship between radiometric properties and the state of coherence and related questions is still growing, although some of the current development seems to parallel analogous work in the field of wave propagation and scattering in random media. For instance, the very popular Gauss-correlated source with Gaussian intensity profile (see, e.g. [1.81]) was extensively investigated by JAKEMAN and PUSEY [1.82,83], and JAKEMAN and McWHIRTER [1.84] as a model for the scattering of light by a deep random phase screen. Another example is the likewise popular quasihomogeneous source correlation that appeared already in TATARSKII's book [1.85] (see also [1.115]).

On the other hand, a number of substantial new contributions to the field have been made recently, on which our attention is focussed here. WALTHER [1.86] settled the question of uniqueness of the definition of generalized radiance (see also

[1.116]). The same author [1.87] discussed the propagation of the generalized radiance through lenses and showed that in a large and significant class of cases the generalized radiance is, except for transmission losses, invariant along light rays traversing a lens. The question of partial coherence in electron microscopy was further pursued by HAWKES [1.80] and FERWERDA [1.88]. The problem of diffuse reflectance and coherence was studied by BALTES et al. [1.89] and HOENDERS et al. [1.90] with emphasis on the radiometric properties of microfacet model scatterers. BALTES and FERWERDA [1.91,92] reinvestigated the generalized Van Cittert-Zernike theorem for partially coherent sources showing not only an intensity but also a phase profile.

The interaction of partial coherence and diffraction is further illustrated in a paper by CARPENTER and PASK [1.93]. As is demonstrated by BOIVIN [1.94], the "diffraction errors", i.e., deviations from geometrical optics due to partial coherence, in radiometry can be reduced by means of toothed apertures. GRAY [1.95] described a method by which optical diffusing surfaces of known statistical properties may be found using photoresist material. These surfaces have a smoothly modulated profile and are thus amenable to theoretical modelling. The recent experimental work by DE SANTIS et al. [1.81] is concerned with a special case of the Gauss-correlated source with Gaussian intensity profile.

While most of the studies on radiometry and coherence are made in terms of the scalar approximation, LEADER [1.96] attempted the corresponding generalization for the electromagnetic vector field. He also established far-zone range criteria for quasihomogeneous sources [1.97]: for example, the far-zone condition for the radiant intensity is that the distance from the source has to be large compared with the product of the source correlation length and the source diameter, divided by the wavelength. Further publications by LEADER [1.98-100] aim at the similarities between coherence theory and scattering theory.

Finally, it is noticed that the "radiometric relation" (between source correlation and radiant intensity) can be rewritten in terms of the spatially averaged source field correlation [1.101] (as introduced by ROSS [1.102] in scattering theory) and, with a different obliquity factor, in terms of the correlation of the source variables rather than that of the source field variables [1.103].

1.2.3 A Moment Problem

A quantum optical moment problem was examined by BALTES et al. [1.104] who constructed mathematical examples for sub-Poissonian radiation fields, i.e., fields whose photon statistics shows a smaller variance than that of the Poissonian distribution belonging to the coherent state. As was shown by MANDEL [1.105], resonance fluorescence may provide a physical realization of such a field.

1.3 Construction of Lambertian Scatterers

In this section, the difficulty of the inverse scattering problem for random objects is demonstrated by two futile attempts to construct a Lambertian diffusor.

1.3.1 Lambertian Source Correlation

The classical description of diffuse emission or reflection [1.106,107] usually starts with the oldest empirical model, namely Lambert's radiant intensity

$$I(\vartheta) \propto \cos\vartheta \quad , \tag{1.1}$$

where ϑ denotes the emission or scattering angle. As first noticed by WALTHER [1.108], the inversion of the scalar radiometric relation for statistically isotropic or "circularly diffuse" sources,

$$I(\vartheta) \propto \cos^2\vartheta \int \rho \, d\rho \, J_0(k\rho \sin\vartheta) W(\rho) \quad , \tag{1.2}$$

with (1.1) leads to the source field correlation

$$W(\rho) \propto (k\rho)^{-1} \sin k\rho \quad , \tag{1.3}$$

which corresponds to that of a blackbody with sufficiently large aperture. In (1.2), J_0 denotes the Bessel function of order zero, $\rho = |\underline{r}_1 - \underline{r}_2|$ the distance between two positions, $\underline{r}_1$ and $\underline{r}_2$, in the source plane , and k the wave number of the considered Fourier component of the radiation field. High spatial frequencies (above k) have been ignored in the result (1.3) and will be ignored throughout this section.

We notice that the obliquity factor $\cos^2\vartheta$ in (1.2) is crucial. PASK [1.109] replaced $\cos^2\vartheta$ by the obliquity factor $(1+\cos^2\vartheta)/2$, which he derived from the corresponding vector field relationship. He then found that the Lambertian source (1.1) goes along with the correlation

$$W(\rho) \propto \sum_{n=1}^{\infty} (-1)^{n+1}\left[1\cdot 3\cdot 5\cdot \ldots (2n-1)\right](k\rho)^{-n} j_n(k\rho) \tag{1.4}$$

rather than with (1.3), where j_n denotes the spherical Bessel function of order n.

1.3.2 Random Scatterer Models

The question now is how an "effective source field" characterized by, say, (1.3) can be realized by illuminating a fluctuating scatterer (random phase object) with appropriate statistical properties. Assuming perpendicular illumination by an expanded coherent beam and assuming that the profile of the scattering filter or surface is simply impressed on the transmitted or reflected wave front, we would find that the requested correlation $W(\rho)$ is directly that of the complex transmission or reflexion function of the scatterer, viz.,

$$W(\rho) \propto \langle \exp[-i\Phi(\underline{r}_1)]\exp[i\Phi(\underline{r}_2)]\rangle \quad , \qquad (1.5)$$

where $\Phi(\underline{r}_1)$ denotes the random phase function. We now have to find a model distribution of $\Phi(\underline{r}_1)$ which yields the desired correlation. One difficulty is again the obliquity factor. In the literature, for the reflexion of perpendicular incident radiation, one finds the factors $(1+\cos\vartheta)^2$ and 1 without being able to say whether the first or the second is preferable [1.110]. This question can be postponed by introducing a general obliquity factor denoted by $f(\cos\vartheta)$.

As a first candidate we examine random scatterers described by the facet or micro-area model (see, e.g. [1.82,90,111]) which goes back to BOUGUER [1.112]. The random surface is described in terms of the probability distribution $P(\underline{m})$ of slopes $\underline{m}$ of the facets, whose average diameter ξ is meant to play the role of the correlation length of the phase correlation $\langle\Phi(\underline{r}_1)\Phi(\underline{r}_2)\rangle$. In the case of circularly diffuse scatterers, it is sufficient to consider the isotropic distribution $P(\underline{m})$, where $m = |\underline{m}|$. In the simplest version of the model one assumes $(k\xi)^2 \gg 1$ and finds the relation [1.90]

$$I(\vartheta) \propto f(\cos\vartheta)P(k\xi \sin\vartheta) \quad . \qquad (1.6)$$

By formal inversion of this result one would find that the slope distribution required for the Lambertian scatterer should obey the condition

$$P(m) \propto [1-(m/k\xi)^2]^{\frac{1}{2}}/f([1-(m/k\xi)^2]^{\frac{1}{2}}) \quad , \quad m < k\xi \quad . \qquad (1.7)$$

Unfortunately, the evaluation of (1.7) strongly depends on the choice of the obliquity factor. For instance, the obliquity factor $f(\cos\vartheta) = \cos^2\vartheta$ would not lead to a well-behaved probability distribution, whereas the factor $f(\cos\vartheta) = \mathrm{const}$ would.

Another surface model is due to BECKMANN [1.113] who started from a Gaussian statistics (variance σ) of the relief height $\zeta(\underline{r}_1)$ and some autocorrelation function

$$C(\rho) = \langle\zeta(\underline{r}_1)\zeta(\underline{r}_2)\rangle/\langle\zeta^2\rangle \quad . \qquad (1.8)$$

Considering only the diffuse component, (1.5) and (1.8) are, under certain restrictions, related by [1.114]

$$W(\rho) = W(0)\ \exp\left\{-k^2\sigma^2[1-C(\rho)]\right\} \qquad (1.9)$$

whence

$$C(\rho) = 1 + (k\sigma)^{-2} \log[W(\rho)/W(0)] \quad . \tag{1.10}$$

For $f(\cos\vartheta) = \cos^2\vartheta$, we may try (1.3) and hence would obtain

$$C(\rho) = 1 + (k\sigma)^{-2} \log[(k\rho)^{-1} \sin k\rho] \quad , \tag{1.11}$$

which is not a well-behaved height correlation. A similarly ill-behaved result is obtained for $f(\cos\vartheta) = 1$ with $W(\rho) = (k\rho)^{-3/2} J_{3/2}(k\rho)$ instead of (1.3).

In general, the radiant intensity of the diffuse component in the case of a statistically isotropic Gaussian surface, reflection, and perpendicular incident reads [1.110]

$$\begin{aligned} I(\vartheta) \propto\ & f(\cos\vartheta) \exp[-k^2\sigma^2(1+\cos\vartheta)^2] \\ & \int \rho\, d\rho\, J_0(k\rho \sin\vartheta)\left\{\exp[k^2\sigma^2(1+\cos\vartheta)^2 C(\rho)]-1\right\} \quad . \end{aligned} \tag{1.12}$$

In view of the ϑ dependence in the exponent under the integral, we do not see a straightforward analytical inversion of the above integral transform of $C(\rho)$ in order to recover $C(\rho)$ when $I(\vartheta)$ is given. As it looks, the construction of a Lambertian scatterer, if possible at all, is an unsolved problem.

1.4 Organization of this Volume

The material is divided into 7 chapters. Both random (Chaps.3 and 4 and part of Chap.2) and deterministic (Chaps.5-7, part of Chap.2) inverse problems are investigated. While most of the book makes use of the scalar approximation, Chap.7 stresses the electromagnetic (vector) aspect of inverse scattering. A discussion of inverse problems should perhaps start with the properties that can be measured. In optical physics we can distinguish the following classes of measurements. *First-order* properties are quantities derived from the two-point-correlation function of the electromagnetic field and include the distribution of the average intensity, the degree of polarization, and the modulus of the first-order degree of coherence. *Second- and higher-order* properties are of interest in the case of random scatterers. They are derived from the corresponding n^{th} order correlations of the scattered field. Examples are the average square intensity, the degree of second-order coherence (Hanbury-Brown-Twiss experiment), and higher moments of the photon statistics. Higher-order properties are examined in Chaps.3 and 4, while Chaps.2, and 5-7 are concerned with first-order properties, mainly the intensity pattern of the scattered field.

In Chap.2, ROSS, FIDDY, and NIETO-VESPERINAS describe the classical theory relating the far- and near-zone spatial intensity distribution to the structural features of both deterministic and random scatterers. After formulating the general theory, the authors consider *weak scatterers* and use the *Born approximation*. Their discussion includes the three-dimensional absorbing anisotropic medium, and the role of analyticity and the theory of entire functions is stressed. Statistical properties of random media are described in terms of correlation functions such as *K correlations*. The chapter is introduced by a summary of the underlying epistemological questions.

In Chap.3, JAKEMAN and PUSEY present an exposition of the relationship between the results of photon-counting experiments and the statistical properties of scattering media. The pertinent inverse problem can be divided in two steps: first, the *inverse detection problem* of retrieving the statistical properties of the field that has propagated to the detector, and secondly the *determination of the statistical features of the scatterer*. As regards the detection problem, the authors describe the two principal types of measurement (single interval statistics and photon-correlation spectroscopy) together with the effect of temporal integration and other instrumental effects. The main part of the chapter is devoted to scattering theory and experiments with *strongly scattering random media*. Both the deep random phase screen situation and the extended region of medium producing multiple scattering are considered. The photon statistics of the scattered field is described in terms of *K distributions* whose origin is discussed in terms of random walk and population statistics.

Chapter 4 by SELLONI is wholly devoted to the *detection problem* with emphasis on quantum theory and microscopic models. After reviewing previous detection models, the author describes very recent work on the *open-system theory of photodetection*, including a first-principle theory of thermal noise. Also presented is a summary of other effects that disturb the transfer of information from the received field to the photoelectric signal. The chapter ends with an overview of statistical methods and the theory of open systems.

While the previous volume [1.1] includes a comprehensive chapter on the question of uniqueness in inverse problems, a systematic account of the *problem of stability* is missing hitherto. Chapter 5 by BERTERO, DE MOL, and VIANO fills this void. The chapter starts with a description of ill-posedness, numerical instability, and the role of prior knowledge. Regularization theory is reviewed with emphasis on Miller's theory, Tikhonov's method, and optimum filtering procedures. Applications include inverse problems in Fourier optics, inverse diffraction, an inverse scattering problem for perfectly conducting bodies, inverse scattering problems in the Born approximation, and object reconstruction from projections.

In view of the existing review literature on radiation transfer and remote sounding [1.10-14], Chap.6 by GOULARD and EMMERMAN is focussed on recent results in one special problem of this field, namely three-dimensional *combustion diagnostics* by optical absorption using the *multiple angular scanning* technique, which is analogous to X-ray tomography.

In Chap.7, BOERNER examines problems of electromagnetic inverse scattering. The author stresses *vector* wave diffraction and *polarization* effects. Not only the inverse optical problems, but also the related topics in microwave physics are discussed: the methods of radar-cross-section and target-polarization-scattering are described and the inverse scattering theories valid in the various electromagnetic frequency regimes are presented. Moreover, the chapter includes an overview of polarization utilization in vector holography and interferometry.

Acknowledgements. It is my pleasure to acknowledge the contributors to this volume — M. Bertero, W-M. Boerner, Christine De Mol, P.J. Emmerman, M. Fiddy, R. Goulard, E. Jakeman, M. Nieto-Vesperinas, P. Pusey, G. Ross, Annabella Selloni, and G.A. Viano — for their enthusiasm and their patience with the editor's suggestions. The authors who wrote most of the text have worked hard to produce high-quality manuscripts. Besides the contributors, discussions with J.C. Dainty, H.A. Ferwerda, H.K.V. Lotsch, D. Maystre, J.G. McWhirter, J.-F. Moser, L. Narducci, and B. Steinle influenced the direction of this volume. Many colleagues kindly provided literature hints and manuscripts of articles prior to publication: in particular I would like to thank R.H.T. Bates, R. Barakat, A.J. Devaney, A.F. Fercher, B. Gopinath, P.W. Hawkes, P.C. Sabatier, and H.G. Schmidt-Weinmar. Landis & Gyr R & D directors H.J. Vonarburg and H. Lienhard generously provided time and support necessary for editing this volume. A special thanks to my wife Gabriella Baltes for her competent help in the less palatable part of the editorial work.

References

1.1 H.P. Baltes (ed.): *Inverse Source Problems in Optics*, Topics in Current Physics, Vol.9 (Springer, Berlin, Heidelberg, New York 1978)

1.2 W.-M. Boerner: "State of the Art Review on Polarization Utilization in Electromagnetic Inverse Scattering", Tech.Rpt.78-3. Communications Laboratory, University of Illinois, Chicago (1978)

1.3 N. Bleistein, J.K. Cohen: "Survey of Recent Progress on Inverse Problems", Tech.Rpt. MS-R-7806. Denver Research Institute, University of Denver, Colorado (1978)

1.4 V.H. Weston: "Electromagnetic Inverse Scattering", in *Electromagnetic Scattering* (Academic Press, New York 1978) pp.289-313

1.5 H.P. Baltes: "Inverse Problems in Optics", in Proc. Intern. Conf. Lasers '78 (STS Press, McLean, Virginia 1979) pp.716-722

1.6 P.C. Sabatier (ed.): *Applied Inverse Problems*, Lecture Notes in Physics, Vol.85 (Springer, Berlin, Heidelberg, New York 1978)

1.7 L. Colin: *Mathematics of Profile Inversion*, NASA Technical Memorandum X-62.150 (Ames Research Center, Moffett Field, Calif. 1972) Chap.6: Electromagnetic Scattering

1.8 K. Chadan, P.C. Sabatier: *Inverse Problems in Quantum Scattering Theory* (Springer, Berlin, Heidelberg, New York 1977)

1.9 A.N. Tikhonov, V.Y. Arsenin: *Solutions of Ill-Posed Problems*, translation ed. by F. John (Winston/Wiley, Washington DC/New York 1977)

1.10 A. Ishimaru: "The Beam Wave Case and Remote Sensing", in *Laser Beam Propagation in the Atmosphere*, ed. by J.W. Strohbehn, Topics in Applied Physics, Vol.25 (Springer, Berlin, Heidelberg, New York 1978) pp.129-170

1.11 A. Ishimaru: *Wave Propagation and Scattering in Random Media*, Vol.2 (Academic Press, New York 1978) pp.493-512
1.12 A. Deepak (ed.): *Inversion Methods in Atmospheric Remote Sounding* (Academic Press, New York 1978)
1.13 E.D. Hinkley (ed.): *Laser Monitoring of the Atmosphere*, Topics in Applied Physics, Vol.14 (Springer, Berlin, Heidelberg, New York 1976)
1.14 A.L. Fymat, V.E. Zuev (eds.): *Remote Sensing of the Atmosphere: Inversion Methods and Applications* (Elsevier, Amsterdam 1978)
1.15 T.S. Huang (ed.): *Picture Processing and Digital Filtering*, 2nd ed., Topics in Applied Physics, Vol.6 (Springer, Berlin, Heidelberg, New York 1979)
1.16 A. Rosenfeld (ed.): *Digital Picture Analysis*, Topics in Applied Physics, Vol.11 (Springer, Berlin, Heidelberg, New York 1976)
1.17 D. Casasent (ed.): *Optical Data Processing, Applications*, Topics in Applied Physics, Vol.23 (Springer, Berlin, Heidelberg, New York 1977)
1.18 W.O. Saxton: *Computer Techniques for Image Processing in Electron Microscopy* (Academic Press, New York 1978)
1.19 P.W. Hawkes (ed.): *Computer Processing of Electron Microscope Images*, Topics in Current Physics, Vol.13 (Springer, Berlin, Heidelberg, New York 1980)
1.20 B. Saleh: *Photoelectron Statistics*, Springer-Series in Optical Sciences, Vol.6 (Springer, Berlin, Heidelberg, New York 1978)
1.21 R. Barakat, J. Blake: *Theory of Photoelectron Counting Statistics: an Essay*, Phys. Rep. *60*, 225-340 (1980)
1.22 R.J. Oliver: Adv. Phys. *27*, 387-435 (1978)
1.23 A.M.J. Huiser: "Fundamental Problems in the Evaluation of Electron Micrographs", Ph.D. Thesis, State University, Groningen (1979)
1.24 P. Van Toorn: "Proposals for the Solutions of the Phase Problem in Electron Microscopy", Ph.D. Thesis, State University, Groningen (1979)
1.25 M. Nieto-Vesperinas: "Statistics of Amorphous Absorbing Media and Object Wave Determination Using the Theory of Entire Functions", Ph.D. Thesis, Queen Elizabeth College, University of London (1978)
1.26 A. Selloni: "Microscopic Models of Photodetection", Ph.D. Thesis, Ecole Polytechnique Fédérale, Lausanne (1979)
1.27 C. De Mol: "Sur la Régularisation des Problèmes Inverses Linéaires", Ph.D. Thesis, Université Libre, Bruxelles (1979)
1.28 J.G. McWhirter, E.R. Pike: J. Phys. A. *11*, 1729-1745 (1978)
1.29 J.G. McWhirter, E.R. Pike: Phys. Scr. *19*, 417-425 (1979)
1.30 J.G. McWhirter: Opt. Acta *27*, 83-105 (1980)
1.31 A. Roger, D. Maystre: Opt. Acta *26*, 447-460 (1979)
1.32 A. Roger, D. Maystre: "A Method for Inverse Problems in Optics. Application to Diffraction Gratings", to be published
1.33 A.L. Fymat: Appl. Opt. *18*, 126-130 (1979)
1.34 G.E. Shaw: Appl. Opt. *18*, 988-993 (1979)
1.35 A. Cohen, J. Cooney, G. Raviv, N. Wolfson: Appl. Opt. *18*, 2466-2469 (1979)
1.36 S.R. Powers, D.J. Somerford: Opt. Commun. *26*, 313-317 (1978)
1.37 V. Pollak: Opt. Acta *25*, 929-936 (1978)
1.38 R.H.T. Bates: Optik *51*, 161-170 (1978)
1.39 R.H.T. Bates: Optik *51*, 223-234 (1978)
1.40 P.J. Napier, R.H.T. Bates: Astron. Astrophys. Suppl. *15*, 427-430 (1974)
1.41 R.H.T. Bates: Mon. Not. R. Astr. Soc. *142*, 413-428 (1969)
1.42 R.H.T. Bates, P.J. Napier: Mon. Not. R. Astr. Soc. *158*, 405-424 (1972)
1.43 R.M. Lewitt, R.H.T. Bates: Optik *50*, 19-33 (1978)
1.44 R.M. Lewitt, R.H.T. Bates, T.M. Peter: Optik *50*, 85-109 (1978)
1.45 R.M. Lewitt, R.H.T. Bates: Optik *50*, 189-204 (1978)
1.46 R.H.T. Bates, T.M. Peters: New Zealand J. Sci. *14*, 883-896 (1971)
1.47 A.E. McKinnon, M.J. McDonnell, P.J. Napier, R.H.T. Bates: Optik *44*, 253-272 (1976)
1.48 R.H.T. Bates, P.J. Napier, A.E. McKinnon, M.J. McDonnell: Optik *44*, 183-201 (1976)
1.49 R.H.T. Bates, P.T. Gough: IEEE Trans. C-*24*, 449-456 (1975)
1.50 R.H.T. Bates, M.J. McDonnell, P.T. Gough: Proc. IEEE *65*, 138-143 (1977)
1.51 R.H.T. Bates, M.O. Milner, G.I. Lund, A.D. Seagar: Opt. Commun. *26*, 22-26 (1978)

1.52 G.L. Rogers: Opt. Commun. *30*, 1-3 (1979)
1.53 P.J. Napier, R.H.T. Bates: Proc. IEEE *120*, 30-34 (1973)
1.54 R.H.T. Bates: Int. J. Eng. Sci. *9*, 1107-1121 (1971)
1.55 P.J. Napier, R.H.T. Bates: Int. J. Eng. Sci. *9*, 1193-1208 (1971)
1.56 R.H.T. Bates, W.M. Boerner, G.R. Dunlop: Opt. Commun. *18*, 421-423 (1976)
1.57 R.H.T. Bates: Arch. Rational Mech. Anal. *38*, 123-130 (1970)
1.58 P. Van Toorn, A.M.J. Huiser, H.A. Ferwerda: Optik *51*, 309-326 (1978)
1.59 G. Ross, M.A. Fiddy, M. Nieto-Vesperinas, M.W.L. Wheeler: Optik *49*, 71-80 (1977)
1.60 M.A. Fiddy, H.A. Greenaway: Opt. Commun. *29*, 270-272 (1979)
1.61 W.D. Montgomery: Opt. Lett. *2*, 120-121 (1978)
1.62 D. Psaltis, D. Casasent: Appl. Opt. *17*, 1136-1140 (1978)
1.63 S.R. Robinson: J. Opt. Soc. Am. *68*, 87-92 (1978)
1.64 A.J. Devaney, R. Chidlaw: J. Opt. Soc. Am. *68*, 1352-1354 (1978)
1.65 M.A. Fiddy, A.H. Greenaway: Nature *276*, 421 (1978)
1.66 J.C. Dainty: "The Role of Entropy in the Inverse Problem", presented at OPTICS '78, 20-23 September 1978, University of Bath, England
1.67 Y. Ohtsuka: Opt. Lett. *1*, 133-134 (1977)
1.68 J.R. Fienup: Opt. Lett. *3*, 27-29 (1978)
1.69 B.J. Hoenders, H.P. Baltes: Lett. Nuovo Cimento *25*, 206-208 (1979)
1.70 B.J. Hoenders, H.P. Baltes: J. Phys. A*13*, 995-1006 (1980)
1.71 A.J. Devaney: J. Math. Phys. *20*, 1687-1691 (1979)
1.72 H.P. Baltes, B.J. Hoenders: Phys. Lett. *69A*, 249-250 (1978)
1.73 B. Steinle, H.P. Baltes: J. Opt. Soc. Am. *67*, 241-247 (1977)
1.74 D. McGuire: Opt. Commun. *29*, 17-21 (1979)
1.75 J.K. Cohen, N. Bleistein: "The Singular Function of a Surface and Physical Optics Inverse Scattering", Tech. Rpt. MS-R-7906 (Department of Mathematics, University of Denver, Denver, Colorado 1978) and Wave Motion *1*, 153-161 (1979)
1.76 G.B. Gillman: Opt. Commun. *29*, 261-264 (1979)
1.77 A.F. Fercher, H. Bartelt, H. Becker, E. Wiltschko: Appl. Opt. *18*, 2427-2439 (1979)
1.78 B. Gopinath: J. Math. Phys. *17*, 1099-1104 (1976)
1.79 A. Roger, D. Maystre, M. Cadilhac: J. Optics (Paris) *9*, 83-90 (1978)
1.80 P.W. Hawkes: Optik *50*, 353-370 (1978)
1.81 P. De Santis, F. Gori, G. Guattari, C. Palma: Opt. Commun. *29*, 256-260 (1979)
1.82 E. Jakeman, P.N. Pusey: J. Phys. A *8*, 369-391 (1975)
1.83 P.N. Pusey, E. Jakeman: J. Phys. A *8*, 392-410 (1975)
1.84 E. Jakeman, J.G. McWhirter: J. Phys. A *9*, 785-797 (1976)
1.85 V.I. Tatarskii: "Locally Homogeneous Fields with Smoothly Varying Mean Characteristics", in *The Effects of the Turbulent Atmosphere on Wave Propagation* (Israel Program for Scientific Translations, Jerusalem 1971) §7
1.86 A. Walther: Opt. Lett. *3*, 127-129 (1978)
1.87 A. Walther: J. Opt. Soc. Am. *68*, 1606-1610 (1978)
1.88 H.A. Ferwerda: "Coherence of Illumination in Electron Microscopy", in *Image Processing and Coherence in Physics*, Workshop Les Houches, 12-13 March (1979) to be published
1.89 H.P. Baltes, B. Steinle, E. Jakeman, B. Hoenders: Infrared Phys. *19*, 461-464 (1979)
1.90 B.J. Hoenders, E. Jakeman, H.P. Baltes, B. Steinle: Opt. Acta *26*, 1307-1319 (1979)
1.91 H.P. Baltes, H.A. Ferwerda: Lett. Nuovo Cimento *27*, 541-543 (1980)
1.92 H.P. Baltes, H.A. Ferwerda, A.S. Glass, B. Steinle: "Retrieval of Structural Information from Far-Zone Intensity and Coherence of Scattered Radiation", Opt. Acta (in press)
1.93 D.J. Carpenter, C. Pask: Opt. Acta *24*, 939-948 (1977)
1.94 L.P. Boivin: Appl. Opt. *17*, 3323-3328 (1978)
1.95 P.F. Gray: Opt. Acta *25*, 765-775 (1978)
1.96 J.C. Leader: Opt. Acta *25*, 395-413 (1978)
1.97 J.C. Leader: J. Opt. Soc. Am. *68*, 1332-1338 (1978)
1.98 J.C. Leader: "Equivalent Source Coherence of Laser-Illuminated Rough Surfaces", submitted to Opt. Acta

1.99 J.C. Leader: J. Opt. Soc. Am. *66*, 183 (1976)
1.100 J.C. Leader: "Similarities and Distinctions between Coherence Theory Relations and Laser Scattering Phenomena", preprint (1979)
1.101 E. Wolf, W.H. Carter: J. Opt. Soc. Am. *68*, 953-964 (1978)
1.102 G. Ross: Opt. Acta *25*, 57-66 (1977)
1.103 E. Wolf: J. Opt. Soc. Am. *68*, 1597-1605 (1978)
1.104 H.P. Baltes, A. Quattropani, P. Schwendimann: J. Phys. A *12*, L35-L37 (1979)
1.105 L. Mandel: Opt. Lett. *4*, 205-207 (1979)
1.106 G. Kortüm: *Reflectance Spectroscopy* (Springer, Berlin, Heidelberg, New York 1969) Chap.2
1.107 D.E. Barrick: "Rough Surfaces", in *Radar Cross Section Handbook*, ed. by G.T. Ruck (Plenum Press, New York 1970) Chap.9
1.108 A. Walther: J. Opt. Soc. Am. *58*, 1256-1259 (1968)
1.109 C. Pask: Opt. Acta *24*, 235-240 (1977)
1.110 W.T. Welford: Opt. Quantum Electron. *9*, 269-287 (1971)
1.111 L.E. Estes, L.M. Narducci, R.A. Tuft: J. Opt. Soc. Am. *61*, 1301-1306 (1971)
1.112 M. Bouguer: *Traité d'Optique sur la Gradation de la Lumière*, ed. by Abbé de la Caille (Guerin & Delatour, Paris 1760) pp.161-228
1.113 P. Beckmann: "Part I - Theory", in *The Scattering of Electromagnetic Waves from Rough Surfaces*, ed. by P. Beckmann, A. Spizzichino (Pergamon Press, London 1963)
1.114 P.J. Chandley: Opt. Quantum Electron. *8*, 329-333 (1976)
1.115 A. Papoulis: J. Opt. Soc. Am. *64*, 779-788 (1974)
1.116 A.T. Friberg: J. Opt. Soc. Am. *69*, 192-198 (1979)

2. The Inverse Scattering Problem in Structural Determinations

G. Ross, M. A. Fiddy, and M. Nieto-Vesperinas

With 9 Figures

The aim of this contribution is to formulate, and to attempt to construct a solution to, the inverse scattering problem. In Sect.2.1, the fundamental epistemological foundations of this task are outlined. Section 2.2 presents the direct scattering problem, and the mathematical equipment required for treating the inverse problem is given in Sect.2.3.

Two types of inverse scattering problems may be distinguished. The first, named the deterministic problem, consists of obtaining information about the detailed structure of a scatterer. The second, known as the statistical problem, is concerned with the retrieval of overall descriptors of the morphology. Both aspects are discussed only for a one-dimensional situation. In this case, the deterministic problem is reduced to the phase problem, investigated in Sect.2.4. The statistical problem is studied in Sect.2.5.

2.1 Philosophical Background

> *But how will you look, Socrates, for something when you don't in the least know what it is? How on earth are you going to set up something you don't know as the object of your search? Or even, supposing, at the best, that you hit upon it, how will you know that it is the thing you didn't know?*
>
> Plato, Meno 80d.

When an electromagnetic field passes through a medium with a certain structure, the inhomogeneity acts as a perturbation. Its effect is a distortion, or corrugation, of the propagating wave front, which is observed as a change in the distribution of energy: this is the scattering phenomenon, our most powerful source of information on the physical world surrounding us.

The modification of the energy repartition is manifest in both the spatial and the temporal frequency spectrum. In general, when concerned with structural determinations, one tends to disregard dynamic effects (allowing for them if necessary) and time-invariant interactions alone between radiation and matter are considered.

Given a source and a medium with a certain structure, the task of obtaining the field and intensity as functions of position is known as the direct source problem in scattering, or the direct scattering problem.

The inverse source problem in scattering or the inverse scattering problem is far more important, particularly in structural studies. It consists of determining the inhomogeneity of a material from a knowledge of the spatial distribution of radiation, usually assuming that the source is given.

One can, in fact, distinguish two inverse scattering problems. The first is to establish the morphology of the medium, in other words to obtain the local values of a function defining its structure: the inhomogeneity distribution function, providing this description, is an intensive magnitude, such as density, refractive index, permittivity. This task will be referred to as the *deterministic problem.*

In many instances, however, a detailed knowledge of the morphology is not required. For various applications, an overall description of the structure may be more relevant. The second inverse scattering problem consists of determining such global characteristics. Although not exclusively, this is mainly the situation when the scattering medium may be regarded, in some sense, as random: various statistical averages of the inhomogeneity distribution function, providing the overall description of the morphology, must be obtained. This task will be referred to as the *statistical problem.*

The direct scattering problem is an exercise in applied mathematics. Although a completely general answer has not (as yet) been formulated, the task is, at least in principle, amenable to a solution. For many particular instances, expressions have been obtained from basic premises and no conceptual difficulties or fundamental obstacles are likely to occur when treating any specific situation.

In contrast, once the theoretical structure provided by the direct counterpart is established, the inverse scattering problem acquires a fundamental epistemological character. The theory of the direct situation must be formulated before the inverse problem can be tackled, but the latter cannot be solved simply by applying the deductive process in reverse direction: strictly speaking, the inverse problem cannot be solved at all.

Solving an epistemological problem consists of being able to put forward an answer to the question 'what can we know, what do we *really* know, and how can we know it?'

The root problem of epistemology is the question 'how can you seek what you do not know?' formulated by PLATO in Meno [2.1]. It is the question of heuristics, of the raison d'être of discovery. Indeed, it seems likely that the first to have been aware of inverse problems in general, of inverse scattering problems in particular and of their epistemological nature was Plato: this is the essence of his famous and brilliant cave metaphor, formulated in the Republic Book VII [2.1]. Of course, the fundamental difficulties of epistemology were known much earlier: one can probably say that they constitute the earliest expression of intellectual pursuit and their formulation marked the dawn of rational inquiry, in other words of civilization as we know it, in the works of Protagoras, Gorgias and – first of all – Xenophanes.

The basic issue in epistemology is the problem of the foundation of knowledge: it is the controversy between dogmatists – who claim that we can know – and sceptics – who claim that we either cannot know or at least cannot know that we can know and when we can know.

The classical sceptical argument is based on the infinite regress, used to show that it is hopeless to try to find foundations for knowledge. This argument is pervasive and persuasive in scattering theories. For example, we shall see that it is indeed impossible to obtain, in inverse scattering problems, information about the scatterer, either in the deterministic or in the statistical situation. In both classes of the inverse problem we obtain information only about the *registration* of objects by a wave, in other words about the perturbation imprinted by an object on a propagating field; this perturbation is known in scattering theories as the object wave. To establish the relationship between the object and the object wave would, in general and without some supplementary knowledge about the object, involve an infinite regress: apart from trivial cases, which correspond to mathematical abstractions rather than physical situations, the object cannot be obtained from the object wave, even if the latter were determined. This is, again, more than can ever be achieved: as will be shown later, complete knowledge of the object wave also involves an infinite regress and hence cannot be reached.

The first aim of this paper is to present the inverse scattering problem as deeply imbedded in general epistemology and only to be understood in this context.

The theory of knowledge, or epistemology, should be identified, according to POPPER [Ref.2.2, Part I] with the theory of scientific method. Its task is the analysis of the method or procedure peculiar to empirical science: its task is to watch the observations, procedures and limitations of the equipment and to point these out before investigators seek to use the methods. Epistemology may be thus described as a theory of the empirical method – a theory of what is usually called 'experience': it is a theory about the range and limits of knowing and about what happens beyond these limits.

Eddington made the distinction between two kinds of epistemological knowledge, i.e., knowledge derived by epistemological study of the procedure of observation [2.3]. The first is derived from an examination of the methodological tools used in the experiment. The second is inferred because of the innermost modality by which experience of external objects is possible.

The first kind of epistemological constraint, described by what BALTES [2.4] entitled 'detection theory' is not fundamental. In inverse scattering problems it is illustrated by our inability to observe the phase of an electromagnetic field at high frequencies, e.g., those corresponding to the visible spectrum. This aspect of methodology, called by POPPER [Ref.2.2, Sect.30] the 'epistemological theory of experiment' is prominent in the deterministic problem: were we able to measure directly the phase, the task would be considerably simplified. But it is not inconceivable that some day, perhaps very soon, a direct method for recording the phase will become available. This major difficulty of the deterministic problem will then disappear. It is, however, doubtful that the measurement of the phase could ever be achieved for *all* frequencies. As physics is understood today, beyond a certain threshold – which we may be able to push all the time further towards higher frequencies – a direct recording of the phase will not be available. Thus, Eddington's distinction, although possibly convenient, may be misleading. Perhaps the first kind of épistemological knowledge ought to be regarded as subsumed into the second.

The second kind of knowledge derived by epistemological study of the procedure of observation is constituted by certain basic categories: these are the fundamental ways in which we *necessarily* perceive and think about the appearances we are presented with. The categories are applied to any material given by experience: all possible perceptions and therefore everything that can come to empirical consciousness, that is, all appearances of nature must be subject to the categories [Ref.2.5, B164].

Kant, as BERLIN [2.6] pointed out, was the first thinker to draw a clear distinction between questions of fact and questions about the patterns in which these facts presented themselves to us – patterns that were not themselves altered however much the facts themselves, or our knowledge of them, might alter. Kant was the first to draw the crucial distinction between facts – the data of experience – and the categories in terms of which we sensed and imagined and reflected about them. In his doctrine of our knowledge of the external world, he taught that the categories through which we saw it were identical for all sentient beings, permanent and unalterable; indeed this is what made our world one, and communication possible.

The role of Kant's categories, by which experience of external objects is possible, is of fundamental importance for inverse problems in general and inverse scattering problems in particular.

In this contribution, we shall discuss both the deterministic and statistical aspects largely for the ideal situation of a noiseless record: this will be the assumption we shall make for tracing the path towards obtaining a solution. The effect of noise will not be systematically investigated in this paper, in which we shall confine ourselves only to the idealized situation (from this point of view) of basic principles.

For this approach, and with this assumption, adopted for our treatment of inverse scattering problems, the most important category belongs to the group entitled by Kant categories of quality: it is the concept of limitation. Every experience, in conformity with the given forms of intuition, is enclosed within limits [Ref.2.5, A509, B537].

The categories describing the fundamental limitation of an experiment are the very reason why the inverse scattering problem exists, in both its deterministic and statistical aspects. As we shall see, the deterministic problem – apart from the limitation of the detection system, which prevents a recording of the phase – is due to the finite bounds of the interval in the space of measurement. This imposes constraints upon the modality in which we can construct the object wave. The statistical problem exists not only for the same motive, but also because the interval in the object space is limited: the problem of obtaining meaningful averages from only a finite realization of the object wave is a fundamental one in statistics.

Moreover, this essential constraint manifests itself in scattering experiments in the form of source geometry, determining the coherence characteristics of the field, and as the finite extent of the aperture with which any measurement is performed, both of which affect the observation.

Knowledge of the external world is had via observation: once the design of the experiment is established, however, certain characteristics, certain patterns, which any observation will present – and which will be discovered *a posteriori* – can be foreseen *a priori*, simply because the pre-established plan of observation will be employed.

Thus, we may distinguish knowledge of the physical universe derived through experience, by study of the results of observation, as *a posteriori* knowledge and knowledge derived by epistemological study of the procedure of observation as *a priori* knowledge.

KANT [Ref.2.5, B3] defines 'pure *a priori* knowledge' as knowledge absolutely independent of all experience. We are using here '*a priori*' in a more relaxed sense, following EDDINGTON [Ref.2.7, Chap.II §IV] and accepting that *a priori* knowledge cannot be regarded as independent of observational experience altogether. Deduction of laws of nature from epistemological considerations implies antecedent observational experience: we regard *a priori* knowledge only as knowledge we have – or may have – of the physical universe prior to the actual observation, prior to the experiment in which we are interested, but of course, not prior to the development of a plan of observation.

The theory of *a priori* knowledge claims that much observationally derived knowledge can be predicted or inferred from an examination of the methodological tools of the physicist, together with an analysis of certain fundamental categories: the categories are thus concepts which prescribe laws *a priori* to appearances and therefore to nature, the sum of all appearances [Ref.2.5, B163]. However, the most understanding can achieve *a priori* is to anticipate the form of a possible experience in general: we can know *a priori* of things only what we ourselves put into them [Ref. 2.5, BXVIII]. KANT [Ref.2.5, A12] entitles 'transcendental' all knowledge which is occupied not so much with objects as with the mode of our knowledge of objects, in so far as this mode of knowledge is to be possible *a priori*. Thus, categories have only transcendental meaning.

To be sure, special laws, as concerning those appearances which are empirically determined, cannot in their specific character be *derived* from the categories, although they are one and all subject to them. To obtain any knowledge whatsoever of the special laws, we must resort to experience; but it is the *a priori* laws that alone can instruct us in regard to experience in general, and as to what it is that can be known as an object of experience [Ref.2.5, B165].

Thus, we are led to the conclusion formulated by KANT [Ref.2.5, BXX], essential when seeking a solution to the inverse scattering problem, that our representation of things, as they are given to us, does not conform to these things as they are 'in themselves' but that these objects, as appearances, conform to our mode of representation.

In inverse problems in general, and inverse scattering problems in particular, ontology is reduced to epistemology [Ref.2.3; 2.8, Chap.18 VI], the physical world itself is defined as that which the tools of science are capable of describing. EDDINGTON [Ref.2.9, p.328] was right in saying that, in the end, what we comprehend about the universe is precisely that which we put into the universe to make it comprehensible. This agrees with Kant's general idea [Ref.2.5, A125] that we can only make sense of the world by imposing some structure originating from the mind upon it.

We are thus conducted to the idea that the observer's presence – in the guise of the plan of observation adopted – is ineluctable and ubiquitous in the final result: this important conclusion can be traced through the whole history of epistemology, from what EDDINGTON [Ref.2.7, Chap.II] defined as 'selective subjectivism' (in the best possible sense of this expression) right back to Protagoras' famous dictum, that 'man is the measure of all things'.

Thus, we are led to the conclusion that our awareness of the categories makes us realize the necessity of an infinite regress in order to attain an 'objective' result. This is equivalent to saying, as Hume pointed out [Ref.2.2 Appendix VII], that we get involved in an infinite regress if we appeal to experience in order to justify *any* conclusion concerning unobserved instances. But, as KANT [Ref.2.5, A512, B540, A523, B551] stressed, this is impossible: 'the regress does not proceed to the infinite, as if the infinite could be given, but only indeterminately far, in order to give that empirical magnitude which first becomes actual in and through this very regress.' POPPER [Ref.2.10, Chap.10.5] adopts the same view: 'the idea that our theories should describe the structural properties of the world ... is hard to think out fully without getting involved in an infinite regress.'

Thus, strictly speaking, epistemological problems like inverse scattering situations cannot be solved. Goethe put it best exactly one hundred and fifty years ago [2.11, 1.9.1829]: 'Kant has unquestionably done the best service, by drawing the limits beyond which human intellect is not able to penetrate and leaving at rest the insoluble problems'.

This conclusion is the main raison d'être of scepticism. It arises because of the supposal that knowledge involves certainty and suggests the necessity of an infinite regress in order to obtain the 'correct' answer; since this can never be carried out, it leads to the defeatist credo summed up, perhaps better than anyone else, by LUCAN [Ref.2.12, II 656-8] 'Nil actum credens, cum quid superesset agendum' (Reckoning that nothing has been achieved if there should be anything left to be done).

We reject both the sceptic and the dogmatic view: in doing so, we follow Hume [see Ref.2.10, 2X] and abandon the idea that knowledge involves certainty. This leads to the realization that our attempts to see and find the truth are not final, but open to improvement. Our conclusion is that we are approximating the truth by stages. We shall use this conclusion in an optimistic way, by a complete re-appraisal of our aims in general and in inverse scattering problems in particular. This is the point of view we adopt here, formulated, in the most poignant way, by POPPER [Ref. 2.10, Introduction XVII]: we realize that the best we can do is to grope for truth without ever achieving it. We shall use the theory of knowledge as a set of defence works against scepticism of the very possibility of knowledge. If the categories are the cause of there being an inverse scattering problem, the *a priori* knowledge provided by them will be seen to be the key to obtaining not a solution, but an acceptable representation of it.

We can never reach complete knowledge of the object wave (let alone the object): we can obtain an encoding of it, in terms we ourselves create, given the *a priori* knowledge of the plan of observation.

To show that this is possible and how it can be accomplished is the main aim of this contribution.

The direct scattering problem is the theoretical structure which must be formulated before any attempt can be made to proceed to the inverse problem. Section 2.2 is devoted to this endeavour. The direct problem will be solved for the most general situation of a three-dimensional absorbing anisotropic medium. In Sect.2.3 the mathematical equipment best suited to make use of the *a priori* epistemological knowledge, which will be applied to treating the inverse problem, is presented: it is the theory of entire functions of exponential type. Section 2.4 discusses the application of this approach for obtaining a solution of the deterministic inverse scattering problem, in terms of the encoding offered by the mathematical tools reflecting the categories. Section 2.5 studies the application of the same method to the statistical situation: as yet, for both inverse aspects a general treatment is available only for a one-dimensional case. The contribution ends with a discussion and conclusions (Sect.2.6).

2.2 The Direct Scattering Problem

Our intellect does not draw its laws from nature, but tries – with varying degrees of success – to impose upon nature laws which it freely invents.

Popper, Conjectures and Refutations, 8.1

Before any attempt could be made to construct a solution – in the broad sense defined in Sect.2.1 – to the inverse scattering problem, in both its deterministic and statistical aspect, the theory corresponding to the direct problem must be formulated. For this, the first task consists in adopting a description for the inhomogeneity of a material.

2.2.1 Description of the Medium

In scattering theories, the models of the medium fall into two broad categories: the perturbed continuum model and the model of an assembly of discrete scatterers. Both models may be used to describe deterministic structures or objects whose morphology displays no obvious regularity.

The continuum model characterizes the medium by some intensive magnitude, such as refractive index or density. For obvious reasons, it is convenient to assign to the medium parameters that appear and act as perturbations in the equations governing the propagation of the type of radiation considered for the direct problem and used for the inverse problem: permittivity, permeability and conductivity for electromagnetic waves, density and pressure for acoustic waves, or potentials for de Broglie waves. It is convenient to assume that the variation in the parameters is about mean values which correspond to the values of an idealized homogeneous medium.

The discrete model considers the scattering function of a single scattering element (known as form factor). The scattering problem is solved for a distribution of scatterers. This distribution is itself the parameter which represents the inhomogeneous medium.

The choice of the model is governed by the problem investigated, scale of inhomogeneity studied, wavelength of the radiation employed.

Only the first model, the perturbed continuum, will be treated here: a general theory for the discrete model is not yet available. Since we shall consider electromagnetic scattering, the object will be described by the macroscopic parameters characterizing a medium in Maxwell's equations. We shall assume that the conductivity is negligibly small and that the permeability μ is uniform throughout the medium and practically equal to that of the vacuum. In this case the only magnitude which, by its inhomogeneity, affects the propagation of the electromagnetic wave and produces its scattering, is the electric permittivity ε. Although Maxwell's equations consider only real permittivities and predict absorption only in media with nonzero conductivity, we shall consider here nonconductive media with complex permittivity, based on classical theories such as Lorentz's (see, e.g. [Ref.2.13, Sect.2.3]). Only 'true' absorption will be taken into account, i.e., the conversion of electromagnetic energy into internal energy. The assumption is also made throughout that a quantum approach is not required.

We shall assume the medium locally anisotropic; this implies that the permittivity

$$\underline{\underline{\varepsilon}}(\underline{r}) = \underline{\underline{\varepsilon}}_R(\underline{r}) + i\underline{\underline{\varepsilon}}_A(\underline{r}) \tag{2.1}$$

is a second-rank tensor, $\underline{r}$ being the positional vector of a generic point in the scattering medium, describing its oriented distance from an arbitrary origin O. It may be shown [Ref.2.13, Sects.14.1 and 14.6.1] that, owing to conservation of energy, the permittivity tensor is symmetric, i.e., only six of the nine components of the tensor are independent. It is, of course, possible to transform this tensor, by choosing an appropriate coordinate system, so that it may be represented by a diagonal matrix: the three values are known as the principal values of the tensor; they can be said to define the axes of an ellipsoid. We may thus describe a complex permittivity by a refractive ellipsoid and an absorptive ellipsoid; in general, their principal axes do not coincide. To avoid a highly complicated problem, we shall present the direct scattering theory for the situation in which at least orthorhombic symmetry is present and thus the principal axes of the two ellipsoids coincide in direction.

In general, the medium as a whole is anisotropic; the overall anisotropy is characterized by two permittivity ellipsoids having as axes the principal values of the real and imaginary parts of the average permittivity

$$\langle\underline{\underline{\varepsilon}}\rangle = \langle\underline{\underline{\varepsilon}}_R\rangle + i\langle\underline{\underline{\varepsilon}}_A\rangle = \lim_{V\to\infty} \frac{1}{V}\int_V d^3\underline{r}\left[\underline{\underline{\varepsilon}}_R(\underline{r})+i\underline{\underline{\varepsilon}}_A(\underline{r})\right] \quad . \tag{2.2}$$

A general solution to the scattering problem for materials with overall anisotropy is not yet available. Here, we shall put forward the treatment of the direct problem only for media which may present local anisotropy, but are overall isotropic. For this situation, the principal axes of the medium as a whole may be arbitrarily chosen; at each point of the primary space we describe the structure of the scattering medium by the inhomogeneity distribution tensor

$$\delta\underline{\underline{\varepsilon}}(\underline{r}) = \delta\underline{\underline{\varepsilon}}_R(\underline{r}) + i\delta\underline{\underline{\varepsilon}}_A(\underline{r}) = \underline{\underline{\varepsilon}}_R(\underline{r}) + i\underline{\underline{\varepsilon}}_A(\underline{r}) - \left(\langle\varepsilon_R\rangle + i\langle\varepsilon_A\rangle\right)\underline{\underline{\mathcal{I}}} \quad , \tag{2.3}$$

where $\underline{\underline{\mathscr{I}}}$ is the unit tensor; the components of $\delta\underline{\underline{\varepsilon}}(\underline{r})$ are

$$[\delta\varepsilon(\underline{r})]_{pq} = [\delta\varepsilon(\underline{r})]_{qp} \quad .$$

For the formulation of the direct scattering theory in this contribution, we shall restrict the discussion to quasi-monochromatic fields: the spectral density of the radiation will be assumed to be uniform with respect to the frequency and restricted to a narrow interval, i.e.,

$$\nu_0 - \Delta\nu \leq \nu \leq \nu_0 + \Delta\nu \quad ; \quad \frac{\Delta\nu}{\nu_0} << 1 \quad .$$

This hypothesis enables one to simplify the problem, by accepting both a harmonic description for the temporal dependence and partial coherence, in a spatial context [2.14].

2.2.2 The Scattered Fields

With the assumptions made, Maxwell's equations for the electric and magnetic fields at a generic point belonging to the irradiated volume may be written, for a time dependence effectively given by exp(-iωt), $\omega = 2\pi\nu$, as

$$\nabla \times \underline{E}(\underline{r}) = \frac{i\omega}{c}\,\mu\underline{H}(\underline{r}) \quad , \tag{2.4}$$

$$\nabla \times \underline{H}(\underline{r}) = -\frac{i\omega}{c}\,[\underline{\underline{\varepsilon}}(\underline{r})\underline{E}(\underline{r})] \quad , \tag{2.5}$$

and

$$\nabla \cdot [\underline{\underline{\varepsilon}}(\underline{r})\underline{E}(\underline{r})] = 0 \quad . \tag{2.6}$$

Using the well-known identity (for orthogonal coordinates)

$$\nabla \times \nabla \times \underline{E} = \nabla(\nabla\cdot\underline{E}) - \nabla^2\underline{E} \tag{2.7}$$

one obtains from (2.3-5)

$$\nabla^2\underline{E}(\underline{r}) + \langle k\rangle^2\underline{E}(\underline{r}) = -\frac{\omega^2}{c^2}\,\mu[\delta\underline{\underline{\varepsilon}}(\underline{r})\underline{E}(\underline{r})] + \nabla[\nabla\cdot\underline{E}(\underline{r})] \quad , \tag{2.8}$$

where

$$\langle k\rangle = \langle k_R\rangle + i\langle k_A\rangle = \frac{\omega}{c}\,(\mu\langle\varepsilon\rangle)^{\frac{1}{2}} = \frac{2\pi}{\langle\lambda\rangle} = \frac{2\pi}{\lambda_0}\langle n\rangle \quad , \tag{2.9}$$

is the average wave number in the medium with average refractive index $<n>$. The wavelength of the electromagnetic field in vacuum is λ_0 and $<\lambda>$ is the average wavelength in the medium.

Equation (2.8) is an inhomogeneous vector Helmholtz equation, i.e.,

$$\left[\nabla^2 + <k>^2\right]\underline{E}(\underline{r}) = \underline{\gamma}(\underline{r}) \quad , \tag{2.10}$$

where $<k>^2 = <k_R>^2 + <k_A>^2$ and the source term $\underline{\gamma}(\underline{r})$ is

$$\underline{\gamma}(\underline{r}) = -\frac{\omega^2}{c^2}\mu[\delta\underline{\underline{\varepsilon}}(\underline{r})\underline{E}(\underline{r})] + \nabla[\nabla\cdot\underline{E}(\underline{r})] \quad . \tag{2.11}$$

The direct scattering problem consists of solving this equation, in other words of obtaining an expression for the field $\underline{E}(\underline{L})$ at an observation point determined by the positional vector $\underline{L}$. The observation point is taken to be outside the inhomogeneous irradiated volume V (which is assumed to be surrounded by homogeneous material of refractive index $<n>$). In this case [Ref.2.15, Chap.13]

$$\underline{E}(\underline{L}) = \underline{E}_0(\underline{L}) + \underline{E}_s(\underline{L}) \quad , \tag{2.12}$$

where $\underline{E}_0(\underline{L})$ is the solution of the homogeneous equation associated with (2.10) and represents the incident field. The scattered field $\underline{E}_s(\underline{L})$ is given by

$$\underline{E}_s(\underline{L}) = -\frac{1}{4\pi}\int_V d^3\underline{r}\ \underline{\underline{G}}(\underline{L},\underline{r})\underline{\gamma}(\underline{r}) \quad . \tag{2.13}$$

In writing the solution (2.12) it has been assumed that the volume V is finite and thus the surface integral which is usually added to (2.12) vanishes: the surface of integration can be taken beyond the limits of the irradiated volume V.

$\underline{\underline{G}}(\underline{L},\underline{r})$ is the Green dyadic

$$\underline{\underline{G}}(\underline{L},\underline{r}) = \frac{\exp(i<k>|\underline{L}-\underline{r}|)}{|\underline{L}-\underline{r}|}\underline{\underline{\mathscr{I}}} \quad . \tag{2.14}$$

The solution of (2.10) in the form (2.13) is thus

$$\underline{E}(\underline{L}) = \underline{E}_0(\underline{L}) + \frac{<k>^2}{4\pi<\varepsilon>}\int_V \frac{\exp(i<k>|\underline{L}-\underline{r}|)}{|\underline{L}-\underline{r}|}[\delta\underline{\underline{\varepsilon}}(\underline{r})\underline{E}(\underline{r})]d^3\underline{r}$$

$$-\frac{1}{4\pi}\int_V \frac{\exp(i<k>|\underline{L}-\underline{r}|)}{|\underline{L}-\underline{r}|}\{\nabla[\nabla\cdot\underline{E}(\underline{r})]\}d^3\underline{r} \quad . \tag{2.15}$$

A simple calculation, which is a straightforward generalization of the treatment presented in [2.16] yields

$$\underline{E}(\underline{L}) = \underline{E}_0(\underline{L}) + \frac{<k>^2}{4\pi<\varepsilon>} \int_V \frac{\exp(i<k>|\underline{L}-\underline{r}|)}{|\underline{L}-\underline{r}|} [\delta\underline{\underline{\varepsilon}}(\underline{r})\underline{E}(\underline{r})] d^3\underline{r}$$

$$- \frac{1}{4\pi<\varepsilon>} \nabla^L \int_V [\delta\underline{\underline{\varepsilon}}(\underline{r})\underline{E}(\underline{r})] \nabla \frac{\exp(i<k>|\underline{L}-\underline{r}|)}{|\underline{L}-\underline{r}|} d^3\underline{r} \quad , \tag{2.16}$$

where ∇^L operates on $\underline{L}$.

By making the assumption $|\underline{L}| > |\underline{r}|$, $|\underline{L}-\underline{r}|$ may be expanded in a convergent series

$$|\underline{L}-\underline{r}| = L\left(1 + \frac{r^2-2L(\underline{r}\cdot\underline{m})}{2L^2} - \frac{[r^2-2L(\underline{r}\cdot\underline{m})]^2}{8L^4} + \frac{[r^2-2L(\underline{r}\cdot\underline{m})]^3}{16L^6} - \dots\right) \tag{2.17}$$

where $L = |\underline{L}|$ and $\underline{m} = \underline{L}/L$.

The truncation of this series after the second term, i.e., the neglect of second- and higher-order terms, is known as the Fresnel approximation [2.17]. Although not essential to the treatment put forward in Sects.2.4,5 for obtaining a solution to the inverse problem, the application of this simplifying hypothesis leads to an elegant result: the expression of the scattered electric field in the Fresnel region is

$$\underline{E}_s(\underline{m}) = \frac{<k>^2}{4\pi<\varepsilon>} \frac{\exp(i<k>L)}{L}$$

$$\times \int_V d^3\underline{r} \left(\underline{m}\times\left\{\exp\left(\frac{i<k>r^2}{2L}\right)[\delta\underline{\underline{\varepsilon}}(\underline{r})\underline{E}(\underline{r})]\times\underline{m}\right\}\right)\exp(-i<k>\underline{m}\cdot\underline{r}) \quad . \tag{2.18}$$

The expression $\underline{U}(\underline{r}) = \delta\underline{\underline{\varepsilon}}(\underline{r})\underline{E}(\underline{r})$, known as the *object wave*, represents the field generated at each point of the primary space: it describes the perturbation of the passing wave by an inhomogeneous structure. In other words, the object wave is the encoding of the inhomogeneity distribution function by the propagating field.

The particular design of the experiment, for which the observation is made on a surface of constant L, is sufficiently important to be distinguished by replacing L with d, which describes the constant distance of the surface from the origin of the primary space.

We thus define

$$\underline{u}_d(\underline{r}) = \underline{U}(\underline{r}) \exp\left(\frac{i<k>r^2}{2d}\right) \tag{2.19}$$

as the effective object wave. With this

$$\underline{E}_s(\underline{m}) = \frac{<k>^2}{4\pi<\varepsilon>} \frac{\exp(i<k>d)}{d} \int_V d^3\underline{r}\left\{\underline{m}\times\left[\underline{u}_d(\underline{r})\times\underline{m}\right]\right\}\exp(-i<k>\underline{m}\cdot\underline{r}) \quad . \tag{2.20}$$

Of course, by taking into account higher order terms in the expansion (2.17), a more accurate representation of the scattered field at a finite distance d from the object may be obtained. Expression (2.20) states, essentially, that the scattered field is the Fourier transform of the effective object wave. To emphasize this relationship, we define

$$\hat{\underline{u}}_d(\underline{m}) = \underline{m} \times \int_V d^3\underline{r}\left[\underline{u}_d(\underline{r})\times\underline{m}\right]\exp(-i<k>\underline{m}\cdot\underline{r}) \quad , \qquad (2.21)$$

where the integral is taken over the *finite* irradiated volume V of the scatterer; the importance of this function will become clear in Sect.2.3.

From Maxwell's equations, the scattered magnetic field in the Fresnel region may be easily obtained

$$\underline{H}_s(\underline{m}) = \frac{<k>^2}{4\pi(<\varepsilon>\mu)^{\frac{1}{2}}}\frac{\exp(i<k>d)}{d}\int_V d^3\underline{r}\left[\underline{m}\times\underline{u}_d(\underline{r})\right]\exp(-i<k>\underline{m}\cdot\underline{r}) \qquad (2.22)$$

reflecting the same Fourier relationship.

When L (or d) $\to\infty$, $r^2/2L$ becomes negligibly small in (2.18,19) and the Fraunhofer approximation is obtained

$$\underline{E}_s(\underline{m}) = \frac{<k>^2}{4\pi<\varepsilon>}\frac{\exp(i<k>L)}{L}\int_V d^3\underline{r}\left\{\underline{m}\times[\underline{U}(\underline{r})\times\underline{m}]\right\}\exp(-i<k>\underline{m}\cdot\underline{r}) \qquad (2.23)$$

$$\underline{H}_s(\underline{m}) = \frac{<k>^2}{4\pi(<\varepsilon>\mu)^{\frac{1}{2}}}\frac{\exp(i<k>L)}{L}\int_V d^3\underline{r}[\underline{m}\times\underline{U}(\underline{r})]\exp(-i<k>\underline{m}\cdot\underline{r}) \quad . \qquad (2.24)$$

The Fraunhofer approximation may be applied when the condition

$$\frac{|<k>|r_{max}^2}{2L} << 1 \qquad (2.25)$$

is fulfilled; it is more exact than the Fresnel approximation, since fewer assumptions require justification. Of course, it may be obtained directly from (2.16).

The expressions (2.23,24) or (2.20,22) for the fields may be regarded as integral equations, on the right side of which the field and the perturbation appear together. As will become clear in Sects.2.4,5, for obtaining a solution to the inverse problem, it is necessary to separate their effects and thus express the scattered field in terms of the inhomogeneity distribution tensor and the incident field. To achieve such an aim, an iteration is required: this is the Liouville-Neumann method of successive substitutions.

The first step in this sequence of approximations consists of replacing on the right-hand side of the expressions for the scattered fields $\underline{E}(\underline{r})$ by $\underline{E}_0(\underline{r})$, the incident wave. This is known as the first Born approximation, valid when the medium does not distort the phase of the passing wave and modifies its amplitude only slightly; a fundamental requirement for its applicability is that the imaginary part of $\delta\underline{\underline{\varepsilon}}(\underline{r})$ is negligible. There are no necessary conditions available for the validity of this approximation; sufficient conditions have been put forward by FRISCH [2.18]. Thus, $\underline{U}(\underline{r}) = \delta\underline{\underline{\varepsilon}}(\underline{r})\underline{E}_0(\underline{r})$ in the first Born approximation.

The second Born approximation consists of introducing into the right-hand side of (2.20-24) the expression for the field obtained in the first Born approximation. We are thus faced with an infinite regress; obviously, the task cannot be fulfilled for the general situation of a three-dimensional scatterer, although an adequate representation may be obtained if the Liouville-Neumann series converges sufficiently fast. This is a fundamental problem in scattering theories, epistemologically inevitable. The only situation when this difficulty can be circumvented and the first Born approximation gives an exact result is a one-dimensional primary space, with a judiciously chosen polarisation of the incident field. More exactly, if the direction of propagation of the incident field is perpendicular to the object and its polarisation is parallel to the object, no secondary, and hence higher order waves, may propagate along the object and therefore only single scattering processes may take place.

2.2.3 Expression for the Intensity

A fundamental characteristic of the scattered radiation is the intensity, given by the time-averaged Poynting vector. For quasi-monochromatic fields, this is equal to the real part of the complex Poynting vector

$$\{\underline{S}\} = \frac{c}{8\pi}\,\mathrm{Re}\left\{\underline{E}_s \times \underline{H}_s^*\right\}$$

and hence the intensity in the Fraunhofer space is given by

$$I_s(\underline{m}) = |\{\underline{S}\}| = \frac{<k>^4 c}{128\pi^3 <\varepsilon>^{3/2}\mu^{\frac{1}{2}}}\,\frac{\exp(-2<k_A>L)}{L^2}\,\mathrm{Re}\left\{\int_V d^3\underline{\rho}\,\exp(i<k_R>\underline{m}\cdot\underline{\rho})\right.$$

$$\left.\cdot\int_V d^3\underline{r}\,\exp(<k_A>\underline{m}\cdot\underline{r})\exp(<k_A>\underline{m}\cdot(\underline{r}+\underline{\rho}))\left\{[\underline{U}(\underline{r})\times\underline{m}]\cdot[\underline{m}\times\underline{U}^*(\underline{r}+\underline{\rho})]\right\}\right\} \quad , \qquad (2.26)$$

i.e.,

$$I_s(\underline{m}) = \frac{<k>^4 c}{128\pi^3 <\varepsilon>^{3/2}\mu^{\frac{1}{2}}}\,\frac{\exp(-2<k_A>L)}{L^2}\,\mathrm{Re}\left\{\int_V d^3\underline{\rho}\,\exp(i<k_R>\underline{m}\cdot\underline{\rho})g(\underline{\rho})\right\} \quad , \qquad (2.27)$$

where $\underline{\rho} = \underline{r}_i - \underline{r}_j$, $\underline{r}_i$ and $\underline{r}_j$ denoting the positioned vectors of two generic points in the medium, and

$$g(\underline{\rho}) = \int_V d^3\underline{r}\,\exp(<k_A>\underline{m}\cdot\underline{r})\exp\left[<k_A>\underline{m}\cdot(\underline{r}+\underline{\rho})\right]\left\{\left[\underline{U}(\underline{r})\times\underline{m}\right]\cdot\left[\underline{m}\times\underline{U}^*(\underline{r}+\underline{\rho})\right]\right\} \quad . \qquad (2.28)$$

Thus, we can define the equivalent of (2.21), viz.,

$$\hat{g}(\underline{m}) = \mathrm{Re}\left\{\int_V d^3\underline{\rho}\,\exp(i<k_R>\underline{m}\cdot\underline{\rho})g(\underline{\rho})\right\} \quad . \qquad (2.29)$$

As will become clear in Sect.2.5, in order to retrieve the overall properties of anisotropic media, it is necessary to examine the depolarisation properties of the medium: this means that the scattered intensity must be examined through an analyser (see also Chap.7). It may be shown [2.19] that the expression for the analysed intensity is, apart from irrelevant factors, given by

$$I_A(\underline{s}) \propto \int\limits_V \chi_A(\underline{\rho}) \exp(-i <k> \underline{\rho}\cdot\underline{s}) d^3\underline{\rho} \quad , \tag{2.30}$$

where $\underline{s} = \underline{m}_0 - \underline{m}$, $\underline{m}_0$ being the unit vector describing the propagation direction of the incident field. This expression is valid only in the first Born approximation; in it

$$\begin{aligned} \chi_A(\underline{\rho}) = & \left\langle \left[\delta\varepsilon(\underline{r}_i)\right]_{13}\left[\delta\varepsilon(\underline{r}_j)\right]_{13}\right\rangle \sin^2\Phi_s \sin^2\theta_A \\ & + \left\langle \left[\delta\varepsilon(\underline{r}_i)\right]_{23}\left[\delta\varepsilon(\underline{r}_j)\right]_{23}\right\rangle \cos^2\Phi_s \sin^2\theta_A \\ & + \left\langle \left[\delta\varepsilon(\underline{r}_i)\right]_{33}\left[\delta\varepsilon(\underline{r}_j)\right]_{33}\right\rangle \cos^2\theta_A \quad , \end{aligned} \tag{2.31}$$

where Φ_s is the scattering angle and θ_A is the angle between the direction of polarisation of the incident field and the axis of the analyser; the sharp brackets indicate volume average.

The direct theory presented here, constitutes the foundation on which a solution to the inverse problem will be sought. In its deterministic form, the aim of the inverse scattering problem is to retrieve the inhomogeneity distribution tensor — or function when the medium is isotropic. For the statistical aspect, overall averages of the inhomogeneity distribution tensor — or function — are required.

Neither solution can be directly obtained from (2.27) or (2.30).

For the deterministic problem, the main steps required in order to obtain a solution are

1) obtaining the scattered field from the observable intensity,
2) the construction, for the general case, of a three-dimensional object wave from a two-dimensional measured record,
3) extracting the inhomogeneity distribution tensor from the Liouville-Neumann series.

Posed in this way, the deterministic problem presents too many difficulties to be amenable to a solution, even in the broad sense attached to this word in Sect.2.1. Inroads have been made, however, and the key to obtaining a solution is now available, at least for a one-dimensional scatterer. This will be the only situation investigated in Sect.2.4.

The statistical problem, too, is not directly amenable to a solution. For media that may be described as random, the first task involved consists of obtaining meaningful ensemble averages from finite records. In addition to this, for all media, the following tasks must be fulfilled:

1) to separate the effect of perturbation and field in the convolution square (2.28) appearing in (2.29),
2) to derive an expression for the scattered intensity for media presenting overall anisotropy.

Again, these problems have not been solved for a general situation. In Sect.2.5, an approach will be put forward only for media presenting overall isotropy — i.e., treating effectively a one-dimensional problem — and only for the case when the statistics of the field may be regarded as independent from those of the medium, i.e., for the range of validity of the first Born approximation.

Both the deterministic and statistical aspects require knowledge of the source, i.e., of the distribution of the incident field. Obviously, this is an inverse problem in itself.

Before attempting the solution to the inverse problem, however, the mathematical equipment best suited for this will be presented.

2.3 Analytic Description and Properties of Scattered Fields

We have found a strange foot-print on the shores of the unknown. We have devised profound theories, one after another, to account for its origin. At last, we have succeeded in reconstructing the creature that made the foot-print. And Lo, it is our own!

A.S. Eddington, Space, Time and Gravitation.

The expression (2.21) for the scattered field obtained in Sect.2.2 leads to some very important observations. On the one hand, it provides the theoretical structure on which any attempt to reach a solution to the inverse problem must be based. On the other hand, its specific form directs us to establishing a particular description of the field. This description also offers the mathematical formalism in terms of which it will be possible to construct the solution to the deterministic problem in Sect.2.4 and to the statistical problem in Sect.2.5.

2.3.1 Entire Functions of Exponential Type

Essentially, (2.21) represents a finite three-dimensional Fourier transform: in this section, we shall examine the properties of functions expressed in this way. So far, these properties have been consistently studied only for a one-dimensional situation. In several dimensions, the main features of the one-dimensional model remain valid but, of course, the mathematical equipment becomes more involved.

Thus, in one dimension, the finite Fourier transform is written

$$\hat{u}_d(x) = \int_a^b u_d(r)\ \exp(-i\langle k\rangle rx)dr \quad , \tag{2.32}$$

where $x = \underline{r}\cdot\underline{m}/r$ and

$$u_d(r) = U(r)\ \exp\left(\frac{i\langle k\rangle r^2}{2d}\right) \tag{2.33}$$

is the effective object wave, given by the object wave $U(r)$ multiplied by a quadratic phase factor, introduced by applying the Fresnel approximation. As $d\to\infty$, $u_d(r)\to U(r)$, i.e., in the Fraunhofer space the effective object wave becomes the object wave itself and $\hat{u}_d(x)\to\hat{u}(x)$. In order to avoid the cumbersome obliquity factor suggested by (2.21) we have assumed, in writing (2.32) that, for the one-dimensional situation discussed, the direction of propagation of the incident field is perpendicular to the object and the direction of polarisation of the incident field is perpendicular to both. The scattered field is considered only in the plane determined by the direction of propagation of the incident field and the object and thus, in the treatment we adopt, the fields will be written as scalars.

The interval (a,b) represents the aperture of the system, i.e., the width of the object or that of the beam, whichever is smaller. It introduces in the space of observation, x, the *a priori* epistemological knowledge; it describes the methodological constraints which will be reflected in the general properties and aspect of the function $\hat{u}_d(x)$ and revealed in any *a posteriori* observation.

The form of (2.32) imposes a certain characteristic behaviour on $\hat{u}_d(x)$, and its consequences can be fully appreciated only if one makes appeal to the powerful concept of analyticity. For this, one must consider $\hat{u}_d(z)$ in the complex plane $z = x + iy$, rather than its restriction to the real axis, as in (2.32).

The effect of a complex $<k>$ would be to rotate the axes of the complex plane z thus defined by an angle equal to $\tan^{-1}(<k_A>/<k_R>)$. In order to avoid this complication, we shall consider $<k>$ real; the extension to complex values should not present any difficulties.

An important theorem, formulated originally by PALEY and WIENER [Ref.2.20, Theorem X], describes the analytic properties of $\hat{u}_d(z)$. We quote here the version put forward by BOAS [Ref.2.21 §6.8]:

Theorem 1: $\hat{u}_d(z)$ is an entire function of exponential type b and its restriction on the real axis belongs to $L^2(-\infty,\infty)$ if and only if it is given by

$$\hat{u}_d(z) = \int_a^b u_d(r)\exp(-i<k>zr)dr \quad , \tag{2.34}$$

where $u_d(r) \in L^2(a,b)$ and $\infty > b \geqq |a| \geqq 0$.

A particular situation is sufficiently important to justify a separate theorem [Ref.2.22 III.1]:

Theorem 2: $\hat{u}_d(z)$ is an entire function of exponential type b and belongs to $L^1(-\infty,\infty)$ on the real axis if and only if

$$\hat{u}_d(z) = \int_a^b u_d(r)\exp(-i<k>zr)dr \quad , \tag{2.35}$$

where $u_d(a) = u_d(b) = 0$, $\infty > b \geqq |a| \geqq 0$, and the function obtained by extending $u_d(r)$ to be 0 outside (a,b) has an absolutely convergent Fourier series on any larger interval $(a-\varepsilon, b+\varepsilon)$, $\varepsilon > 0$. Since $\hat{u}_d(x) \in L^1(-\infty,\infty)$, it follows that $u_d(r)$ is continuous.

An entire function is a function that is analytic everywhere in the finite complex plane. The growth properties of entire functions are described by the order τ and type σ of the function

$$\tau = \limsup_{|z|\to\infty} (\ln\ln[\max|\hat{u}_d(|z|)|]/\ln|z|) \tag{2.36}$$

and

$$\sigma = \limsup_{|z|\to\infty} (\ln[\max|\hat{u}_d(|z|)|]/|z|^\tau) \quad , \tag{2.37}$$

where the supremum means the least upper bound.

The rate of growth of the function in the complex plane, as $|z|\to\infty$, is such that

$$|\hat{u}_d(z)| \leqq \exp(\sigma|z|^\tau) \quad . \tag{2.38}$$

Functions of exponential type b have unit order and finite type b: their growth is limited by that of a simple exponential – hence the name.

In general, the growth of $|\hat{u}_d(z)|$ is anisotropic. A descriptor of this direction-dependent growth is the Phragmén-Lindelöf (or indicator) function, $h(\theta)$. For a function of order 1,

$$h(\theta) = \limsup_{|z|\to\infty} \frac{\ln|\hat{u}_d(z)|}{|z|} \quad , \tag{2.39}$$

where $z = |z| \exp(i\theta)$.

The indicator function may be identified with a supporting function, i.e., a function defining the domain in which $\hat{u}_d(z)$ does not vanish [2.23]. As such, its value for any direction θ may be found from the indicator diagram, which is a non-empty, bounded, closed set of points, the projection of which onto the direction $\arg z = \theta$ is the support function.

For entire functions of exponential type b, the indicator diagram is the line segment (-ib,-ia) [2.23] and

$$h(\theta) = \frac{b-a}{2} |\sin\theta| \quad .$$

Thus, the maximum value of $h(\theta)$ lies along the imaginary axis and the minimum rate of growth occurs in the direction of the real axis.

Therefore, entire functions of exponential type b, which from now on will be referred to as functions of class E, are defined by the following three conditions: the class E consists of all functions $\hat{u}_d(z)$ that are entire, for which, along the real axis,

$$\int_{-\infty}^{\infty} |\hat{u}_d(x)|^2 dx < \infty$$

and for which there exist constants b and K such that

$$|\hat{u}_d(z)| \leqq K \exp(b|z|) \quad ,$$

i.e.,

$$|\hat{u}_d(z)| = O[\exp(b|z|)] \quad .$$

From the definition of $h(\theta)$, it follows that along the real axis $h(0) = h(\pi) = 0$ and hence $\hat{u}_d(x)$ has a rate of growth lower than that of an exponential function. Thus, along the real axis,

$$|\hat{u}_d(x)| < K \exp(b|x|) \quad , \quad \text{i.e.,} \quad |\hat{u}_d(x)| = o[\exp(b|x|)] \quad . \tag{2.40}$$

As, on the real axis, $\hat{u}_d(x) \in L^2(-\infty,\infty)$, the growth properties describe a decrease towards zero as $|x| \to \infty$; expressions (2.40) state that the decrease of $|\hat{u}_d(x)|$ as $|x| \to \infty$ must be slower than that of a negative exponential function.

Therefore, the statement

$$|\hat{u}_d(x)| = O[\exp(-b|x|)]$$

not being sufficiently precise, we seek to establish the conditions required for

$$|\hat{u}_d(x)| = O\{\exp[-\beta(x)]\}$$

to describe the growth rate of our function of class E along the real axis. These conditions have been established by the Ingham-Levinson theorem. We give it here in the form put forward by KAWATA [Ref.2.24 §8.3]:

Theorem 3: I) Suppose that $\beta(x)/x$ is a nonincreasing function of $x > 1$ such that $\beta(x)/x \to 0$ as $x \to \infty$ and

$$\int_1^\infty \frac{\beta(x)}{x^2}\, dx < \infty \quad . \tag{2.41}$$

Then there exists a function $u_d(r)$ which vanishes outside a certain interval (a,b), and for which its Fourier transform $\hat{u}_d(x)$ satisfies

$$\hat{u}_d(x) = O\{\exp[-\beta(x)]\} \tag{2.42}$$

for large $|x|$.

II) Let $\beta(x)$ be a nonnegative function for $x > 1$. If there is a function $u_d(r)$ which is not identically zero and which vanishes outside a certain interval (a,b), and for which its Fourier transform $\hat{u}_d(x)$ satisfies (2.42), then (2.41) holds.

For example, a possible expression for $\beta(x)$ is

$$\beta(x) = x^\alpha$$

with $0 \leq \alpha < 1$. The importance of this particular form will become clear in Sect.2.5.

2.3.2 Distributions of Zeros for Functions of Class E

It is well known that entire functions of exponential type are the simplest generalization of polynomials. Like ordinary algebraic polynomials, functions of class E may be described everywhere by their roots or zeros $z_{j,d}$, to within a linear phase factor [2.25], i.e.,

$$\hat{u}_d(z) = z^{q_d} \exp[i(a+b)z]B_d \prod_{j=-\infty}^{\infty}\left(1 - \frac{z}{z_{j,d}}\right) \quad , \tag{2.43}$$

where B_d is a scaling constant and q_d the order of a zero at the z origin. When $a = -b$ and there is no zero at the z origin

$$\hat{u}_d(z) = B_d \prod_{j=-\infty}^{\infty}\left(1 - \frac{z}{z_{j,d}}\right) \tag{2.44}$$

the expression being known as a Hadamard product. If q identical factors appear in (2.44), the corresponding zero is said to be of order q. Since $\hat{u}_d(z)$ is analytic, its roots must be isolated zeros; otherwise the function would be zero everywhere.

The indicator diagram provides information on the density of zeros in any angular interval in the complex plane. The density of zeros is defined by

$$D = \lim_{|z|\to\infty} \frac{N(|z|)}{|z|} \quad ,$$

where $N(|z|)$ is the number of zeros $z_{j,d}$ for which $|z_{j,d}| \leq |z|$. It is well known [2.21] that functions having a line segment indicator diagram have a finite nonvanishing density of zeros only along a line parallel to, or coincident with, the real axis.

To be more precise, all the roots of $\hat{u}_d(z)$, except possibly for a set of zero density, lie in arbitrarily small angles containing the normals to the sides of the indicator diagram. The density of the set of roots inside each of these angles is given by the length of the indicator diagram. Thus, for $\hat{u}_d(z)$ defined by (2.34), the density of zeros in the complex plane is asymptotically proportional to (b-a).

In any physical situation, the density of zeros at some finite $|z|$ will be approximately given by this value. Thus, the zeros of $\hat{u}_d(z)$ lie 'near' the real axis and tend to be distributed 'regularly' along the real axis: $\hat{u}_d(z)$ is a function of class A or of completely regular growth [2.23].

The behaviour of $\hat{u}_d(z)$ is therefore uniquely described by the zero distribution of the function of class E which may be defined in the complex plane associated with each distance d. From (2.33) it follows that, for any d, the location of the complex zeros determines uniquely the conjugate function U(r) in the primary space. It is important to understand how this takes place, in other words to establish the precise way in which the information on the object wave in the primary space is encoded by the zeros of the associated function of class E describing the scattered field at each d.

For simplicity and convenience, we shall consider only the particular situation for which $d \to \infty$, i.e., we shall examine the encoding of information by zeros in the Fraunhofer space. It is a straightforward matter to apply our treatment to any finite distance d in the Fresnel region.

On the level of abstraction, we may regard the *a priori* epistemological knowledge representing the aperture of the system as defined by two Dirac delta functions $(a/|a|)\delta(r-a)$ and $(b/|b|)\delta(r-b)$. The Fourier transform of these 'aperture edge' markers is a sine function, having all zeros real and equidistant. Depending on the relative values of a and b, a unimodular linear phase factor multiplies the Hadamard product, as expressed by (2.43); this is inessential since, as will be seen later, it is not observable. The unimodular linear phase factor becomes real when $a = -b$.

This particular distribution of zeros will be referred to as the fundamental lattice of zeros. It consists of an equidistant array of real zeros, with unit cell $x_j - x_{j-1}$ given by $1/(b-a)$, i.e., in inverse proportion to the zero density. Obviously, this definition of the fundamental lattice is nonphysical and the model assumed does not satisfy the requirements for the applicability of Theorem 1: the two Dirac edge markers are not $L^2(a,b)$ and, of course, the sine function in the conjugate space is not $L^2(-\infty,\infty)$.

Any information within the aperture (a,b) is encoded by the movement of at least one zero from its fundamental lattice position.

Whilst the zeros in the complex plane associated with the Fraunhofer space encode information about U(r) defined in the interval (a,b), their presence is a direct consequence of the existence of the finite aperture (a,b). Thus, whilst the detailed information in (a,b) is encoded by the specific location of zeros, the general pattern in which this information is recorded is that of a 'regular' distribution of zeros, which is a perturbation of the fundamental lattice: this is the *a priori* epistemological knowledge associated with the scattering phenomenon.

The perturbation of the fundamental lattice is subject to severe constraints. The distribution of zeros is always such that they asymptotically become equidistant on lines parallel to, or on, the real axis. Movement of zeros too far from the lattice positions results in an apparent change of support, i.e., of the interval (a,b) in the primary space. Also, inserting additional zeros in $\hat{u}(z)$ can lead to the function collapsing everywhere to zero (corollary to Carlson's theorem [2.21]).

Thus, all information about the object wave U(r) resides in the displacement of a certain set of zeros of $\hat{u}(z)$ from the fundamental lattice locations. As an illustration, Fig.2.1 shows the zero distribution of $\hat{u}(z)$, for $U(r) = \exp(-r^2)$, $a = -b$.

For certain classes of U(r), particular symmetries are displayed by the distribution of zeros of $\hat{u}(z)$. This results in specific characteristics being exhibited by $\hat{u}(x)$. Obviously, similar features of $u_d(r)$ lead to the same zero symmetries in the complex plane associated with the Fresnel space d (i.e., considered at the finite distance d in the Fresnel region), where $\hat{u}_d(z)$ is defined.

Three important situations can be distinguished.

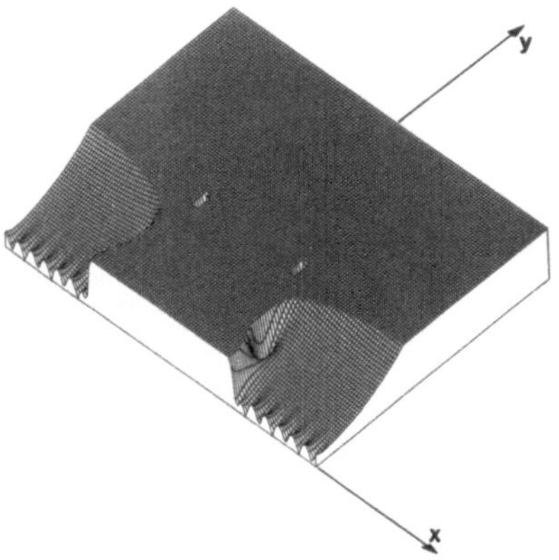

Fig. 2.1. The distribution of zeros in the complex plane of the finite Fourier transform of a Gaussian. The amplitude of the Fourier transform represented has been truncated to facilitate the observation of the zeros

I) A Hermitian $U(r)$, i.e., $U(r) = U^*(-r)$, is associated with a $\hat{u}(z)$ having zeros symmetrically distributed about the real x axis: the function $\hat{u}(x)$ takes only real values. At any real zero of odd order, $\hat{u}(x)$ changes sign. If every real zero is of even order, or if there are no real zeros, $\hat{u}(x)$ has everywhere nonnegative values, being zero only at the even-order real zeros.

II) A real $U(r)$ results in a symmetry of the zeros of $\hat{u}(z)$ about the y axis, $\hat{u}(x)$ being Hermitian. If, in addition to this, $U(r)$ is nonnegative, no zero occurs at any finite value of y along the imaginary axis; the reciprocal is not necessarily true.

III) A real and even $U(r)$ corresponds to zeros of $\hat{u}(z)$ symmetric about both the x and y axes, and hence a real and even $\hat{u}(x)$.

Under certain circumstances, $\hat{u}(z)$ has only real zeros. The necessary conditions which the corresponding $U(r)$ must fulfil are not known, although sufficient conditions have been put forward [Ref.2.26, Appendix]. Apart from trivial cases, namely $U(r)$ being simple and monotonic in behaviour, the only interesting situation [2.26] occurs when $U(r)$ and hence $\hat{u}(x)$ are prolate spheroidal functions. Some examples of functions having randomly distributed real zeros are discussed by VOELCKER and REQUICHA [2.27].

2.3.3 Encoding of Information by Zeros

Let us examine now the specific way in which a zero of the function of class E in the Fraunhofer space codes information in the conjugate primary space. For this, it is convenient to express the distances in z space in terms of the separation between any two successive zeros of the fundamental lattice. We refer to this interval as one unit. To make things as simple as possible, we will discuss only the coding of real-valued information in the primary space. This implies, as seen above, that the distribution of zeros is symmetric about the y axis. The extension to the encoding of complex information is immediate. For convenience, we shall also assume that $a = -b$, thus rendering unnecessary to take into account the effect of the unimodular phase factor multiplying the Hadamard product.

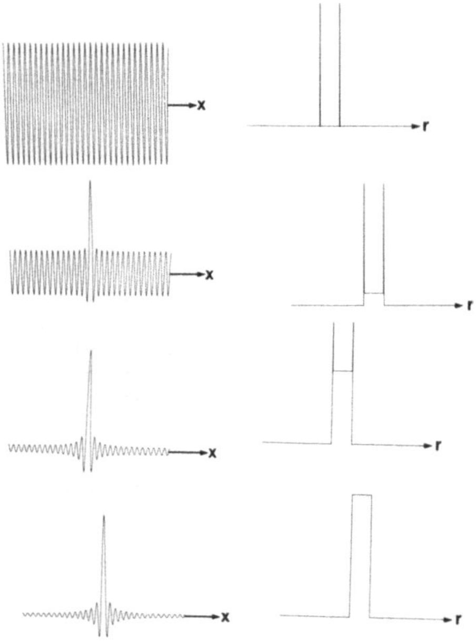

Fig. 2.2. The effect on the real x axis of moving the zero from the origin of the fundamental lattice along the y axis is shown on the left. The modulus of the Fourier transform of the function thus defined is shown alongside, on the right. The transition from sin(x) to sinc(x) is thus illustrated

Starting with the fundamental zero lattice, which encodes the aperture width in the r space, we study the consequences of displacing zeros from their equidistant $x_{|j|}$ locations. Let us first investigate the effects of moving the zero at the origin along the imaginary axis; this is illustrated in Fig.2.2.

As can be seen, this displacement induces a constant or dc level, i.e, a plane wave, within (a,b). On the x axis, a nonzero minimum is introduced in $|\hat{u}(x)|$ at the origin. When the zero is moved to infinity, the situation becomes physical; the simplest possible signal is introduced within the aperture. The Dirac delta functions locating the aperture edges have now disappeared. The aperture becomes encoded not by nonphysical markers, but by the signal itself, confined to the interval (a,b).

Any dc level introduced in (a,b) renders the edge markers immediately redundant: it automatically defines the aperture in the primary space and translates the zero at the origin along the y axis to infinity. The distribution of zeros thus obtained, corresponds, as can be easily seen from the Hadamard product, to sinc(z) in the Fraunhofer space. This is associated with the rectangular function in the primary space. Such a distribution of zeros will be known as the sinc lattice and is the simplest physically realisable: in this case, the function defined in the object space is $L^2(a,b)$ and that obtained in the conjugate space is $L^2(-\infty,\infty)$: Theorem I may be applied. This corresponds, therefore, to the simplest instance in which a function of class E may be used to describe the scattered field. We shall consider the sinc lattice as the basic zero distribution, since this contains the lowest level of information in the object space. We may now regard this as the *a priori* epistemological knowledge in the scattering experiment.

Maintaining the symmetry about the y axis, let us now examine the effect of moving a pair of zeros from $x_{|j|}$ on the sinc lattice. Calculating the new Hadamard product, and performing the required inverse Fourier transform, it is easy to see that the information thus introduced in the interval (a,b) always consists of a harmonic having j complete cycles added to the pre-existing dc level.

In order to define this harmonic, it is convenient to imagine the complex plane displaying, for each $|j|$, a network consisting of two sets of curves: contours of constant phase at any given point in (a,b), chosen, for example, at $r=a$, and contours of constant maximum amplitude. The position of a zero, therefore, specifies the intersection of a contour of equal phase and a contour of equal maximum amplitude. Any position for a zero is thus coding an amplitude and phase at $r=a$ for the associated j^{th} harmonic defined in (a,b).

The contours of constant amplitude will not be discussed here. For the purpose of this contribution it is sufficient to say that the amplitude depends upon the distance of the zero from its lattice position. A simple calculation shows that the family of curves of constant phase is the set of arcs of circles with centres on the y axis, and which pass through the points $x_{|j|}$, of equation

$$x^2_{|j|} + y^2_{|j|} - 2|j| \cot\Phi \, y_{|j|} - |j|^2 = 0 \quad .$$

Each arc joining the points $x_{\pm j}$ encodes a cosine with phase Φ at $r=a$.

For example, by moving the pair of symmetric zeros on the arcs of the circles of infinite radius between the two lattice positions, i.e., on the segment of real axis $|x| < |x_{|j|}|$, an even harmonic is encoded, with $\Phi = 0$. If $|x| > |x_{|j|}|$, the even harmonic has the phase $\Phi = \pm\pi$.

An odd harmonic, i.e., a pure sine, in (a,b) is encoded by moving the zeros on the circle

$$x^2_{|j|} + y^2_{|j|} = j^2 \quad .$$

If the zeros are on the arc in the upper half plane the phase $\Phi = -\pi/2$ and if in the lower half plane $\Phi = \pi/2$.

In general, the phase Φ varies from 0 to $-\pi$ for $y_{|j|} \geqq 0$ and from 0 to $+\pi$ for $y_{|j|} \leqq 0$. The phase of the harmonic changes from Φ to $-\Phi$ when the pair of zeros is moved to the complex conjugate position.

This encoding of the information in the r space is shown in Figs.2.3 and 2.4, which illustrate the harmonics introduced by the zero displacements indicated.

In general, a translation of the jth zero pair from the lattice position to a new location in the complex plane, generates in (a,b) a harmonic having j complete cycles, but not necessarily odd or even. The phase of the harmonic depends on the ratio of the real and imaginary parts of the displacement.

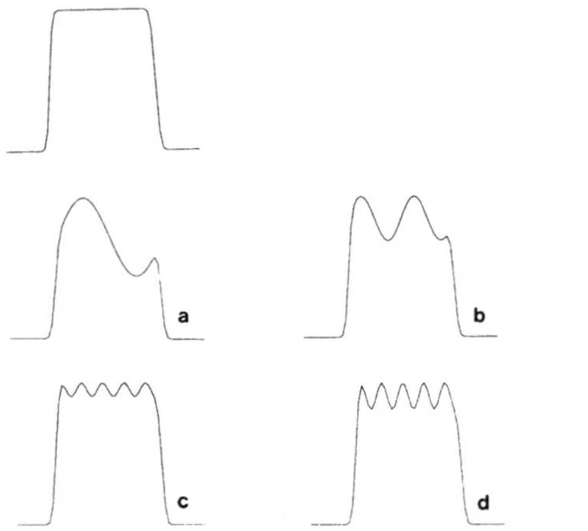

Fig. 2.3a-d. The change in the information encoded in the r space, from the original rect, as the zero pairs are moved from their position on the sinc lattice x_j on the circle $x_j^2 + y_j^2 = j^2$ for (a) $j = 1$, $\tan^{-1}(y_j/x_j) = \psi_0$; (b) same as a), but $j = 2$; (c) same as a), but $j = 5$; (d) $j = 5$, $\tan^{-1}(y_j/x_j) = 2\psi_0$

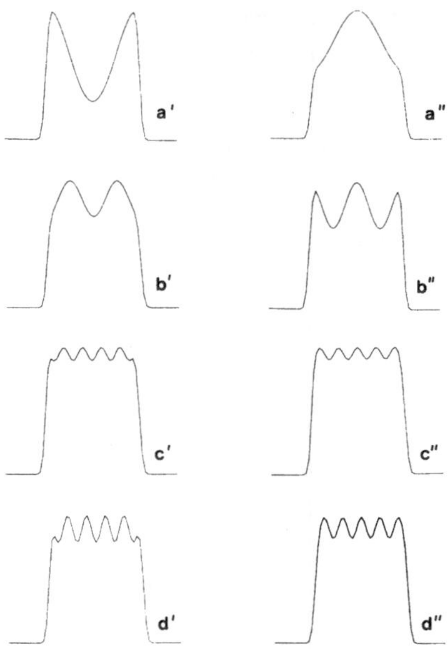

Fig. 2.4a-d. The change in the information encoded in the r space, from the original rect, as zero pairs are moved from their position on the sinc lattice x_j to $x_j + \Delta x_j$ for (a') $j=1$, $\Delta x_j=\delta<0$; (a") $j=1$, $\Delta x_j=\delta>0$; (b') $j=2$, $\Delta x_j=\delta<0$; (b") $j=2$, $\Delta x_j=\delta>0$; (c') $j=5$, $\Delta x_j=\delta<0$; (c") $j=5$, $\Delta x_j=\delta>0$; (d') $j=5$, $\Delta x_j=2\delta<0$; (d") $j=5$, $\Delta x_j=2\delta>0$

If the displacement of a zero, from the fundamental lattice position, away from $z = 0$, exceeds a few units, the rate of increase of the amplitude of the associated harmonic decreases. This can be seen from the behaviour of the corresponding factor in the Hadamard product: if a zero is moved from its lattice position along these directions, the information in the r space becomes essentially invariant for a displacement beyond a certain distance. This defines an area about the lattice position, within which the effect of each zero is concentrated.

For various applications, it is useful to manipulate the zeros by modifying the signal in the primary space. One such instance will be given in Sect.2.5, in which it will be shown that a suitable modification of the zero configuration is essential for constructing a solution to the statistical problem. The most common manipulation consists in the multiplication of the object wave by a function $w(r)$ having an envelope which decreases in amplitude towards the ends of the interval (a,b), thus affecting selectively the harmonics of which $u_d(r)$ is constituted, the relative amplitude of the lower harmonics being increased with respect to that of the higher harmonics. The convolution of $\hat{u}_d(x)$ with $\hat{w}(x)$, the Fourier transform of $w(r)$, has the effect of moving the zeros of $\hat{u}_d(z)$ away from the real axis, this displacement

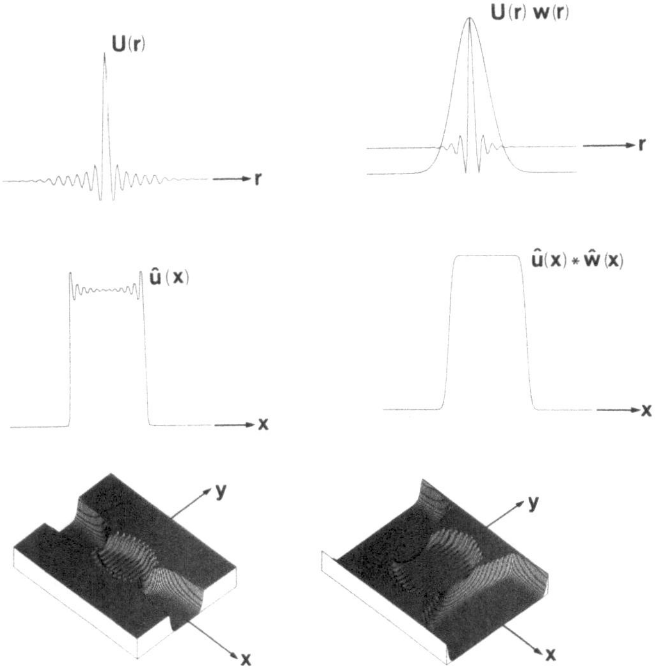

Fig. 2.5. The modification of U(r) by a Gaussian envelope, and the resultant smoothing of $\hat{u}(x)$; the corresponding zero distribution of $\hat{u}(z)$ is also shown

being larger for small j, because of the weighting of the lower order harmonics. The consequence of this on the x axis is a smoothing process. For the particular case in which the multiplying function goes to zero at the ends of the interval (a,b), Theorem 2 is satisfied, the function defined in the z space $\in L^1(-\infty,\infty)$ and thus converges as a simple limit towards the infinite Fourier transform, the requirement $w(a)=w(b)=0$ being maintained while $a\to-\infty$, $b\to\infty$. The density of zeros increases as this limit is taken; hence they cannot stay in the vicinity of the real axis, since then the convergence would be possible only as a limit in the mean and not as a simple limit. The smoothing effect of the modifying function w(r), showing the effect of displacing the zeros away from the real axis, is illustrated in Fig.2.5.

It is of fundamental importance to extend the mathematical equipment presented here to two and three dimensions. An n-dimensional Fourier transform of a function which is confined to a bounded domain remains an entire function of exponential type in each of its variables. Functions of several complex variables no longer have zeros that are isolated points: they display 'lines' of zeros. To be more precise, the function will be zero on a (n-1)-dimensional line. An analogue to the Hadamard product may be written [2.28,29] but it is far more complicated to interpret.

Of course, the description of the encoding of information in the r space by the complex zeros of a function of class E applies to all complex planes associated with values of d for which the Fresnel approximation is valid. In each complex plane, a suitably perturbed lattice of complex zeros encodes all the information about the scattered field at the distance d, and hence about the object wave U(r). The density of zeros is invariant with respect to d, since it depends only on the width of the interval (a,b). The information can be seen to propagate, as d increases, along lines of zeros in a four-dimensional complex space $[x,y,\hat{u}_d(z),d]$.

The Fresnel approximation, defined by the truncation of the series (2.17) after the second term, is not sufficiently rigorous. Indeed, in the term of second order which was neglected, there is a contribution of the same order as $r^2/2L$ which was introduced, namely $x^2r^2/2L$. By taking this into account, the quadratic phase factor in (2.19) should be written $\exp[i<k>r^2(1-x^2)/2L]$. This is known as the wide-angle Fresnel approximation [2.30]: it is easy to see that the Fresnel approximation as introduced in Sect.2.2 implies a paraxial situation. The application of the wide-angle Fresnel approximation does not alter any of the conclusions reached in this section, except that its effect consists of a curvature of the complex plane. The complex plane becomes a complex surface; the more terms taken in the binomial expansion (2.17), the more pronounced the curvature of the complex surface. Therefore, it appears that the functions of class E used for describing the scattered field, are defined in a complex plane in the Fraunhofer space and in complex surfaces of increasing curvature as we approach the primary space. Hence, the propagation of information takes place along lines of zeros defined in a four-dimensional *curved* complex space. The scatterer disturbs the geometry of the complex space in which the propagation of information is described, this distortion being more severe nearer the object. We conjecture that the zeros of the sinc lattice propagate along the geodesics of this space.

This model offers a physical interpretation to GABOR's treatment [2.31] of the transmission of information. The denumerable infinity of isolated zeros may be understood from Gabor's quantal point of view: each zero can be associated with a degree of freedom or logon [2.32], each zero codes one harmonic in the r space and thus labels a 'bound state' of U(r). The effective disc of influence of a zero in the complex plane may be related to Gabor's idea of information cell. The propagation of information via these cells at each d may be associated with the concept of tubes of information [2.33].

A function of class E may be described not only by its zeros, but also in terms of its values taken at any denumerable infinity of points in the complex plane. A particular case is when the points are taken on the real axis, where they may or may not be equidistant. The reconstruction of $\hat{u}(x)$ in terms of its values on the real axis was put forward by SHANNON [2.34]. In fact, Shannon's interpolation formula may be easily derived from the Hadamard product [2.35]. The description in terms of complex zeros is conceptually more general and is indispensible for tackling various problems. Of greatest importance, as will be seen in Sects.2.4,5, is the insight offered for the solution of inverse scattering problems.

The model put forward in this section is based on the concept of *a priori* knowledge of the fundamental category in scattering situations: the finite interval (a,b) in the primary space. This leads to the transcendental knowledge that a 'regular' distribution of zeros is defined in a Fourier conjugate space: the exact location of the zeros is the *a posteriori* factual knowledge. We thus describe the encoding of information in the object space in terms of concepts we ourselves create, namely the complex zeros of a function of class E. The properties of the distributions of zeros will be seen to provide the key to constructing a solution to the inverse scattering problem, in both its deterministic and statistical aspects.

2.4 The Deterministic Problem

The gods did not reveal, from the beginning,
All things to us; but in the course of time,
Through seeking we may learn, and know things better.

But as for certain truth, no man has known it,
Nor will he know it; neither of the gods
Nor yet of all the things of which I speak.
An even if by chance he were to utter
The final truth, he would himself not know it;
For all is but a woven web of guesses.

Xenophanes, DK B 18 and 34
(Translated by K. Popper)

2.4.1 Limitations of Measurements

If in any experimental situation $\hat{u}_d(x)$ can be measured for a known d, (2.32) suggests that the object wave U(r) may be retrieved by a simple inverse Fourier transform

$$U(r) = \exp\left(-\frac{i\langle k\rangle r^2}{2d}\right)\int_{-\infty}^{\infty} \hat{u}_d(x)\exp(i\langle k\rangle rx)dx \quad . \tag{2.45}$$

Unfortunately, things are not as simple as that. First, the integral in (2.45) can be performed only over $x \in (-X,X)$: the interval of measurement is always finite, the maximum value of X being 1. Given $\hat{u}_d(x) = 0$, $|x| > X$, it follows from Theorem 1 that the object wave can only be reconstructed as a function of class E. The larger X, the closer an approximant to the object wave the determined function will be. More exactly, the quality of our encoding of the object wave depends on the number of harmonics from which it is comprised. This is equal to the number of zeros, for a given (a,b), in the interval of measurement, namely $2X(b-a)\langle k\rangle$.

Secondly, and more important, $\hat{u}_d(x)$ is not an observable magnitude at high (e.g., optical) frequencies. For this region of the electromagnetic spectrum, the field characteristic that can be recorded is the flux of energy density, given by the modulus of the Poynting vector, i.e., the intensity. Essentially, and ignoring factors irrelevant for the present discussion,

$$\hat{g}_d(x) = \hat{u}_d(x)\hat{u}_d^*(x) = \int_{-R}^{R} g_d(\rho)\exp(i\langle k\rangle \rho x)d\rho \quad , \tag{2.46}$$

where $R = b - a$ and

$$g_d(\rho) = \int_0^{R-|\rho|} u_d(r)u_d^*(r+\rho)dr \tag{2.47}$$

is the convolution square of the effective object wave.

Theorem 1 renders possible the analytic continuation of $\hat{g}_d(x)$ into the complex plane

$$\hat{g}_d(z) = \int_{-R}^{R} g_d(\rho) \exp(i<k>\rho z)dz \tag{2.48}$$

and establishes that $\hat{g}_d(z) \in E(-\infty,\infty)$. The integrability properties of $\hat{u}_d(x)$ and $\hat{g}_d(x)$ may be different. The intensity is always $\in L^1(-\infty,\infty)$, as may be easily seen from Hölder's inequality [Ref.2.24, §1.10]: if $pq \geqq 0$, $1/p + 1/q = 1$, $f(x) \in L^p(-\infty,\infty)$ and $g(x) \in L^q(-\infty,\infty)$, then $f(x)g(x) \in L^1(-\infty,\infty)$. It is, in fact, obvious that $g_d(-R) = g_d(R) = 0$ and hence Theorem 2 could be invoked, leading to the same conclusion, namely that $\hat{g}_d(x) \in L^1(-\infty,\infty)$.

Clearly, the inverse Fourier transform of $\hat{g}_d(x)$ provides only information about $g_d(\rho)$, from which alone the object wave U(r) cannot be retrieved since, in general, the integral equation (2.47) is not soluble. The intensity is related only to $|\hat{u}_d(x)|$, the phase information is not directly available. This is the phase problem: for the one-dimensional situation considered here and with a judiciously chosen polarisation of the incident field, it is the only fundamental difficulty presented by the deterministic aspect of the inverse scattering problem.

2.4.2 The Phase Problem

Being a function of class E, $\hat{g}_d(z)$ is also characterized everywhere by its regular distribution of zeros. As can be seen from (2.46), $\hat{g}_d(z)$ has the zeros of both $\hat{u}_d(z)$ and $\hat{u}_d^*(z^*)$. From the Hadamard product (2.44) it follows, since

$$\left(1 - \frac{z^*}{z_{j,d}}\right)^* = \left(1 - \frac{z}{z_{j,d}^*}\right)$$

that the zeros of $\hat{u}_d^*(z^*)$ are in complex conjugate position to those of $\hat{u}_d(z)$. Hence, the zeros of $\hat{g}_d(z)$ are always symmetric about the x axis: this conclusion is also consistent with $g_d(\rho)$ being at least Hermitian, if not only real and even. A Hadamard product may be written for $\hat{g}_d(z)$, containing the factors corresponding to both $\hat{u}_d(z)$ and $\hat{u}_d^*(z^*)$; obviously, the unimodular phase factor applied to it is real.

On the real axis, the intensity will always display maxima and minima. The minima occur at, or near, the x_j coordinate of each pair of complex zeros at z_j. Between them, maxima in intensity are observed. Both maxima and minima are only apparent, since, from the theorem of maximum modulus [2.36], no maxima are permitted in a region of analyticity, and all minima must be zero.

This highly oscillatory behaviour of the intensity, with an average frequency given by b-a, is the well-known speckle effect [2.37]. However much the object waves U(r) may differ from each other, the scattered intensity $\hat{g}_d(x)$ will always display the same general pattern: an oscillatory behaviour, with the frequency imposed by the pre-established aperture. This characteristic is the *a priori* epistemological knowledge which is manifest in every observation, determined by the fundamental categories; the factual, *a posteriori* knowledge is the detailed shape of $\hat{g}_d(x)$.

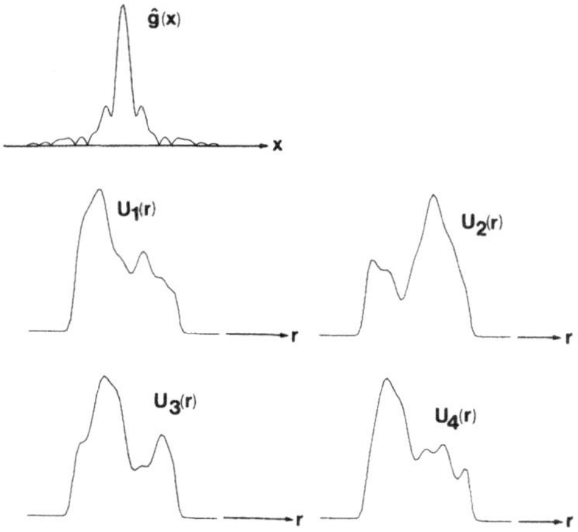

Fig. 2.6. An intensity profile corresponding to a function $\hat{u}(z)$ having three pairs of complex zeros symmetric about the y axis and the four distinct associated real U(r)'s, the other four real ones being the mirror image of those displayed

Since $\hat{g}_d(z)$ has its roots symmetric about the real axis, $\hat{g}_d(x)$ is invariant with respect to a zero in $\hat{u}_d(z)$ being flipped from $z_{j,d}$ to $z^*_{j,d}$. Hence, if N complex zeros are present in the interval of measurement, 2^N different functions $\hat{u}_d(z)$ can be generated, all compatible with the same observed $\hat{g}_d(x)$. An illustration is given in Fig.2.6. Of course, any real zeros are seen as zeros on the x axis and hence their location is immediate. Since real zeros coincide with their complex conjugate, they are not a source of ambiguity in the deterministic problem which, if all zeros of $\hat{g}_d(z)$ are real, can be solved without difficulty.

The first modalities for obtaining the phase information associated with the observable $|\hat{u}_d(x)|$ have been sought by using the analytic properties of $\hat{u}_d(z)$ for a particular, and in fact nonphysical, situation.

It has been known for a long time (e.g. [2.38]) that, if the lower limit of integration, a, in (2.34) is nonnegative, then, even if $b \to \infty$, $\hat{u}_d(z)$ is regular in the upper half of the complex plane. The restriction imposed on the lower limit a has a special significance. In temporal processes, the requirement $a \geqq 0$ is known as the 'causality condition'. This terminology is also used in spatial problems, although the meaning of the expression is clearly lost. With this constraint on a, $\hat{u}_d(x)$ is said to be a 'causal transform' [2.39]. The causality condition is determined by the fundamental assumption that no effect can precede its cause, i.e., in a scattering context, that the scattered wave cannot occur prior to the incident wave reaching the scatterer. A consequence of this is that a Hilbert transform may be written, relating, on the real axis, the real and imaginary parts of $\hat{u}_d(x)$, i.e.,

$$\mathrm{Re}\{\hat{u}_d(x)\} = -\frac{1}{\pi} P \int_{-\infty}^{\infty} \frac{\mathrm{Im}\{\hat{u}_d(x')\}}{x'-x} dx' \quad ,$$

$$\mathrm{Im}\{\hat{u}_d(x)\} = \frac{1}{\pi} P \int_{-\infty}^{\infty} \frac{\mathrm{Re}\{\hat{u}_d(x')\}}{x'-x} dx' \quad . \qquad (2.49)$$

Thus, if one is able to measure either the real or the imaginary part of the scattered field, the complex function $\hat{u}_d(x)$ (and hence its phase) can be calculated; the object wave U(r) may then be determined from (2.45), albeit with an integration performed only over the finite interval (-X,X). The procedure could require a shift of origin in the primary space, to insure that the support (a,b) lies entirely in the positive half of the r axis.

However, it is not possible to determine the real (or imaginary) part of $\hat{u}_d(x)$ although, in some situations, an estimate of it may be obtained [2.40]. In general, only $|\hat{u}_d(x)|$ is recorded in a direct measurement. The obvious development of the theory consisted in formulating a relation between the observed modulus and the required phase. Since

$$\log \hat{u}_d(z) = \log|\hat{u}_d(z)| + i[\varphi_d(z) \pm 2n\pi] \qquad (2.50)$$

and $\log \hat{u}_d(z)$ has the same analytic properties as $\hat{u}_d(z)$ except where $\hat{u}_d(z) = 0$, the following expression, known as a logarithmic Hilbert transform, may be written [2.41]:

$$\varphi_d(x) = \frac{x}{\pi} P \int_{-\infty}^{\infty} \frac{\log|\hat{u}_d(x')|}{x'(x'-x)} dx' \quad . \qquad (2.51)$$

Since $\hat{u}_d(x) \in L^2(-\infty,\infty)$, $|\hat{u}_d(x)| \to 0$ as $|x| \to \infty$, and thus its logarithm cannot be $L^2(-\infty,\infty)$. The 'subtraction' of the expression, consisting of applying the factor x' to the denominator in the integrand, is required to render the function transformed $L^2(-\infty,\infty)$.

A Hilbert transform is calculated by taking an integral along a semicircular contour of indefinitely large radius in the upper half of the complex plane. The condition for its validity is that $\log|\hat{u}_d(z)|$ should not have any singularities within this contour: these occur where $\hat{u}_d(z) = 0$. Consequently, the correct $\varphi_d(x)$ is obtained from (2.51) only when $\hat{u}_d(z)$ has a zero free half plane, and, obviously, this must be known to be the case. Of course, it is possible to see directly from the intensity profile whether $\hat{u}_d(z)$ has only real zeros. As discussed in Sect.2.3, however, these situations are known to be exceptional. There is no criterion available to enable us to establish either *a priori*, or by examining the observed $|\hat{u}_d(x)|$, whether $\hat{u}_d(z)$ has a zero-free half plane, other than all zeros being real. Thus, the zeros of the function describing the field are the effective cause of the phase problem.

2.4.3 Solutions to the Zero Problem

Two methods present themselves for overcoming the difficulty introduced by the presence of zeros. The first consists of translating all zeros into the lower half plane, thus satisfying the condition required for the validity of the logarithmic Hilbert transform. Historically, this was the first approach adopted. The second way is to attempt to locate the zeros. Once this is done, the contribution from those in the upper half of the complex plane could be added to the phase given by (2.51).

The possibility of translating all zeros of $\hat{u}_d(z)$ into the lower half plane is offered by the application of Rouché's theorem [2.36], which states:

Theorem 4: If two functions $\hat{r}_d(z)$ and $\hat{u}_d(z)$ are regular in a region $\mathscr{C}$ of the complex plane and if $|\hat{u}_d(z)| < |\hat{r}_d(z)|$ at every point of the boundary of $\mathscr{C}$, then this is a sufficient but not necessary condition for $\hat{r}_d(z)$ and $\hat{r}_d(z) + \hat{u}_d(z)$ to have the same number of zeros in $\mathscr{C}$.

The function $\hat{r}_d(z)$ is termed a reference function if it is chosen as having no zeros in the upper half plane and satisfies Rouché's theorem in this half plane. Hence the behaviour of $\hat{r}_d(z) + \hat{u}_d(z)$ is dominated, at least from this point of view, by $\hat{r}_d(z)$. The reference function $\hat{r}_d(z)$ may be constant, in which case $\hat{r}_d(z) = \hat{r}$. The desired phase associated with $|\hat{u}_d(x)|$ may be easily obtained by applying a Hilbert transform to $\log|\hat{u}_d(x)+\hat{r}|$ as indicated in [2.41].

The addition of a reference function (or constant) to the scattered field is a holographic procedure. Indeed, it is well known that holography is a method of phase retrieval, although it has not been generally recognized that, from the point of view of the theory of analytic functions, this is carried out by a translation of all zeros into one half plane.

Holography, however, presents several serious drawbacks. It may be counterproductive if all complex zeros are already in the lower half plane; as said above, this cannot be established from $|\hat{u}_d(x)|$. Furthermore, $\hat{r}$ should not be smaller than the value required to move all zeros; at the same time, $\hat{r}$ should not be much larger than the minimum value required, because the sensitivity of the measurement would be reduced. But, not knowing which zeros are in the upper half plane, it is impossible to determine *a priori* the value of $\hat{r}$ satisfying both requirements: Rouché's theorem offers only a sufficient and not a necessary condition. In principle, one could establish the magnitude of $\hat{r}$ which ensures the translation of all complex zeros into one half plane. As the level of $\hat{r}$ is increased, if all zeros move together, either away from the real axis, or towards it, $\hat{u}_d(z)$ has a zero free half plane. If some of the zeros move towards the real axis and some away from it, the zeros in $\hat{u}_d(z)$ are not restricted to one half plane only. The trend in the displacement of zeros would be observed by following the change in maxima and minima displayed by the intensity on the real axis. To say that such a test could be tedious, is an euphemism.

In addition to this, given the desirability of having available for measurement as large as possible an interval, it is useful to record the information over a 2π angular range: in a holographic situation, such a requirement is experimentally demanding. Besides, for certain types of radiation (X-rays, electrons, neutrons) an appropriate reference function may be difficult to obtain experimentally; X-ray workers, for example, know that methods based on the use of an isomorphous derivative are by no means straightforward.

Thus, in order to solve the deterministic problem, one is effectively reduced to the task of locating the zeros of $\hat{u}_d(z)$.

Of course, the application of a method based on the Hilbert transform is valid even for $b \to \infty$, and for this situation, which is nonphysical, one cannot define a density of zeros. In fact, it seems that nothing is known about the properties of the distribution of zeros associated with a function defined on a semi-infinite interval. In any physically realisable situation, b will be finite and hence $\hat{u}_d(z)$ will be of class E, with the established properties of zeros. Therefore, it seems only too reasonable to make full use of this, rather than rely on theories based on nonphysical models.

Let us assume that N complex zeros of $\hat{u}_d(z)$ are present in the z space: 2^N object waves U(r) may then be defined, all compatible with the observed $|\hat{u}_d(x)|$. Several methods have been suggested for solving this ambiguity and selecting the actual U(r), i.e., for effectively establishing to which half plane each zero does belong [2.41-43]. All of them consist, essentially, of determining the 2^N possible functions U(r) and comparing them with some *a priori* information about the object wave in the r space, e.g., the presence of an interval within (a,b) on which $U(r) \equiv 0$ [2.42], thus retrieving the correct solution. Such methods work satisfactorily if the number N of complex zeros is not too large. If N exceeds, say, 10, the time required to compute the 2^N possibilities may become prohibitively long.

Other algorithms put forward [2.44-46], not relying explicitly on the analytic properties of the field, consist of a process of successive approximations from an initial distribution of arbitrary phases. How fast the iteration sequence converges, if at all, will obviously depend on the amount of information available *a priori* and on the proximity of the original guess to the actual solution. An excellent and detailed discussion of such methods may be found in [2.43].

Thus, the need remains to find a method for locating the zeros in such a way that the 2^N ambiguity ceases to be a handicap. For this, we shall make use of the properties of zeros of functions of class E discussed in Sect.2.3. In fact, it is obvious that if the zeros can be located, $\hat{u}_d(z)$ may be directly obtained from the Hadamard product (2.44), thus rendering the application of a Hilbert transform unnecessary.

2.4.4 Zero Location

It is a relatively simple matter to locate the zeros of $\hat{g}_d(z)$. Any real zeros are seen on the intensity profile along the real axis: their position is directly established. Complex zeros may be located by a straightforward process of analytic continuation, from the observation on the real axis, into the complex plane. This is achieved by taking the inverse Fourier transform of $\hat{g}_d(x)$ given by (2.46), thus obtaining $g_d(\rho)$, or rather its approximant of class E, given the fact that the interval of measurement is finite. Subsequently, $g_d(\rho)$ is multiplied by $\exp(-\langle k \rangle \rho y)$, yielding

$$\hat{g}_d(x+iy) = \int_{-R}^{R} g_d(\rho) \exp(-\langle k \rangle \rho y) \exp(i \langle k \rangle \rho x) dx \qquad (2.52)$$

for each y, regarded as a parameter. In this way, the complex plane is swept and the zeros of the intensity are located. Obviously, it is sufficient to explore only one half of the complex plane, i.e., either the lower or the upper semi-plane, given the symmetry of the zeros about the x axis. Thus, for all complex zeros of $\hat{u}_d(z)$ present in the interval of measurement, one can obtain $x_{j,d}$ and $|y_{j,d}|$. For each $|y_{j,d}|$ there are two values of $y_{j,d}$ which correspond to the same observed intensity: one of them belongs to $\hat{u}_d(z)$ and the other to $\hat{u}_d^*(z^*)$. If N complex zeros are present in one half plane, the ambiguity in solving the phase problem is 2^N. In other words, 2^N different functions $u_d(r)$ generate the same intensity $\hat{g}_d(x)$. This is the fundamental ambiguity present in the deterministic problem. In practice, there could be hundreds of complex zeros in the interval of measurement: the ambiguity may be considerable.

Thus, the phase and hence the first step of the deterministic problem is reduced to the task of establishing the sign associated with each $|y_{j,d}|$, i.e., of deciding to which half plane each of the zeros of $\hat{u}_d(z)$ belongs. If this can be achieved, introducing the correct $z_{j,d}$ into the Hadamard product (2.44) provides the true $\hat{u}_d(x)$.

Of course, if $u_d(r)$ is known to be Hermitian – or can be made so – $\hat{u}_d(z)$ will have zeros at $z_{j,d}$ and $z_{j,d}^*$. The zeros of $\hat{g}_d(z)$ are those of $\hat{u}_d(z)$, but the order of each is doubled. Once the zeros in the intensity are located, $\hat{u}_d(z)$ is reconstructed from the Hadamard product, in which the zeros are introduced with the appropriate order. Therefore, for this particular class of effective object waves, a solution to the phase problem can always be constructed in a straightforward manner, up to a maximum harmonic determined by the interval of measurement.

Such situations, however, which consist of highly specific properties of the object wave, resulting in a distribution of zeros which does not lead to an ambiguity in solving the deterministic problem, are very rare indeed. The fundamental desideratum remains of being able to establish with speed and ease the sign of $y_{j,d}$'s.

For this, it is obvious that we must make use of the modality in which the information is propagated from the object to the Fraunhofer space along the lines of zeros. Let us examine how the sign of the $y_{j,d}$'s – and hence U(r) – can be easily determined from two measurements of intensity at d_1 and d_2, without any knowledge of the object wave and without a reference function.

In order to outline the ideas involved, let us assume that, using the location of the N zeros in intensity (in one half plane) at d_1, we generate the 2^N possible distributions of zeros. For each of them, $\hat{u}_{d1}(x)$ may be calculated by a Hadamard product and, subsequently, for each $\hat{u}_{d1}(x)$, U(r) – or rather its approximant of class E – may be obtained by a Fourier transform from

$$U(r) = \exp\left(-\frac{i<k>r^2}{2d_1}\right)\int_{-X}^{X} \hat{u}_{d_1}(x)\, \exp(i<k>rx)dx \quad .$$

For each of the 2^N object waves U(r) thus determined, one can calculate the corresponding

$$\left|\hat{u}_{d_2}(x)\right| = \left|\int_a^b U(r)\exp\left(\frac{i<k>r^2}{2d_2}\right)\exp(-i<k>rx)dr\right| \quad .$$

The presence of the quadratic phase factor $\exp(i<k>r^2/2d_2)$ ensures that not only will the 2^N functions $\hat{u}_{d_2}(x)$ be distinct, but also that their moduli will be different. By comparing the observed $|\hat{u}_{d_2}(x)|$ with each of the 2^N calculated functions, one can identify the correct U(r). In principle, the method is simple and can be easily applied, involving only two conventional scattering measurements, both in the Fresnel region, for two finite distances d, or one of them in the Fraunhofer region.

The main obstacle in the application of this approach to practical situations as a routine method is that, like other methods put forward above, it requires the calculation of 2^N possible solutions. Thus, the method will work satisfactorily when N is not too large, but these situations are exceptional. It is essential to avoid the overwhelming drawback introduced by the 2^N calculations required to obtain the possible moduli $|\hat{u}_{d_2}(x)|$, while, at the same time, preserving the simplicity and soundness of the approach. For this, it seems logical to make use of the specific way in which each zero in the z space codes information in the r space.

The first step of the modified method is the same and consists of determining x_j and $|y_j|$ from $|\hat{u}_{d_1}(x)|$ in the manner discussed above. For convenience and accuracy, let us assume that d_1 describes the Fraunhofer space and $d_2 = d$ represents a finite distance defining a Fresnel space; for simplicity we assume real, weak scatterers.

The interval (a,b) in the primary space establishes the sinc lattice associated with the given experiment. The zeros, when on the sinc lattice positions, define the zeroth approximation of the object wave, as a constant in the interval (a,b).

The first approximation of the object wave consists of describing it by its first harmonic. For this, the pair of zeros with $j = 1$ is moved from the lattice position to $x_1 + i|y_1|$ and, subsequently, to $x_1 - i|y_1|$. For both distributions of zeros, the two functions $\hat{u}(x)$ may be calculated from a Hadamard product. The corresponding first approximation to the object wave consists, therefore, of two possible harmonics, having one complete cycle each within the interval (a,b), and equal amplitudes, but phases at $r = a$ equal to Φ_1 and $-\Phi_1$ respectively. The two corresponding Fresnel intensities are calculated [introducing higher order terms from (2.17), if necessary] and compared with the measured intensity. The better match determines the sign of y_1 and hence the phase of the first harmonic component of U(r).

The next step consists of moving z_2 from the lattice position to $x_2 + i|y_2|$ and subsequently to $x_2 - i|y_2|$. The same procedure is carried out, obtaining two possible second approximations for the object wave. They consist of the first harmonic, whose phase is correct, and two possible second harmonics, whose phases at $r = a$ depend on x_1, y_1 and x_2, $\pm y_2$. The Fresnel intensities are calculated and again compared with the measured intensity. The better fit determines the sign of y_2, and hence the phase of the second harmonic component of U(r). The procedure is repeated for z_3, z_4, ..., z_N.

The method for assessing which of the two intensities is better at each step in the sequence of successive approximations consists of subtracting each of the calculated from the measured intensity. The areas between the measured distribution of intensity and the two calculated ones are determined: the lower corresponds to the better fit. A calculation of the standard deviation provides the same information.

The fundamental advantage of this method, besides its simplicity, is the dramatic decrease in the number of operations required, from 2^N to merely 2N, assuming N pairs of zeros symmetric with respect to the imaginary axis in the interval of measurement, i.e., N zeros either side of the z origin. Thus, the method is amenable to being applied in practice when hundreds of complex zeros are present in (-X,X). A solution is constructed, having a resolution given by the highest frequency harmonic, whose corresponding zero can be located.

The extension of the method to complex object waves is immediate; for N complex zeros located, the number of steps in the computing procedure is reduced to N, if the object wave is known *a priori* to be either real or purely imaginary, and to 2N for the general situation of a complex U(r). We believe that this is a fundamental lower limit: it is equal to the amount of structural information (logon content) in the r space.

In principle, the determination of the sign of each y_j could be simplified further still by continuation from the Fraunhofer to the Fresnel space for both possibilities at every step, without calculating two object waves each time. This could be accomplished by following the propagation of information between the object and Fraunhofer space along specific lines of zeros.

So far, only simulated experiments have been carried out on noise-free data. They have all been successful: the only difficulty encountered in assessing the sign of y_j occurs when $|y_j|$ is very small, but then the difference introduced in U(r) by the zero flip is negligible. All objects were assumed to be weak scatterers.

In every practical experiment, noise will be present, affecting the location of zeros and hence the whole procedure. The effect of noise is to introduce a region of uncertainty about each zero and hence an error in the amplitude and phase of each harmonic from which U(r) is built. The precise effect of noise levels is under investigation. Obviously, the problem will be common to all methods of phase retrieval. If the data are very accurate, it should be possible to locate not only the zeros within the interval of measurement, but also, using the same process of analytic continuation, at least some zeros outside it. This may provide the key to reconstructing the object wave up to a spatial frequency corresponding to the zero with the highest j that can be thus located.

A point to consider is that not all zeros are necessarily of first order; however, if a higher order zero is present and treated as a simple zero, the Hadamard product will generate an intensity different from the observed one.

Figure 2.7 illustrates the steps in the method. In the example given, we start from a distribution of intensity in the Fraunhofer space and locate the associated zeros in one half plane. We have chosen a function $\hat{u}(z)$ displaying 50 complex zeros on either side of the origin and distributed symmetrically about the imaginary axis, in order to code a real object wave. The 2^N ambiguity corresponding to this number is of the order of 10^{15}. The observable Fraunhofer and Fresnel intensities are dis-

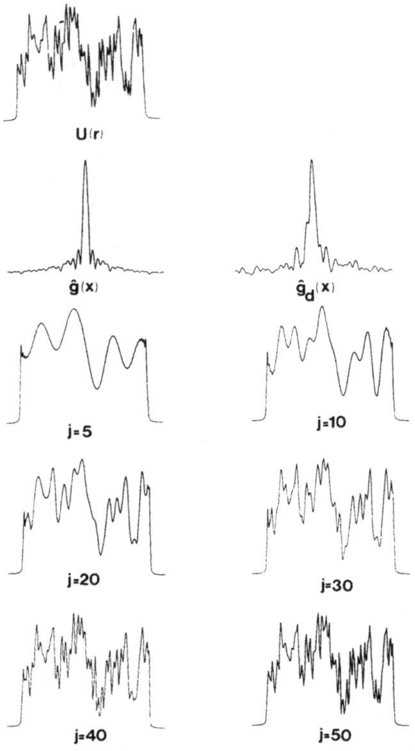

Fig. 2.7. An object, U(r), and its Fraunhofer and Fresnel intensity. Six reconstructions of the object are shown, corresponding to the location of the first 5, 10, 20, 30, 40 and 50 zeros

played, as well as six successive steps in the approximation process corresponding to the first 5, 10, 20, 30, 40 and 50 zeros being located. The whole process requires only 100 calculations compared to 10^{15}, taking only 120 seconds on the CDC 6600 computer and representing functions by 2048 points.

In the given illustration, the number of complex zeros considered was established beforehand to be the total number of complex zeros for that function. Hence, the final reconstruction is identical to the defined object wave. In reality, of course, the total number of complex zeros is never known *a priori* and there will almost certainly be complex zeros beyond $|X|$, perhaps an indefinitely large number of them. As such, in general, the object wave U(r) cannot be obtained with complete accuracy and, even if in some instances this may be so, we would not know it. The best we can do is to construct the object wave from higher and higher harmonics, by locating as many zeros as possible. Thus, the solution will be more and more similar to the object wave, without ever being identical to it, except fortuitously and without us realizing it. If the algorithm used for obtaining U(r) could involve an infinite Fourier transform of $\hat{u}_d(x)$, the interval (a,b) would be exactly obtained. In practice, the numerical calculation of the Fourier transform requires $\hat{u}_d(x)$ to be confined to an interval of finite extent which, for the purposes of the algorithm, may or may not be equal to the interval of measurement (-X,X). Thus, an approximant of class E is obtained to describe the constructed object wave.

The method would work, in principle, with any nonidentically zero degree of coherence of the incident field. However, the degree of coherence acts as an envelope on $g_d(\rho)$ and, depending on its shape and width, may attenuate $g_d(\rho)$ near the ends of the interval (-R,R). The result of this, as discussed in Sect.2.3, would be a movement of the zeros away from the real axis, this effect being stronger for lower values of j. This is inconvenient, because more of the complex plane must be explored. Also, the further a zero from the real axis, the more susceptible its location would be to the noise inherent in the system. Hence, as high as possible a degree of coherence of the incident field is desirable: ideally, $g_d(\rho)$ should not be distorted at all.

2.4.5 Extensions of the Method

The construction of the object wave in this way for a one-dimensional situation provides the solution — in the broad sense we attach to this word — to the deterministic inverse scattering problem. The only condition required for this is an adequate design of the experiment, namely the use of a direction of polarisation parallel to the object; this renders multiple scattering impossible. Thus, the first Born approximation may be invoked and the inhomogeneity distribution function is proportional to the obtained object wave.

One can regard a two-dimensional scatterer as a collection of one-dimensional strips and apply the above procedure to each of them. We hope, however, to present a rigorous approach to this situation. For a two-dimensional situation, an additional difficulty must be overcome. In order to retrieve the inhomogeneity distribution function from the object wave, it is necessary, as suggested by (2.18), to know the values of the field at each point of the object space: for a strong scatterer, the object wave may bear no resemblance at all to the inhomogeneity distribution function. This leads to the realization that we are faced with the problem of an infinite regress. The relationship between the object wave and the inhomogeneity distribution function is a simple one only if the first Born approximation may be invoked. In this case, the field at every point of the primary space is the incident field. If the first Born approximation is not sufficiently accurate, it is necessary to expand the object wave in a Liouville-Neumann series as discussed in Sect.2.2, from which, if it converges sufficiently fast, the inhomogeneity distribution function may be, in principle, obtained. The second Born approximation is already quite involved.

The use of the Liouville-Neumann expansion — or of its first (Born) approximant — requires knowledge of the incident field, i.e., of the source. If the incident field is not given, the repartition of its modulus may be easily obtained and the phase distribution could be retrieved by applying the procedure presented above. In practice, one could arrange, in most cases, an adequately high coherence of the incident field, so that the coherence factor deviates only negligibly from a constant value in the interval (-R,R). This difficulty may be thus circumvented and, for the range of validity of the first Born approximation, the inhomogeneity distribution function is proportional to the object wave constructed. Again, the conclusion is reached that as high as possible a degree of coherence is desirable.

For a three-dimensional situation, obviously, the effect of multiple scattering is much more important. Besides, in this case, solving the deterministic problem involves an additional task. This consists in obtaining the object wave in a three-dimensional primary space from recorded data in two-dimensional spaces, Fraunhofer or Fresnel [2.47]. Traditional approaches to solving this difficulty rely on exploring a three-dimensional Fourier conjugate space by varying the direction of the incident wave vector and recording the scattered field in the same plane; such methods necessarily involve redundant information. Procedures for solving this problem have been put forward [2.48-57]. It is worth investigating whether this task could not be considerably simplified by using the analytic properties of the scattered field. Of course, the problem is even more complicated when the medium is locally anisotropic.

Some of the difficulties involved in these, and related, tasks, are examined by HOENDERS [2.58] and their synoptic presentation is given by BALTES [2.4].

The presence of zeros in the function describing the scattered field was seen to be the cause of the phase problem: the knowledge of their properties provides the key to its solution. The *a priori* knowledge of the categories to which our observation is subjected, enabled us to put forward a representation of the object in terms of concepts we ourselves create.

2.5 The Statistical Problem

> *And do you know also that although they make use of the visible forms and reason about them, they are thinking not of these, but of the ideals which they resemble; not of the figures which they draw, but of the absolute square and the absolute diameter, and so on .... the forms which they draw or make, and which have shadows and reflections in water of their own, are converted by them into images, but they are really seeking to behold the things themselves which can only be seen with the eye of the mind.*
>
> Plato, Republic VI 510

For many practical applications, a detailed knowledge of the inhomogeneity distribution function — or tensor — is not necessary. For example, in the study of the relationship between the morphology and properties of a material, one would not expect the macroscopic mechanical, electrical or optical characteristics to depend on the detailed structure, but only on parameters describing the global aspects of morphology.

2.5.1 Overall Characterization of the Medium

Any finite realisation of a medium, any finite sample studied, is deterministic. The problem of structural averages is particularly relevant when, for some reason, the material in which one is interested cannot be entirely explored or it would be highly inconvenient to do so. On the level of abstraction, we may regard the problem of obtaining global values as applicable when the medium investigated is assumed to have an infinite extent, out of which, obviously, only a finite section could ever be examined. The concept of an infinite medium makes it relatively easy to describe global properties by mathematical averages.

To simplify the treatment, we consider for the time being a scalar description for $\delta\varepsilon(\underline{r})$. The generalisation to a tensor quantity will be outlined later. It is convenient and usual to classify the models of infinite media into two groups, according to their integrability properties. In group 1 we include all media such that

$$\lim_{V\to\infty} \frac{1}{V}\int_V d^3\underline{r}\,|\delta\varepsilon(\underline{r})|^2 < \infty \quad , \tag{2.53}$$

and in group 2 those for which

$$\int_{-\infty}^{\infty} d^3\underline{r}\,|\delta\varepsilon(\underline{r})|^2 < \infty \quad . \tag{2.54}$$

This classification is somewhat arbitrary, since the second group is a subclass of the first group [Ref.2.59, Sect.21]. The properties of media in group 2 are sufficiently distinct to warrant a separate treatment: their reduction to the features presented by the main group 1 will be examined later. The inhomogeneity distribution function of a medium included in the first group is said to belong to the Wiener class [2.24], i.e., $\delta\varepsilon(\underline{r}) \in S(-\infty,\infty)$ for all directions determined by $\underline{r}$. Sometimes, and perhaps misleadingly in a spatial context, these media are said to be of 'finite power'. Remarkable examples in this group – which do not belong to group 2 as well – are periodic structures and random materials. A medium in the second group is such that $\delta\varepsilon(\underline{r}) \in L^2(-\infty,\infty)$ for all directions determined by $\underline{r}$. Such media are sometimes referred to as being of 'finite energy'.

The problem of establishing global characteristics is particularly relevant for random structures. A stochastic medium, $\delta\varepsilon(\underline{r})$, is statistically determined if one knows its nth order distribution function, i.e.,

$$F(\delta\varepsilon_1,\delta\varepsilon_2,\ldots\delta\varepsilon_n;\underline{r}_1,\underline{r}_2,\ldots,\underline{r}_n) = P\{\delta\varepsilon(\underline{r}_1)\leqq\delta\varepsilon_1,\delta\varepsilon(\underline{r}_2)\leqq\delta\varepsilon_2,\ldots,\delta\varepsilon(\underline{r}_n)\leqq\delta\varepsilon_n\}$$

or the nth order probability density

$$\frac{\partial^n F}{\partial\delta\varepsilon_1\partial\delta\varepsilon_2\ldots\partial\delta\varepsilon_n} = f(\delta\varepsilon_1,\delta\varepsilon_2,\ldots,\delta\varepsilon_n;\underline{r}_1,\underline{r}_2,\ldots,\underline{r}_n) \quad .$$

These descriptors provide, for an arbitrarily high n, the complete overall characterization of the morphology of a random medium.

Posed in this way, the problem has no solution, since it involves an infinite regress: one cannot, in general, describe completely the statistical properties of a material. The only nontrivial case in which the nth order distribution function can be determined from the 1st and 2nd moment, is when the process may be assumed to be Gaussian, i.e., the values of $\delta\varepsilon(\underline{r})$ at different $\underline{r}$ are normally distributed. In this situation, all distribution functions up to any desired n are completely determined in terms of the mean and autocovariance of the medium [2.60]. Strictly speaking, no medium is ever Gaussian, and hence the statistical problem can never be completely solved. Even so, however, the mean and the autocovariance of the medium provide useful information, by assessing the scale of inhomogeneity and, possibly, by characterizing its nature.

The mean and autocovariance are defined by ensemble averages: assuming their ergodicity [2.61], we shall express them as volume averages. The mean has already been defined, by (2.2). The autocovariance is given by

$$G_1(\rho) = \frac{\lim\limits_{V\to\infty} \frac{1}{V} \int\limits_V \delta\varepsilon(\underline{r})\delta\varepsilon^*(\underline{r}+\underline{\rho})d^3\underline{r}}{\lim\limits_{V\to\infty} \frac{1}{V} \int\limits_V |\delta\varepsilon(\underline{r})|^2 d^3\underline{r}} \quad , \tag{2.55}$$

where the denominator defines

$$\left\langle |\delta\varepsilon(\underline{r})|^2 \right\rangle = \lim_{V\to\infty} \frac{1}{V} \int\limits_V |\delta\varepsilon(\underline{r})|^2 d^3\underline{r} \quad .$$

In writing this expression, we have used the hypothesis of wide-sense homogeneity of the medium [2.62], justified by the ergodicity already assumed. We have also stated that $G_1(\underline{\rho}) = G_1(\rho)$, because the medium was assumed to be overall isotropic. This descriptor of average morphology may be applied to all functions of class S, hence the subscript 1 attached to it, denoting the first group of media defined above.

If $\delta\varepsilon(\underline{r}) \in L^2(-\infty,\infty)$, a similar overall description of the medium may be put forward. For this situation we define

$$G_2(\rho) = \frac{\int\limits_{-\infty}^{\infty} \delta\varepsilon(\underline{r})\delta\varepsilon^*(\underline{r}+\underline{\rho})d^3\underline{r}}{\int\limits_{-\infty}^{\infty} |\delta\varepsilon(\underline{r})|^2 d^3\underline{r}} \quad . \tag{2.56}$$

Both $G_1(\rho)$ and $G_2(\rho)$ are characteristic functions [Ref.2.24, Theorems 10.32 and 10.41] [$G_1(\rho)$ must be continuous at $\rho = 0$ for this to be true]: in the study of their common properties we shall drop the subscripts.

Characteristic functions may be expressed in the form [2.63]

$$G(\rho) = \int\limits_{-\infty}^{\infty} \exp(i\langle k\rangle\underline{\rho}\cdot\underline{s})dW(\underline{s}) \quad , \tag{2.57}$$

where $W(\underline{s})$ is a bounded, nondecreasing function such that $W(-\infty) = 0$ and $W(\infty) = 1$, i.e., it is a distribution function. If $W(\underline{s})$ is absolutely continuous, its derivative is known as the spectral density function (or spectral density) $\hat{G}(\underline{s})$ and thus

$$G(\rho) = \int_{-\infty}^{\infty} \exp(i\langle k\rangle \underline{\rho}\cdot\underline{s})\hat{G}(\underline{s})d\underline{s} \quad . \tag{2.58}$$

The spectral density is given by

$$\hat{G}(\underline{s}) = \int_{-\infty}^{\infty} \exp(-i\langle k\rangle \underline{\rho}\cdot\underline{s})G(\rho)d^3\underline{\rho} \quad , \tag{2.59}$$

where the integral is taken as a limit in the norm. Thus, $\hat{G}(\underline{s})$ contains all the information about the overall descriptors of the structure if the medium is Gaussian and valuable information about it in any case.

It is important to establish the correspondence between the idealized quantities introduced above, expressing averages defined for media of infinite extent, and the functions describing a medium in the physical situation, introduced in Sect.2.2. For this, let us examine (2.28): it is arguable whether in this expression the effect of the field may be separated from that of the medium, i.e., they may be regarded as independent. Some authors [2.64,65] consider this assumption correct in any situation. Others [2.66] claim that such an approach is dishonest. This hypothesis may be certainly made within the range of validity of the first Born approximation. With this,

$$g(\rho) = \sin^2\vartheta \; V \left\langle \delta\varepsilon(\underline{r})\delta\varepsilon(\underline{r}+\underline{\rho})\right\rangle_V \left\langle \underline{E}_0(\underline{r})\cdot\underline{E}_0^*(\underline{r}+\underline{\rho})\right\rangle_V \quad , \tag{2.60}$$

where we have taken into account that, in the first Born approximation, $\delta\varepsilon(\underline{r})$ must be assumed real and hence $\langle k_A\rangle = 0$. We shall adopt this hypothesis for the time being; the extension of the treatment – and of the conclusions reached – to absorbing media will be attempted later on. In the obliquity factor, ϑ represents the angle between the direction of polarisation of the incident field and the direction of observation $\underline{m}$. The first average in (2.60) represents the convolution square of the inhomogeneity distribution function over the finite volume examined. The last factor represents the average – over the irradiated volume – of the mutual intensity of the incident field. Let us assume, in order to facilitate the comparison with the characteristic function, that $\vartheta = \pi/2$ and that the incident field is completely coherent. Thus, if the medium is isotropic,

$$g^c(\rho) = \int_V \delta\varepsilon(\underline{r})\delta\varepsilon(\underline{r}+\underline{\rho})d^3\underline{r} \quad , \tag{2.61}$$

where the superscript c indicates complete coherence of the incident field.

The relationship between $g^c(\rho)$ and $G(\rho)$ is now obvious. For $\delta\varepsilon(\underline{r}) \in L^2(-\infty,\infty)$, we define

$$G_2'(\rho) = \lim_{V\to\infty} g^c(\rho) \tag{2.62a}$$

and hence

$$G_2(\rho) = \frac{G_2'(\rho)}{G_2'(0)} \quad , \tag{2.62b}$$

whilst for the more general situation $\delta\varepsilon(\underline{r}) \in S(-\infty,\infty)$,

$$G_1'(\rho) = \lim_{V\to\infty} \frac{1}{V} g^c(\rho) \tag{2.63a}$$

and thus

$$G_1(\rho) = \frac{G_1'(\rho)}{G_1'(0)} \quad . \tag{2.63b}$$

It is well known [2.24] that both limits do exist. On this basis, we shall establish the convergence of the intensity, as described by (2.29) towards the spectral density $\hat{G}(\underline{s})$. This is determined by the displacement of zeros of $\hat{g}(\underline{m})$ which takes place when $V\to\infty$, as will become clear later. With the assumptions made, $\hat{g}^c(\underline{s})$ may be written

$$\hat{g}^c(\underline{s}) = \int_V d^3\underline{\rho}\, \exp(-i\langle k\rangle \underline{\rho}\cdot\underline{s}) g^c(\rho) \quad , \tag{2.64}$$

where $\underline{s} = \underline{m}_0 - \underline{m}$, $\underline{m}_0$ being the unit vector describing the propagation of the incident field.

If $\delta\varepsilon(\underline{r}) \in L^2(-\infty,\infty)$, the relationship between the observed intensity and the spectral density is straightforward. Indeed, in this case, by Plancherel's theorem [2.24]

$$\hat{G}_2'(\underline{s}) = \underset{V\to\infty}{\text{l.i.m.}}\; \hat{g}^c(\underline{s}) \tag{2.65a}$$

and

$$\hat{G}_2(\underline{s}) = \frac{\hat{G}_2'(\underline{s})}{G_2'(0)} \quad , \tag{2.65b}$$

where the limit in the mean is the limit in the norm of the L^2 space and the subscript 2 signifies the second group of media defined above.

If $\delta\varepsilon(\underline{r}) \in S(-\infty,\infty)$, things are more complicated. Indeed, in this case neither (the normalized) $\hat{g}^c(\underline{s})$ nor $\hat{g}^c(\underline{s})/V$ converges towards $\hat{G}_1(\underline{s})$, not even as a limit in the mean: although the mean convolution square of the inhomogeneity distribution function over the finite volume examined, $g^c(\rho)/V$, is bounded, it is not square integrable and hence its infinite Fourier transform cannot be defined. However, it is possible to modify $g^c(\rho)/V$ such that the result of the operation is square integrable. If $\psi(\underline{\rho}) \in L^2(-\infty,\infty)$, then

$$\frac{1}{V} g^c(\rho)\psi(\underline{\rho}) \in L^2(-\infty,\infty) \quad .$$

A very important instance is

$$\psi(\underline{\rho}) = \frac{\sin(<k>\underline{\rho}\cdot\underline{\Delta s})}{<k>\underline{\rho}\cdot\underline{\Delta s}} \tag{2.66}$$

and, with this, observing that the normalizing factor

$$G_1'(0) = \left\langle[\delta\varepsilon(\underline{r})]^2\right\rangle$$

we may write

$$\frac{1}{\left\langle[\delta\varepsilon(\underline{r})]^2\right\rangle} \underset{V\to\infty}{\text{l.i.m.}} \frac{1}{V}\int_V g^c(\rho)\, \frac{\sin(<k>\underline{\rho}\cdot\underline{\Delta s})}{<k>\underline{\rho}\cdot\underline{\Delta s}}\, \exp(-i<k>\underline{\rho}\cdot\underline{s})d^3\underline{\rho}$$

$$= W(\underline{s}+\underline{\Delta s}) - W(\underline{s}-\underline{\Delta s}) = \Delta W(\underline{s}) \quad , \tag{2.67}$$

where $W(\underline{s})$ is defined by (2.57). This may be seen from the inversion formula of the Fourier-Stieltjes transform [Ref.2.24, Sect.4.4]. Expression (2.67) is known as the generalized Fourier transform of $\left\{1/\left\langle[\delta\varepsilon(\underline{r})]^2\right\rangle\right\}(1/V)g^c(\rho)$ as $V\to\infty$ and is always $L^1(-\infty,\infty)$. Hence, in hindsight, the limit in the mean in (2.67) could have been written as a simple limit.

Thus, for $\delta\varepsilon(\underline{r})\in S(-\infty,\infty)$ the normalized relative intensity (per unit volume), i.e., the scattering cross section, does not converge towards the corresponding spectral density. Only the integrated scattering cross section, in an interval however small, converges towards the difference between the values of the spectral distribution function $W(\underline{s})$ corresponding to the ends of this interval. Assuming that $\hat{G}_1(\underline{s})$ may be defined, it is only possible to state that

$$\frac{1}{\left\langle[\delta\varepsilon(\underline{r})]^2\right\rangle} \lim_{V\to\infty} \frac{1}{V}\int_{\underline{s}-\underline{\Delta s}}^{\underline{s}+\underline{\Delta s}} \hat{g}^c(\underline{s}')d\underline{s}' = \int_{\underline{s}-\underline{\Delta s}}^{\underline{s}+\underline{\Delta s}} \hat{G}_1(\underline{s}')d\underline{s}' = \Delta W(\underline{s}) \quad . \tag{2.68}$$

The expression $\left\{1/\left\langle[\delta\varepsilon(\underline{r})]^2\right\rangle\right\}(1/V)g^c(\rho)$ is, as $V\to\infty$, the Fourier-Stieltjes transform of $W(\underline{s})$. This is the only sense in which we can speak of the Fourier transform of a function in $S(-\infty,\infty)$. The meaning of the generalized Fourier transform is that of a locally averaged spectral density, i.e., convolved with an aperture of half-width (or radius) $\underline{\Delta s}$. The multiplication of $\left\{1/\left\langle[\delta\varepsilon(\underline{r})]^2\right\rangle\right\}(1/V)g^c(\rho)$ by $(1/<k>\underline{\rho}\cdot\underline{\Delta s})\sin(<k>\underline{\rho}\cdot\underline{\Delta s})$ is equivalent to the convolution of $\left\{1/\left\langle[\delta\varepsilon(\underline{r})]^2\right\rangle\right\}\times (1/V)\hat{g}^c(\underline{s})$ by a rect function (of appropriate dimensionality) of half-width $\underline{\Delta s}$. Therefore, for media in this group, it is inconceivable to obtain the spectral density at a point $\underline{s}$, even assuming the nonphysical situation of an indefinitely large volume, and must accept an average over an interval of the $\underline{s}$ axis, in the vicinity of $\underline{s}$. Although, on the face of it, this appears an important drawback for functions in $S(-\infty,\infty)$, since it results in a loss of resolution, from a practical point of view such a constraint will always be present, even when $\delta\varepsilon(\underline{r})\in L^2(-\infty,\infty)$. Indeed, whatever the integrability properties of $\delta\varepsilon(\underline{r})$, when a measurement of intensity is performed, only a local average is obtained: this is an inescapable manifestation of the fundamental categories.

For $\delta\varepsilon(\underline{r}) \in L^2(-\infty,\infty)$ one can also write

$$\frac{1}{G_2'(0)} \lim_{V\to\infty} \int_{\underline{s}-\underline{\Delta s}}^{\underline{s}+\underline{\Delta s}} \hat{g}^c(\underline{s}')d\underline{s}' = \int_{\underline{s}-\underline{\Delta s}}^{\underline{s}+\underline{\Delta s}} \hat{G}_2(\underline{s}')d\underline{s}' = \Delta W(\underline{s}) \quad . \tag{2.69}$$

In practice, the most convenient, readily available form for $\psi(\underline{\rho})$ is the sinc-like form already introduced. By comparing (2.61) with (2.60) it is easy to see that an inbuilt factor is available, multiplying the finite convolution square of the inhomogeneity distribution function: it is the volume average of the mutual intensity of the incident field (assuming the validity of the first Born approximation). Indeed, the local average of the measured intensity, expressed by (2.68,69) as the convolution of $\hat{g}^c(\underline{s})$ with an aperture of half-extent $\underline{\Delta s}$, may be generated by placing the aperture *either* in the source space *or* in the space of measurement (Fraunhofer): both are Fourier conjugates of the primary space. Thus, assuming an extended incoherent source of uniform intensity, $\langle \underline{E}_0(\underline{r})\cdot\underline{E}_0^*(\underline{r}+\underline{\rho})\rangle$ is essentially given by $\mathrm{sinc}(\langle k\rangle\underline{\rho}\cdot\underline{\Delta s})$ in one dimension or $(2/\langle k\rangle\underline{\xi}\cdot\underline{\Delta s})J_1(\langle k\rangle\underline{\xi}\cdot\underline{\Delta s})$ in two dimensions, where $\underline{\xi}$ is the component of $\underline{\rho}$ defined by $\underline{m}_0\cdot\underline{\xi}=0$. In the first Born approximation, therefore, with a judiciously chosen polarisation of the incident field,

$$g(\rho) = g^c(\rho)\mu_0(\langle k\rangle\underline{\rho}\cdot\underline{\Delta s}) \quad , \tag{2.70}$$

where $\mu_0(\langle k\rangle\underline{\rho}\cdot\underline{\Delta s})$ is the complex degree of coherence of the incident field and a factor of proportionality has been omitted. With this,

$$\hat{g}(\underline{s}) = \int_V d^3\underline{\rho}\,\exp(-i\langle k\rangle\underline{\rho}\cdot\underline{s})g(\rho) = \int_{\underline{s}-\underline{\Delta s}}^{\underline{s}+\underline{\Delta s}} \hat{g}^c(\underline{s}')d\underline{s}'$$

and hence

$$\frac{1}{G_1'(0)} \lim_{V\to\infty} \frac{1}{V}\hat{g}(\underline{s}) = \int_{\underline{s}-\underline{\Delta s}}^{\underline{s}+\underline{\Delta s}} \hat{G}_1(\underline{s}')d\underline{s}' = \Delta W(\underline{s}) \tag{2.71a}$$

and

$$\frac{1}{G_2'(0)} \lim_{V\to\infty} \hat{g}(\underline{s}) = \int_{\underline{s}-\underline{\Delta s}}^{\underline{s}+\underline{\Delta s}} \hat{G}_2(\underline{s}')d\underline{s}' = \Delta W(\underline{s}) \tag{2.71b}$$

depending on the integrability properties of the medium.

2.5.2 Analytical Properties of Overall Descriptors

In Sect.2.3, the analytic description for the field and intensity has been put forward as representing the most adequate mathematical tool in scattering problems. We shall apply now this approach to the statistical problem, thus generalizing the treatment presented so far in this section for the real axis. The observable $\hat{g}(\underline{s})$ is always a function of class E. The spectral density $\hat{G}(\underline{s})$ is never a function of class E, nor is $W(\underline{s})$: in defining these idealized quantities we have ruled out the possibility that $\delta\varepsilon(\underline{r}) \equiv 0$ outside a given interval. The spectral density $\hat{G}(\underline{s})$ and the spectral distribution $W(\underline{s})$ may be entire functions or analytic in certain regions of the complex plane or not analytic anywhere. It is interesting to follow the movement of zeros that occurs, as the scattering volume increases, in the transition from $\hat{g}(\underline{s})$ to $W(\underline{s})$.

No characteristic functions having a rate of growth higher than that of a Gaussian function are known [Ref.2.67, Sect.13.2] with the sole exception of the characteristic function corresponding to a Poisson distribution. In no physical situation can one possibly conceive of a medium which may be described by a spectral density corresponding to a Poisson distribution. We shall assume, therefore, that the rate of growth of all characteristic functions that can be attached to media with physical significance is contained between that of a constant and that of a Gaussian.

It is convenient to classify characteristic functions into three classes, by considering their rate of growth described by

$$G(\rho) = O\left\{\exp\left[-\left(\frac{|\rho|}{\rho_0}\right)^{\alpha}\right]\right\} \quad \text{as} \quad |\rho| \to \infty \ ,$$

class 1: $1 < \alpha \leqq 2$,

class 2: $\alpha = 1$,

class 3: $0 \leqq \alpha < 1$.

Characteristic functions of class 1 have spectral densities which are entire functions. This follows from a well-known theorem [Ref.2.68, Theorem 26]. Presumably owing to the difficulties of expressing the spectral density associated with these functions, the only $G(\rho)$ in this group that has found applications is that corresponding to $\alpha = 2$. For

$$G(\rho) = \exp\left(-\frac{\rho^2}{\rho_0^2}\right)$$

the corresponding $\hat{G}(s)$ in three dimensions is given by

$$\hat{G}(s) = (2\langle k\rangle\rho_0)^3 \exp\left[-\left(\frac{\langle k\rangle\rho_0 s}{2}\right)^2\right] \ ,$$

where ρ_0 describes the scale of inhomogeneity of the medium. For this class of characteristic functions, as V increases, the zeros of $\hat{g}(\underline{s})$ move away from the real axis. This follows from the argument presented in Sect.2.2 and, in fact, Theorem 2 may be invoked asymptotically. The effect of the displacement of the zeros is observed as a smoothing of $\hat{g}(\underline{s})$: we say that it is self-apodised. The zero displacement ensures that, although $\hat{g}(\underline{s})$ will always be of class E, it will approximate better and better the entire function of order greater than 1 $\Delta W(\underline{s})$ which is $L^1(-\infty,\infty)$.

Characteristic functions of class 2 have probably been the most widely used in scattering theories and experiments and for this reason will be discussed in more detail. The behaviour of their spectral density is also described by TITCHMARSH's theorem 26 [2.68]. We give here the version put forward by KAWATA [Ref.2.24, Theorem 11.7.1].

Theorem 5: Let $G(\rho)$ be the characteristic function of a distribution function $W(s)$. If, for some $\rho_0 > 0$,

$$G(\rho) = O\left[\exp\left(-\frac{|\rho|}{\rho_0}\right)\right] \quad \text{as} \quad \rho \to \infty$$

then $W(s)$ is analytic, $-\infty < s < \infty$. Furthermore, there exists a function $W(s+i\sigma)$ which is analytic in the strip $|\sigma| < 1/\langle k\rangle\rho_0$ and on the real axis it coincides with $W(s)$.

Moreover, the points $\sigma = \pm(1/\langle k\rangle\rho_0)$ are singular points of $W(s+i\sigma)$. It can be easily seen that this theorem reinforces Theorem 3.

All characteristic functions belonging to this group may be represented by the family

$$G(\rho) = \frac{1}{2^{\nu-1}\Gamma(\nu)}\left(\frac{|\rho|}{\rho_0}\right)^{\nu} K_\nu\left(\frac{|\rho|}{\rho_0}\right) , \qquad (2.72)$$

where ν is a real number and $K_\nu(|\rho|/\rho_0)$ is the MacDonald function (or the Bessel function of imaginary argument) of order ν. $\Gamma(\nu)$ is the well-known Euler's function. This generality may be even better appreciated by examining the corresponding spectral density. Indeed, from [2.62]

$$\hat{G}(s) = 2^{2(\nu-2)}\frac{\Gamma\left(\nu+\frac{3}{2}\right)}{\Gamma(\nu)}\frac{(2\langle k\rangle\rho_0)^3}{\left(4+4\langle k\rangle^2\rho_0^2s^2\right)^{\nu+3/2}} .$$

It is easy to see that this is the simplest expression of a function analytic in a strip, with singularities at $\sigma = \pm(1/\langle k\rangle\rho_0)$, which become poles of arbitrarily high order $(\nu+3/2)$ when ν is a half-integer.

It is well known [Ref.2.69, Sect.3.7] that MacDonald functions of half-integer order may be expressed analytically by

$$K_{n+\frac{1}{2}}\left(\frac{|\rho|}{\rho_0}\right) = \left(\frac{\pi}{2}\frac{\rho_0}{|\rho|}\right)^{\frac{1}{2}} \exp\left(-\frac{|\rho|}{\rho_0}\right) \sum_{p=0}^{n} \frac{(n+p)!}{(n-p)!}\left(2\frac{|\rho|}{\rho_0}\right)^{-p} ,$$

from which the growth properties of these functions are also clearly understood; in the above, $n = 0,1,2,\ldots$. For $n = 0$, the family of characteristic functions gives

$$G(\rho) = \exp\left(-\frac{|\rho|}{\rho_0}\right)$$

and, besides this, the characteristic function (2.72) has been used for $\nu = 1$, 3/2, 5/2 [2.70,71].

As V increases, the zeros of $\hat{g}(\underline{s})$, owing to the apodising effect provided by a judiciously chosen coherence factor, move away from the real axis. Since $\hat{g}(\underline{s})$ is smoothed by this, i.e., locally averaged, its values tend, as V increases, towards the locally averaged values of $\hat{G}(\underline{s})$.

Characteristic functions belonging to class 3 have a spectral density which is never analytic in a strip of nonzero width, nor even everywhere on the infinite real $\underline{s}$ axis. Two types of behaviour may be distinguished here. The first is represented by the functions $G(\rho) \notin E$. An example of this is

$$G(\rho) = \frac{1}{\left[1+\left(\frac{\rho}{\rho_0}\right)^2\right]^2} \quad . \tag{2.73}$$

The associated spectral density is given by

$$\hat{G}(s) = (2\langle k\rangle\rho_0)^3 \exp(-\langle k\rangle\rho_0|s|) \quad ,$$

which is analytic everywhere on the real axis except at the origin. The second type of behaviour of characteristic functions, of class 3 is represented by $G(\rho) \in E$; then $\hat{G}(\underline{s})$ is identically zero outside a finite interval on the s axis and hence, again, is not analytic everywhere on the real s axis. For these functions, as V increases, the zeros of $\hat{g}(\underline{s})$ could move either away from the real axis, or towards it, or both. In the example (2.73) they move away from the real axis, but, if $G(\rho)$ were periodic, the zeros of $\hat{g}(\underline{s})$ would tend to accumulate on the real axis. For the second type of behaviour, i.e., $G(\rho) \in E$, some of the zeros of $\hat{g}(\underline{s})$ might move away from the real s axis and the rest would tend to accumulate on it, in order to define the interval over which $\hat{G}(\underline{s})$ is nonidentically zero.

2.5.3 Determination of Overall Descriptors from Finite Records

Having examined the way in which $G(\rho)$ characterizes the inhomogeneity of a medium and having studied the relationship between $g^c(\rho)$ and $G(\rho)$ as $V \to \infty$, the most important practical problem, which represents the essence of the statistical inverse scattering problem, is that of assessing $G(\rho)$ from a finite record in the primary space, i.e., a finite scattering volume. In practice, this would be done by obtaining $G(\rho)$ from $\hat{g}(\underline{s})$. For a random medium, this is a fundamental statistical inference task, known as the 'time series' problem [2.72-74].

According to a theorem by BOCHNER [Ref.2.75, Supplement p.295] the existence of the spectral distribution function $W(\underline{s})$ is proved under the general assumption that the limit

$$G'(\rho) = \lim_{V\to\infty} \frac{1}{\Omega(V)} \int_V \delta\varepsilon(\underline{r})\delta\varepsilon(\underline{r}+\underline{\rho})d^3\underline{r} \qquad (2.74)$$

exists for all ρ. The arbitrary positive monotonically increasing function $\Omega(V)$ must satisfy the condition

$$\lim_{V\to\infty} \frac{\Omega(V+1)}{\Omega(V)} = 1 \quad .$$

This theorem has been already implicitly invoked in defining $G_1'(\rho)$ for $\Omega(V) = V$ and $G_2'(\rho)$ for $\Omega(V) = 1$.

Bochner's theorem allows one to introduce several estimators for $G(\rho)$. Amongst them, the most important and commonly used correspond to the following:

estimator 1: $\Omega(V) = V$,
estimator 2: $\Omega'(V) = V'(\underline{\rho})$,

where $V'(\underline{\rho})$ is the intersection of the volume V and its 'ghost' displaced by $\underline{\rho}$, i.e., it represents the convolution square of the shape function describing the scattering volume V. Accordingly, we define the autocovariance estimators

a.e. 1: $\frac{1}{\Omega(V)} g^c(\rho)$,

a.e. 2: $\frac{1}{\Omega'(V)} g^c(\rho)$.

The first is a biased estimator [2.73] of $G'(\rho)$ and the second an unbiased estimator. It is widely accepted [2.76] that the use of the biased estimator is preferable, because it has a nonnegative Fourier transform, unlike the unbiased estimator: this is obviously important when considering intensities. Besides, the biased estimator is consistent, i.e., the bias and variance tend to zero as $V\to\infty$. Further and even more important, the biased estimator is immediately physically available in a scattering context. Hence, we shall discuss only the determination of $G(\rho)$ using the biased estimator.

It is well known [2.73] that

$$\frac{1}{V}\mathscr{E}\left[g^c(\rho)\right] = \frac{V'(\underline{\rho})}{V} G'(\rho) \quad , \qquad (2.76)$$

where $\mathscr{E}$ indicates ensemble average and $V'(\rho)$ was defined above.

The Fourier transform of the estimator $(1/V)g^c(\rho)$ is $(1/V)\hat{g}^c(\underline{s})$ with the obvious property that

$$\frac{1}{V}\mathscr{E}\left[\hat{g}^c(\underline{s})\right] = \int_V \frac{V'(\underline{\rho})}{V} G'(\rho)\exp(-i<k>\underline{\rho}\cdot\underline{s})d^3\underline{\rho} = \hat{G}'(\underline{s}) * \frac{\hat{V}'(\underline{s})}{V} \quad , \tag{2.77}$$

where $\hat{V}'(\underline{s})$ is a sinc^2-like function.

The procedure for obtaining a consistent estimate of $G(\rho)$ requires the use of a function $A(\rho)$ known as a lag window [2.74] or, more appropriately in an optical context, an apodising function. It must be an even function, with $A(0)=1$, preferably having an envelope decreasing monotonically towards the boundary of the volume V. A modified $g^c(\rho)$ is obtained by the operation $(1/V)g^c(\rho)A(\rho)$. In the conjugate $\underline{s}$ space this gives the convolution $(1/V)\hat{g}^c(\underline{s}) * \hat{A}(\underline{s})$ and, by taking an ensemble average, we obtain

$$\frac{1}{V}\mathscr{E}\left[\hat{g}^c(\underline{s})*\hat{A}(\underline{s})\right] = \frac{1}{V}\hat{A}(\underline{s}) * \mathscr{E}\left[\hat{g}^c(\underline{s})\right] \quad . \tag{2.78}$$

From (2.77) we observe that

$$\frac{1}{V}\mathscr{E}\left[\hat{g}^c(\underline{s})\right] \rightarrow \hat{G}'(\underline{s}) \tag{2.79}$$

as $V\rightarrow\infty$, which means that, without the use of the apodiser, the estimator $(1/V)\hat{g}^c(\underline{s})$ will be asymptotically unbiased: of course, it will be inconsistent. The relevance of $\hat{A}(\underline{s})$ is to render this estimate consistent. Given (2.79), we can write

$$\hat{A}(\underline{s}) * \hat{G}'(\underline{s}) = \frac{1}{V}\mathscr{E}\left[\hat{A}(\underline{s})*\hat{g}^c(\underline{s})\right] \quad . \tag{2.80}$$

It may be shown that the narrower $\hat{A}(\underline{s})$ the smaller the bias but the larger the variance of $(1/V)\left[\hat{A}(\underline{s})*\hat{g}^c(\underline{s})\right]$. Conversely, a broad $\hat{A}(\underline{s})$ ensures a small variance but, of course, the bias of $(1/V)\left[\hat{A}(\underline{s})*\hat{g}^c(\underline{s})\right]$ increases. In practice, a compromise is called for. A detailed calculation of the bias and variance of the spectral estimators is given in [2.73].

In scattering situations, an inbuilt apodising function $A(\rho)$ is available: it is the partial coherence of the incident field. Indeed, with

$$A(\rho) = \mu_0(<k>\underline{\rho}\cdot\underline{\Delta s})$$

in the first Born approximation, from (2.70) it is obvious that

$$g^c(\rho)A(\rho) = g(\rho)$$

and it follows that (2.80) may be written

$$\int_{\underline{s}-\underline{\Delta s}}^{\underline{s}+\underline{\Delta s}} \hat{G}'(\underline{s}')d\underline{s}' = \frac{1}{V}\mathscr{E}\int_{\underline{s}-\underline{\Delta s}}^{\underline{s}+\underline{\Delta s}} \hat{g}^{c}(\underline{s}')d\underline{s}' \quad , \tag{2.81}$$

which establishes the correspondence with (2.68,69); we may write (2.81) in the form

$$\int_{\underline{s}-\underline{\Delta s}}^{\underline{s}+\underline{\Delta s}} \hat{G}'(\underline{s}')d\underline{s}' = \frac{1}{V}\mathscr{E}[\hat{g}(\underline{s})]$$

and thus stress the parallel with (2.71).

Hence, from an observed scattered intensity, obtained from a finite scatterer, it is possible to construct an acceptable and useful representation of the spectral density, in the interval of measurement available. The method proposed consists of a judicious movement of the complex zeros away from the real axis, which has the effect of rendering a function of class E more and more similar to an averaged spectral density. On the real axis, this is observed as a smoothing of the speckled intensity: the process is achieved by the manipulation of the partial coherence of the incident radiation.

The partial coherence acts effectively as a spatial filter, rendering it practically impossible to observe scales of inhomogeneity larger than the width of the coherence factor. A judicious apodisation, therefore, implies the use of a coherence factor which is considerably narrower than the maximum extent of the scattering volume, in order to ensure the consistency of the estimate, yet at the same time considerably wider than the maximum scale of inhomogeneity present, in order to reduce the bias. If the extent of the scattering volume is, say, $2000\langle\lambda\rangle$ and the maximum scale of inhomogeneity of the order of a few $\langle\lambda\rangle$, a convenient width of the coherence factor is $250\langle\lambda\rangle$.

So far, we have examined only one manifestation of the fundamental categories in the statistical aspect of the inverse scattering problem, namely the finite character of the record in the primary space. The categories, however, manifest themselves also as a limitation of the available space of measurement. In order to solve the statistical inverse scattering problem, it is not sufficient to achieve an unbiased and consistent representation of the spectral density in a finite interval, it is necessary to obtain an unbiased and consistent estimate of the characteristic function. To accomplish this, means ought to be found for continuing analytically the unbiased consistent estimate of $\hat{G}(\underline{s})$, ideally towards infinite values of $\underline{s}$. This is possible – in principle – only for characteristic functions belonging to class 1 and 2, i.e., with growth at least as fast as that of an exponential function, in which case $\hat{G}(\underline{s})$ is either entire or analytic in a strip about the real axis. For characteristic functions of class 3, it is sometimes possible to continue $\hat{G}(\underline{s})$, for example when the only singularity occurs at the $\underline{s}$ origin. In general, however, the estimators of $\hat{G}(\underline{s})$ corresponding to $G(\rho)$ of class 3 are not analytically continuable.

The practical procedure used so far for achieving this, relies on a trial and error method [2.70,71]. With a judiciously chosen coherence and by the process of analytic continuation indicated, it is possible to detect and measure, using visible light, inhomogeneities on a scale of about 100 Å and even finer. Two illustrations are given, indicating the suitably smoothed intensity (Figs.2.8,9). The experimental

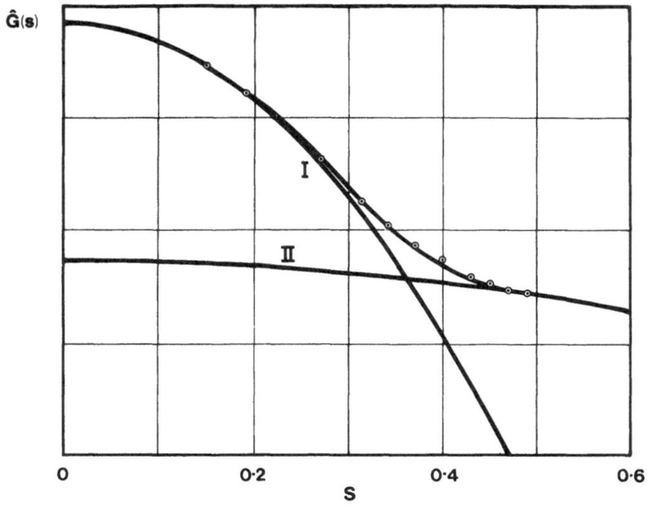

Fig. 2.8

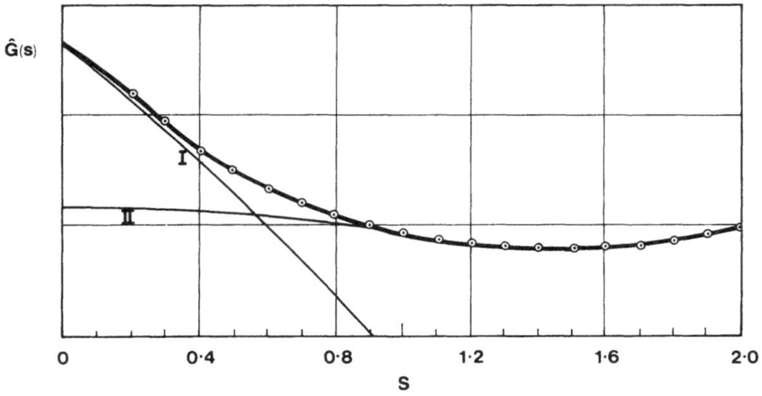

Fig. 2.9

Fig. 2.8. Experimental points of the suitably smoothed intensity, i.e., of the spectral density estimator and the spectral densities (full lines) corresponding to the characteristic functions:
I $G(\rho) = \exp(-\rho^2/\rho_0^2)$, $\rho_0 = 0.727$ µm
II $G(\rho) = (1+|\rho|/\rho_0) \exp(-|\rho|/\rho_0)$, $\rho_0 = 0.055$ µm

Fig. 2.9. Experimental points of the suitably smoothed intensity, i.e., of the spectral density estimator and the spectral densities (full lines) corresponding to the characteristic functions:
I $G(\rho) = 1/[1+(\rho^2/\rho_0^2)]^2$, $\rho_0 = 0.451$ µm
II $G(\rho) = \exp(-\rho^2/\rho_0^2)$, $\rho_0 = 0.032$ µm

points, recording the estimator of $G(\underline{s})$ are matched by superpositions of spectral densities. In Fig.2.8, curve I corresponds to $G(\rho) = \exp(-\rho^2/\rho_0^2)$ for $\rho_0 = 0.727$ µm and curve II to $G(\rho) = (1+|\rho|/\rho_0) \exp(-|\rho|/\rho_0)$ with $\rho_0 = 0.055$ µm. The latter is the generic characteristic function of class 2 given by (2.72) for $\nu = 3/2$. In Fig.2.9, curve I corresponds to $G(\rho) = 1/[1+(\rho^2/\rho_0^2)]^2$ for $\rho_0 = 0.451$ µm and curve II to $G(\rho) = \exp(-\rho^2/\rho_0^2)$ for $\rho_0 = 0.032$ µm.

In making the transition from (2.28) to (2.60), in other words in assuming the independence between field and perturbation necessary for constructing a solution to the statistical problem, the first Born approximation was invoked. This implies real inhomogeneity distribution functions — and hence real characteristic functions. Indeed, the properties of characteristic functions and of the corresponding spectral densities have been outlined above only for nonabsorbing media. For complex permittivities, the characteristic function may be written [2.77]

$$G(\rho) = G_R(\rho) + iG_A(\rho) \quad ,$$

where, if $\delta\varepsilon(\underline{r}) \in S(-\infty,\infty)$,

$$G_R(\rho) = \frac{1}{<|\delta\varepsilon(\underline{r})|^2>}\left[<\delta\varepsilon_R(\underline{r})\delta\varepsilon_R(\underline{r}+\underline{\rho})> + <\delta\varepsilon_A(\underline{r})\delta\varepsilon_A(\underline{r}+\underline{\rho})>\right]$$

$$G_A(\rho) = \frac{1}{<|\delta\varepsilon(\underline{r})|^2>}\left[<\delta\varepsilon_R(\underline{r})\delta\varepsilon_A(\underline{r}+\underline{\rho})> - <\delta\varepsilon_A(\underline{r})\delta\varepsilon_R(\underline{r}+\underline{\rho})>\right] \quad .$$

Obviously, the characteristic function is Hermitian,

$$G(-\rho) = G^*(\rho)$$

and thus $\hat{G}(\underline{s})$ is real. However, the existence of the nonzero imaginary part $G_A(\rho)$ renders $\hat{G}(\underline{s})$ uneven, although it may still be symmetric with respect to a nonzero spatial frequency.

If the medium is Gaussian [2.61],

$$<\delta\varepsilon(\underline{r})\delta\varepsilon(\underline{r}+\underline{\rho})> = 0 \quad ,$$

i.e.,

$$<\delta\varepsilon_R(\underline{r})\delta\varepsilon_R(\underline{r}+\underline{\rho})> = <\delta\varepsilon_A(\underline{r})\delta\varepsilon_A(\underline{r}+\underline{\rho})> \quad ,$$

$$<\delta\varepsilon_R(\underline{r})\delta\varepsilon_A(\underline{r}+\underline{\rho})> = -<\delta\varepsilon_A(\underline{r})\delta\varepsilon_R(\underline{r}+\underline{\rho})> \quad .$$

The determination of theoretical expressions for $G(\rho)$ assuming a certain type of inhomogeneity is not an easy problem. If $\delta\varepsilon(\underline{r})$ may be regarded as describing a Markov process, a straightforward calculation leads to

$$G(\rho) = \exp\left(-\frac{|\rho|}{\rho_0}\right)\exp(i2\pi s_0\rho) \quad ,$$

where the parameter s_0 is due to the correlation between the absorptive and refractive parts of permittivity and acts as a carrier frequency: it represents the spatial frequency in the $\underline{s}$ space with respect to which the spectral density is symmetric. It is not unreasonable, therefore, to assume that, for absorbing media, all characteristic functions may be expressed in the form

$$G(\rho) = |G(\rho)| \exp(i2\pi s_0\rho) \quad ,$$

where $|G(\rho)|$ represents the characteristic function defined for real inhomogeneity distribution functions, with properties studied above. Since for such scatterers (2.60) cannot be written, there is no simple correspondence between $g(\rho)$ and $G(\rho)$: for absorbing media the statistical aspect of the inverse scattering problem has not been solved yet, nor even systematically investigated.

So far, only the situation of a locally isotropic medium has been discussed. In dealing with anisotropic media, exactly the same problems are encountered but, in addition to these, the tasks are more complicated since all the elements of a characteristic matrix ought to be established. For the general case of an inhomogeneity of class $S(-\infty,\infty)$ for any direction in the medium, the three elements of the characteristic matrix — expressed in diagonal form — are [2.19]:

$$G'_{11} = \left\langle \left[\delta\varepsilon(\underline{r}_i)\right]_{11}\left[\delta\varepsilon(\underline{r}_j)\right]_{11}\right\rangle + \left\langle \left[\delta\varepsilon(\underline{r}_i)\right]_{21}\left[\delta\varepsilon(\underline{r}_j)\right]_{21}\right\rangle + \left\langle \left[\delta\varepsilon(\underline{r}_i)\right]_{31}\left[\delta\varepsilon(\underline{r}_j)\right]_{31}\right\rangle$$

$$G'_{22} = \left\langle \left[\delta\varepsilon(\underline{r}_i)\right]_{12}\left[\delta\varepsilon(\underline{r}_j)\right]_{12}\right\rangle + \left\langle \left[\delta\varepsilon(\underline{r}_i)\right]_{22}\left[\delta\varepsilon(\underline{r}_j)\right]_{22}\right\rangle + \left\langle \left[\delta\varepsilon(\underline{r}_i)\right]_{32}\left[\delta\varepsilon(\underline{r}_j)\right]_{32}\right\rangle$$

$$G'_{33} = \left\langle \left[\delta\varepsilon(\underline{r}_i)\right]_{13}\left[\delta\varepsilon(\underline{r}_j)\right]_{13}\right\rangle + \left\langle \left[\delta\varepsilon(\underline{r}_i)\right]_{23}\left[\delta\varepsilon(\underline{r}_j)\right]_{23}\right\rangle + \left\langle \left[\delta\varepsilon(\underline{r}_i)\right]_{33}\left[\delta\varepsilon(\underline{r}_j)\right]_{33}\right\rangle$$

in which, owing to the symmetry of the tensor $\delta\underline{\underline{\varepsilon}}(\underline{r})$,

$$\left\langle \left[\delta\varepsilon(\underline{r}_i)\right]_{mn}\left[\delta\varepsilon(\underline{r}_j)\right]_{mn}\right\rangle = \left\langle \left[\delta\varepsilon(\underline{r}_i)\right]_{nm}\left[\delta\varepsilon(\underline{r}_j)\right]_{nm}\right\rangle$$

where $m,n = 1,2,3$.

In order to determine the characteristic matrix for media exhibiting overall isotropy, it is necessary to find only two averages of the form

$$\left\langle \left[\delta\varepsilon(\underline{r}_i)\right]_{mn}\left[\delta\varepsilon(\underline{r}_j)\right]_{mn}\right\rangle$$

since all such terms with $m \neq n$ are equal between them and all those with $m = n$ are similarly equal between them.

Making appeal to (2.30,31), derived assuming the validity of the first Born approximation, it is obvious that these averages – or, rather, an estimate of them – may be easily obtained by performing two scattering measurements. The first is made with $\theta_A = 0$, i.e., with the polariser parallel to the analyser: this yields the diagonal elements. The second is made with $\theta_A = \pi/2$, i.e., with the analyser perpendicular to the polariser: from this, one obtains the off-diagonal elements of the characteristic matrix. It is customary to refer to these two arrangements by the symbols V_v and H_v, respectively.

In order to solve the statistical inverse scattering problem two main extensions of the treatment put forward are required. The first is to overcome the restrictions imposed by the first Born approximation. This would involve, however, an infinite regress. The second task is to generalize the theory to media which exhibit overall anisotropy.

From the discussion presented, it is clear that an overall statistical description of a scattering medium is not, strictly speaking, possible. The theory put forward, however, does enable one to construct a useful representation of it.

2.6 Conclusions

All the interests of my reason, speculative as well as practical, combine in the three following questions:
1. What can I know?
2. What ought I to do?
3. What may I hope?

Kant, Critique of Pure Reason A805, B833.

The inverse scattering problem is due to unavoidable fundamental limitations inherent in any observation, which act as categories conditioning any measurement. The contribution has attempted to indicate how we can turn these categories to our advantage, by using the *a priori* knowledge they provide. The key to constructing a solution to the inverse scattering problem is offered by the use of the mathematical equipment which is generated by, and describes logically, the fundamental categories through which we perceive physical reality. At best, however, what we can know is only an encoding of the physical reality in terms of concepts we ourselves create.

Analyticity is a topic which holds a fascination for theoretical physicists. Some of the most important contributors to modern physics have wondered what is the significance of analytic functions in our description of the world. CHEW [2.78] was sufficiently interested in this problem to formulate a specific question: 'does analyticity in itself constitute a suitable *a priori* principle for the formulation of theories?' We have put forward a partial answer to this fundamental question in a scattering context, by showing that the description of scattered fields involves *a priori* a particular class of analytic functions, namely entire functions of exponential type. Indeed, by examining the vector Helmholtz equation governing the propagation of the field, it may be shown that its solutions are entire functions. We believe that, for a finite extent of the primary space, these functions are always of order one, on a complex surface with a curvature depending on the distance from the scatterer. This mathematical representation also offers a link with information theory. The indicator function, which is the most important descriptor of functions of class E, may be regarded as the channel capacity [2.79] of the experiment. Each individual zero is a degree of freedom or a logon.

What ought we to do next? First and foremost, it is important to examine further the properties of the distributions of zeros. Any fresh knowledge in this area would contribute to our understanding of encoding and propagation of information and thus to the possible improvement of the experimental design and computational procedure.

By relying on the properties of analytic functions, it is, in principle, possible to continue the data outside the experimentally available region. This would provide the means of reconstructing the object wave with an increased resolution, corresponding to the harmonic associated with the most distant zero that can be located. Ideally, that would be a simple exercise. This is the ultimate aim of obtaining the solution to the inverse scattering problem. However, in the presence of noise, unavoidably affecting any observation, this desideratum cannot be achieved. The noise introduces a region of uncertainty about the exact position of each zero. The extent of this region increases beyond the interval of observation, owing to the effective finite range of influence each zero has.

Amongst the many difficulties still awaiting a rigorous treatment, an important objective for future work is to study in detail the effect of noise on zero location and in particular to establish the acceptable uncertainty in the position of each. It is an impossible effort to locate zeros beyond a certain distance outside the interval of measurement; indeed, even establishing their precise position inside it will be a problem.

Thus, when, at the end of this chapter, we try to fathom what we can hope, the wistful words of Faust, the eternal seeker of solutions to inverse problems, come to mind [Ref.2.80, p.35]:

O happy he, who still can hope
Out of this sea of error to arise!
We long to use what lies beyond our scope,
Yet cannot use even what within it lies.

Acknowledgements. The algorithm for phase retrieval presented in Sect.2.4 has been developed by Mr. H. Moezzi. The authors are greatly indebted to him, as well as to Dr. D.A. Jarvis and Mr. I. Manolitsakis, for kindly allowing them to use as yet unpublished material. A travel grant awarded by the British Council to one of the authors (M. N.-V.) facilitated the collaborative effort involved in the writing of this contribution. The authors are also grateful to Mrs. J. Burns-Moyes and Mrs. A. Baker for their patient and careful preparation of the script.

References

2.1 Plato: *The Collected Dialogues* (Princeton University Press, Princeton 1973)
2.2 K.R. Popper: *The Logic of Scientific Discovery* (Hutchinson, London 1972)
2.3 J.W. Yolton: *The Philosophy of Science of A.S. Eddington* (Martinus Nijhoff, The Hague 1960)
2.4 H.P. Baltes: "Introduction", in *Inverse Source Problems in Optics*, ed. by H.P. Baltes, Topics in Current Physics, Vol.9 (Springer, Berlin Heidelberg, New York 1978)
2.5 I. Kant: *Critique of Pure Reason*, translated by N. Kemp Smith (Macmillan, London 1964)
2.6 I. Berlin: "The Purpose of Philosophy", in *Concepts and Categories*, ed. by I. Berlin (Hogarth Press, London 1978) pp.1-11
2.7 A. Eddington: *The Philosophy of Physical Science* (Cambridge University Press, Cambridge 1939)
2.8 J. Agassi: *Science in Flux* (Reidel, Dordrecht 1975)
2.9 A. Eddington: *Relativity Theory of Protons and Electrons* (Cambridge University Press, Cambridge 1936)
2.10 K.R. Popper: *Conjectures and Refutations* (Routledge and Kegan Paul, London 1963)
2.11 J.P. Eckermann: *Conversations with Goethe* (Dent and Sons, London 1951)
2.12 Marcus Annaeus Lucanus: *The Civil War* (The Pharsalia) (Heinemann, London 1928)
2.13 M. Born, E. Wolf: *Principles of Optics* (Pergamon Press, Oxford 1975)
2.14 M.J. Beran, G.B. Parrent, Jr.: *Theory of Partial Coherence* (Prentice-Hall, Englewood Cliffs 1964)
2.15 P.H. Morse, H. Feschbach: *Methods of Theoretical Physics* (McGraw-Hill, New York 1953)
2.16 G. Ross: Opt. Acta *25*, 57 (1978)
2.17 J.W. Goodman: *Introduction to Fourier Optics* (McGraw-Hill, New York 1968)
2.18 U. Frisch: Ann. d'Astrophys. *29*, 645 (1966); *30*, 565 (1967)
2.19 D.A. Jarvis: *The Characterization of Orientation in Anisotropic Materials by Means of Elastic Light Scattering;* Ph.D. Thesis, Queen Elizabeth College, University of London (1977)
2.20 R.E.A.C. Paley, N. Wiener: *Fourier Transforms in the Complex Domain* (American Mathematical Society, Ann Arbor, Michigan 1934)
2.21 R.P. Boas: *Entire Functions* (Academic Press, New York 1954)
2.22 K. Chadan, P.C. Sabatier: *Inverse Problems in Quantum Scattering Theory* (Springer, Berlin, Heidelberg, New York 1977)
2.23 B.Ja. Levin: *Distribution of Zeros of Entire Functions* (American Mathematical Society, Providence, Rhode Island 1964)
2.24 T. Kawata: *Fourier Analysis in Probability Theory* (Academic Press, New York 1972)
2.25 E.C. Titchmarsh: Proc. Lond. Math. Soc. (2) *25*, 283 (1925)
2.26 G. Ross, M.A. Fiddy, M. Nieto-Vesperinas, M.W.L. Wheeler: Proc. Roy. Soc. (London) *A360*, 25 (1973)
2.27 H.B. Voelcker, A.A.G. Requicha: IEEE Comm. *21*, 933 (1973)
2.28 B.A. Fuks: *Introduction to the Theory of Analytic Functions of Several Complex Variables* (American Mathematical Society, Providence, Rhode Island 1963)
2.29 M. Plancherel, G. Polya: Commentari Mathematici Helvetici *9*, 224 (1936-7)
2.30 G. Ross, M.A. Fiddy, M. Nieto-Vesperinas, I. Manolitsakis: Opt. Acta *26*, 229 (1979)
2.31 D. Gabor: "Light and Information", in *Proc. Symposium on Astronomical Optics and Related Subjects*, ed. by Z. Kopal (North-Holland, Amsterdam 1956) pp.17-30
2.32 G. Ross: Phil. Trans. Roy. Soc. London *A268*, 177 (1970)
2.33 J.T. Winthrop: J. Opt. Soc. Am. *67*, 833 (1977)
2.34 C.E. Shannon: Bell Systems Techn. J. *27*, 623 (1948)
2.35 B.J. Hoenders, H.A. Ferwerda: Optik *37*, 542 (1973)
2.36 A.S.B. Holland: *Introduction to the Theory of Entire Functions* (Academic Press, London 1973)
2.37 G. Ross, M.A. Fiddy: Opt. Acta *25*, 205 (1978)
2.38 H.M. Nussenzveig: *Causality and Dispersion Relations* (Academic Press, London 1972)

2.39 J.S. Toll: Phys. Rev. *104*, 1760 (1956)
2.40 D.L. Misell, R.E. Burge, A.H. Greenaway: J. Phys. D *7*, L27 (1974)
2.41 R.E. Burge, M.A. Fiddy, A.H. Greenaway, G. Ross: Proc. Roy. Soc. London *A350*, 191 (1976)
2.42 A.H. Greenaway: Opt. Lett. *1*, 10 (1977)
2.43 H.A. Ferwerda: "The Phase Reconstruction Problem for Wave Amplitudes and Coherence Functions", in *Inverse Source Problems in Optics*, ed. by H.P. Baltes, Topics in Current Physics, Vol.9 (Springer, Berlin, Heidelberg, New York 1978) Chap.2
2.44 D.L. Misell: J. Phys. D *6*, L6 and 2200 (1973)
2.45 R.W. Gerchberg, W.O. Saxton: Optik *35*, 237 (1972)
2.46 D.L. Misell: Adv. in Optical and Electron Microscopy *7*, 185 (1978)
2.47 G. Ross, G.J. Brownsey: Plastics and Rubber: Materials and Applications *1*, 109 (1976)
2.48 E. Wolf: J. Opt. Soc. Am. *60*, 18 (1970)
2.49 P.D. Rowley: J. Opt. Soc. Am. *59*, 1946 (1969)
2.50 R. Dändliker, K. Weiss: Opt. Commun. *1*, 323 (1970)
2.51 W. Carter: J. Opt. Soc. Am. *60*, 306 (1970)
2.52 A. Klug, R.A. Crowther: Nature *238*, 435 (1972)
2.53 M.V. Berry, D.F. Gibbs: Proc. Roy. Soc. Lond. A*314*, 143 (1970)
2.54 D.W. Sweeney, C.M. Vest: Appl. Opt. *12*, 2649 (1973)
2.55 H.H. Lipson, W.A. Wooster: *Interpretation of X-ray Diffraction Photographs* (Macmillan, London 1960)
2.56 H.G. Schmidt-Weinmar: J. Opt. Soc. Am. *65*, 1059 (1975)
2.57 A.F. Fercher, H. Bartelt, H. Becker, E. Wiltschko: Appl. Opt. *18*, 2427 (1979)
2.58 B.J. Hoenders: "The Uniqueness of Inverse Problems", in *Inverse Source Problems in Optics*, ed. by H.P. Baltes, Topics in Current Physics, Vol.9 (Springer, Berlin, Heidelberg, New York 1978) Chap.3
2.59 N. Wiener: *The Fourier Integral and Certain of its Applications* (Dover, New York 1933)
2.60 A. Papoulis: *Probability, Random Variables and Stochastic Processes* (McGraw-Hill, New York 1965)
2.61 J.L. Doob: *Stochastic Processes* (Wiley, New York 1967)
2.62 G. Ross: Opt. Acta *15*, 451 (1968)
2.63 E. Lukacs: *Characteristic Functions* (Griffin, London 1960)
2.64 R.C. Bourret: Can. J. Phys. *40*, 782 (1962)
2.65 R.C. Bourret: Nuovo Cimento *26*, 1 (1962)
2.66 J.B. Keller: Proc. Symp. Appl. Math. *13*, 227 (1962)
2.67 Yu.V. Linnik: *Decomposition of Probability Distributions* (Oliver and Boyd, London 1964)
2.68 E.C. Titchmarsh: *Introduction to the Theory of the Fourier Integral* (Oxford University Press, Oxford 1948)
2.69 G.N. Watson: *A Treatise on the Theory of Bessel Functions* (Cambridge University Press, Cambridge 1958)
2.70 G. Ross: Opt. Acta *16*, 95 (1969)
2.71 G. Ross, R.L. Addleman: J. Phys. D *6*, 1537 (1973)
2.72 L.H. Koopmans: *The Spectral Analysis of Time Series* (Academic Press, New York 1974)
2.73 G.M. Jenkins, D.G. Watts: *Spectral Analysis and its Applications* (Holden-Day, San Francisco 1968)
2.74 R.B. Blackman, J.W. Tukey: *The Measurement of Power Spectra* (Dover, New York 1958)
2.75 S. Bochner: *Lectures on Fourier Integrals* (Princeton University Press, Princeton 1959)
2.76 E. Parzen: Technometrics *3*, 167 (1961)
2.77 M. Nieto-Vesperinas: "Statistics of Amorphous Absorbing Media and Object Wave Determination Using the Theory of Entire Functions"; Ph.D. Thesis, Queen Elizabeth College, University of London (1978)
2.78 G.F. Chew: *The Analytic S Matrix* (Benjamin, New York 1966)
2.79 N.J. Bershad: J. Opt. Soc. Am. *59*, 157 (1969)
2.80 J.W. von Goethe: *Faust*, translated by W.H. Bruford (Dent & Sons, London 1954)

3. Photon-Counting Statistics of Optical Scintillation

E. Jakeman and P. N. Pusey

With 9 Figures

In contrast to the weak statistical scatterers and their first-order correlation and average scattered intensity studied in Sect.2.5, this chapter aims at strong scatterers and the fluctuations and higher-order correlations of the scattered radiation as explored by photon-counting techniques. The pertinent inverse detection and scattering problems are discussed with emphasis on a new class of distributions – the K distributions – which were recently found to characterize the fluctuations in scattered intensity for a wide class of experimental situations.

3.1 Introductory Remarks

In this chapter we present an exposition of the relationship between the statistical properties of photodetections in light scattering experiments employing photon-counting techniques and the statistical properties of the scattering medium. The inverse problem for such measurements can be divided into two parts. First a connection must be established between the raw data – the photo-electron emission times – and the properties of the received electromagnetic field; this is the inverse detection problem. Secondly, connection must be made between the properties of the received electromagnetic field and those of the scattering medium; this is the inverse scattering problem. The detection problem in photon-counting experiments is a fairly specialised area of physics which has developed rapidly since the advent of the laser (see also Chap.4). Several detailed reviews of the subject are now available, however [3.1-3], and our aim here will be to summarise results on the most important aspects of the problem. The interaction of electromagnetic radiation with matter, on the other hand, is of widespread interest throughout science and work on this subject is not open to brief survey. Instead we shall confine discussion to a limited, though important class of scattering systems which give rise to strong intensity fluctuations or *scintillation* of the scattered electromagnetic field. This choice is motivated by the recent introduction of photon-counting techniques into measurements of optical scintillation [3.4,5]. Although a natural approach to adopt in low intensity situations such as stellar scintillation measurements [3.6] photon-counting would not appear to be appropriate in many experiments involving lasers, for example, where the photon flux may be high. However, in contrast to analogue processing techniques, a photon-counting system offers the possibility of achieving high dynamic range coupled with linear and virtually noiseless signal processing. These are clearly advantages in the measurement of statistical properties for which there is little physical insight, let alone mathematical predictions. The accuracy and reliability of statistical data obtained by photon-counting techniques provide, in these circumstances, a firmer basis for the development of new theoretical models for the scattering interaction and for the scattering medium itself.

In Sect.3.2 we review the salient features of the *inverse detection problem* in photon-counting experiments. The brief description given above is, of course, a gross oversimplification of the actual inversion which must be accomplished. The true nature of the problem is revealed by tracing the flow of information through the detection apparatus. The scattered light of interest falls on the photocathode of a photomultiplier tube; this has a finite area so that there is some spatial averaging of the incident intensity pattern. Photo-electrons are emitted randomly by the photocathode at a rate which is modulated according to the square of the envelope of the incident electromagnetic field and multiplication at subsequent dynodes leads to output pulses of various shapes and sizes associated with each initial event. In addition to photo-electron emissions related to the incident photon flux there will be thermally induced events or "dark counts" within the detector which constitute an additive Poisson distributed noise. Due to the finite response time of certain elements of the detector there will also be a dead time after each pulse during which photo-events will not occur. Selection and shaping techniques are used to convert the detector output into a series of identical pulses or "counts" (these techniques are discussed at length in the literature [3.7] and as they do not affect the statistics of the output pulses they will not be considered further here). One more procedure is undertaken prior to statistical analysis: time is divided into sample intervals or integration times with the number of pulses or counts in each interval representing the basic data presented for analysis. There is thus some time-averaging of fluctuations in the photo-electron emission rate. Four of the processes mentioned above, the square-law detector response, dead times, spatial averaging and temporal averaging can be regarded as instrumental functions requiring deconvolution. The random photo-electron emission process, on the other hand, presents a noise limitation on photon-counting measurements. As we shall see in Sect.3.2, it is not an additive noise process, but depends on the strength of the incident photon flux. This feature means that noise considerations in the post-detection processing of optical signals are usually quite different from those encountered in the case of microwave signals [3.8,9]. In fact the thermally excited detector dark counts are most closely analogous to the additive Gaussian (Johnson) noise which often limits the performance of radar systems. However, dark counts are usually negligible by comparison with signal-induced detector shot-noise in photon-counting experiments.

In addition to the "quantum" noise introduced by the detector as a result of incident signal photons, there will in some experiments be an additive random emission of photo-electrons due to external background radiation or to stray light. Fluctuations in the signal, too, may present a noise problem (as in the fading of radio waves) whilst the finite duration of experimental observations will always provide a limitation on the accuracy with which the value of any statistical quantity can be measured. Both instrumental functions and noise considerations play a role in the design of signal processing equipment to "solve" the inverse detection problem.

Although the detection part of the inverse problem will be considered in Sect.3.2 in fairly general terms, subsequent sections on the *scattering problem* will concentrate on the propagation of waves through layers or extended regions of random media containing refractive index inhomogeneities which are large compared to the wavelength. In particular, developments in discrete scatterer models stimulated by recent photon-counting measurements of scintillation generated by laboratory scattering systems will be reviewed [3.10,11].

The simplest mathematical model for the thin layer configuration is the random phase screen which introduces randomly varying path differences into an incident plane wave. The inverse problem for this model consists of relating the statistical properties of amplitude fluctuations which develop beyond the screen as a result of free propagation, to those of the phase fluctuations introduced initially by the scattering layer. If these are large compared to the wavelength of the incident radiation, the phase screen is said to be "deep" and it is in this situation that the use of discrete scatter approximations can be most easily justified. The reader is referred to an article on random phase screens in the earlier volume of this series for an extensive bibliography of the subject [3.12] but in view of its relevance to the present chapter a brief review of its historical development is appropriate in this introductory section.

The earliest theoretical study of phase screen scattering emanated from the radio frequency area where a model was sought for the scattering of radio waves by fluctuations in the ionosphere [3.13]. The growth of interest in interplanetary scintillation which took place throughout the fifties and sixties promoted further theoretical work including that of MERCIER [3.14], who first showed that with monochromatic plane wave illumination plots of scintillation index (normalised rms intensity) against propagation distance in the Fresnel region beyond the screen could exhibit a high contrast peak if the phase fluctuations introduced by the scattering layer or surface were sufficiently strong. Unlike the weak scattering case, where the Born approximation can be used to effect a simple relationship between properties of the initial wave front and those of the subsequent amplitude fluctuations, the inverse propagation problem for a deep random phase screen is complicated and solution is usually attempted on the basis of an assumed mathematical model for the properties of the initial wave front. Even so, numerical evaluation of the effects of propagation are difficult and although analytical investigation by several authors during the last decade has clarified the mathematical structure of the problem [3.15-19], a number of important questions remain unanswered. This is due in part to the fact that radio frequency measurements are confined to remote natural phenomena so that adequate testing of theoretical models has not been feasible for workers in the area of science most closely associated with phase screen scattering.

Although radio frequency scintillation has stimulated some work on the extended medium scattering problem, this is greatly outweighed by the enormous effort which has been applied to the theoretical and experimental study of scintillation effects arising when light propagates through extended regions of the atmosphere containing refractive index fluctuations generated by turbulent mixing [3.20-23]. These effects limit the performance of laser systems operating over long paths and are therefore of considerable practical importance. As in the radio frequency case mentioned above, laboratory experiments which simulate this kind of scattering in a controllable fashion are not easily contrived. In view of the close relationship between the extended medium configuration and the simple thin layer or screen problem it is perhaps surprising that the vast literature on atmospheric propagation contained, until recently, little reference to early phase screen theory and no reports of laboratory experiments on this system. It is interesting to note that in practice, localised regions of strong turbulence in the atmosphere are far more likely than the homogeneous distribution commonly assumed in theoretical studies of propagation.

From the point of view of testing phase screen scattering theory, experiments at optical frequencies clearly have many advantages. Phase screens which are deep compared to the wavelength of light are easily constructed and strong scintillation effects can be generated in the laboratory without difficulty. The fact that these effects are visible to the naked eye is, perhaps, the greatest advantage, and it is not surprising that the twinkling of starlight is the natural scintillation phenomenon of most long standing interest. Measurement of this phenomenon has been undertaken by many workers over the past three decades using a variety of techniques [3.6]. However, the low power per unit band-width per unit area of starlight has meant that in early experiments the scintillation pattern was invariably averaged over wavelength, space and time in order to obtain sufficient post-detection signal for analogue processing techniques to be viable. Only recently, with the introduction of photon-counting methods, has it become possible to avoid this averaging and establish the true nature of stellar scintillation for the first time [3.4]. Detailed laboratory experiments designed specifically to test the predictions of phase screen theory in the Fresnel region were eventually carried out in 1976. Phase fluctuations in the experiments were generated by a rising plume of hot air, and the scintillation of a scattered laser beam was analysed using photon-counting techniques [3.24]. A new class of distributions – K distributions – were found to characterize the fluctuations in intensity of the scattered light for a wide range of parameter values [3.5]. Subsequently it was discovered that these distributions fitted data from a variety of systems (reviewed in Sect.3.5) involving the scattering of electromagnetic radiation from turbulent media [3.10]. The inclusion of non-Poissonian step number fluctuations in the random walk problem (Sects.3.3 and 3.4) appears to provide a possible explanation for this universality and may lead to new models for the scattering medium itself.

3.2 Photon-Counting Statistics

In this section we consider the inverse detection problem in photon-counting experiments. Several reviews of photon-counting techniques and photon-correlation spectroscopy have now been published [3.1-3] and the reader is referred to these for historical background and bibliography. Here we shall discuss only the "bare bones" of the subject, presenting results for the most part as logical steps in a development governed by the parallel improvement of electronic hardware and signal processing techniques which has occurred over the last two decades. The two principle types of statistical measurement in photon-counting experiments will be discussed first, together with the effect of temporal integration. Other instrumental effects and the problems of noise and finite experiment time will be considered in separate subsections.

3.2.1 Single-Interval Statistics

The term "photon-counting" is something of a misnomer since the actual countable events are current pulses generated by the emission of photo-electrons from the photocathode of the detector. A fully quantum-mechanical treatment of detection involving quantisation of the incident Maxwell field as well as the electron-emission process was first given by GLAUBER [3.25] and several aspects of this problem are discussed in Chap.4. A semi-classical approach, in which the field is not quantised, has more intuitive appeal, perhaps, and gives identical results for the types of optical field discussed in this chapter. It cannot cope, however, with manifestly quantum effects such as antibunching [3.26-30]. According to semi-classical theory, when detection takes place at a single point (i.e., when there is no spatial averaging) the emission of photo-electrons is a doubly stochastic random process governed by [3.31]

$$p(n;T) = \int_0^\infty \frac{(\eta U)^n}{n!} \exp(-\eta U) P(U) dU \quad , \tag{3.1}$$

where p(n;T) is the probability of counting n photo-electron emissions in the time interval T when the integrated intensity

$$U(t;T) = \int_{t-T/2}^{t+T/2} I(t')dt' \tag{3.2}$$

is characterized by the probability density P(U). By intensity we mean here the square of the envelope of a linearly polarized incident field

$$I(t) = |V(t)|^2 \quad , \tag{3.3}$$

where V(t) is the complex analytic signal or positive frequency component of the field. Here η is a constant efficiency factor for the emission process, but may be used to take into account the random sampling effects of pulse losses in the multiplier tube and discrimination at the detector output. The quantity p(n;T) may then be regarded as the distribution of events or counts actually presented for processing in real experiments. We shall neglect dark counts at this stage; these add in a purely random fashion to the signal-induced counts and will be discussed briefly later on (see Sect.3.2.4). We shall also ignore the possibility of unpleasant detector imperfections, such as correlated after-pulsing, which are usually avoided in practice by judicious choice of photo-multiplier tube.

Inversion of (3.1) to obtain properties of U(t) would appear to involve first reconstructing P(U) from p(n;T) (which can be calculated directly from the data). This problem has received scant attention in the literature in spite of the fact that the earliest and simplest photon-counting measurements were of p(n;T). It is solved in principle by the following transformation [3.32]

$$p(U) = \frac{1}{2\pi} \exp(\eta U) \int_{-\infty}^{+\infty} \sum_{n=0}^{\infty} \left(\frac{ix}{\eta}\right)^n p(n;T) \exp(-ixU)dx \quad . \tag{3.4}$$

To our knowledge no distributions of integrated intensity have ever been obtained using (3.4). This is partly due to the fact that in the early days of photon-counting it was soon recognized that two kinds of light were of overriding importance. Statistical models for these types of optical field are well understood and can be used with confidence to calculate the expected properties of p(n;T). The inverse problem then reduces to the fitting of model parameters to give best agreement with experimental data. A second reason why the inversion (3.4) has not been attempted is the existence of a simple relationship between the moments of p(n;T) and those of P(U). In fact it can easily be shown from (3.1) that

$$\begin{aligned} N^{[r]} &= \langle n(n-1)(n-2)\ldots(n-r+1)\rangle = \eta^r \langle U^r \rangle \\ n^{[r]} &= N^{[r]}/\langle n\rangle^r = \langle U^r \rangle / \langle U \rangle^r \end{aligned} \tag{3.5}$$

i.e., the normalized factorial moments of the photon-counting distribution are identical to the normalized moments of the integrated intensity fluctuation distribution (we shall assume throughout that processes are stationary and ergodic so that the angle brackets may be interpreted as time or ensemble averages). Since, for many purposes, a knowledge of the moments is an adequate measure of the distribution itself there seems little purpose in pursuing the difficult route defined by (3.4). The relationship (3.5) has led to the moments of p(n;T) becoming the principal single-interval statistic of interest in photon-counting experiments.

As we have already mentioned, two kinds of optical field are of particular importance. Early photon-counting experiments were carried out at a time when the statistical properties of laser light were of great interest. For an ideal, single mode, amplitude-stabilized laser the intensity is constant ($U=U_0=I_0T$) and

$$P(U) = \delta(U-U_0) \quad . \tag{3.6a}$$

From (3.1) the photon-counts are then Poisson distributed

$$p(n;T) = (n!)^{-1}(\eta U_0)^n \exp(-\eta U_0) \quad ; \quad n^{[r]} = 1 \quad . \tag{3.6b}$$

This result for a coherent field may be contrasted with that for a Gaussian field. In this case the intensity varies with time and two limiting cases can be identified. If we first assume that temporal smoothing is negligible (U=IT) then [3.33,34]

$$P(U) = \exp(-U/U_0)/U_0 \tag{3.7a}$$

and from (3.1) the photon-counts obey a geometrical distribution

$$p(n;T) = (\eta U_0)^n/(1+\eta U_0)^{n+1} \quad ; \quad n^{[r]} = r! \quad . \tag{3.7b}$$

On the other hand, if T is long compared to the fluctuation time of the light then smoothing will occur and U will become constant, again leading to the result (3.6b).

The Gaussian model was first introduced as a description of thermal light, although the result (3.7b) can be derived on the basis of a photon-rate equation approach which at no time involves the Gaussian field hypothesis [3.35,36]. Gaussian light is in fact most frequently encountered in laser light scattering configurations. If the scattered field can be expressed as the sum of many independent randomly phased contributions, the real and imaginary parts of V(t) will be Gaussian distributed, by virtue of the central limit theorem, and the intensity will obey the negative exponential distribution (3.7a). This situation is easily accomplished in practice, indeed it is difficult to avoid, since most scatterers are rough compared to an optical wavelength and contain many inhomogeneities within the scattering volume defined by a typical laser beam. The most familiar manifestation of Gaussian light is of course the speckle-like structure of the far field intensity pattern.

As is well known, a Gaussian variable is completely specified by a knowledge of its spectrum so that for light exhibiting the photon-counting distribution (3.7b) in the short sample (integration) time limit, it is necessary to determine, in addition, only the field spectrum or its Fourier transform: the first-order correlation function

$$g^{(1)}(\tau) = \langle V^*(0)V(\tau)\rangle / \langle |V|^2\rangle \quad . \tag{3.8}$$

The decay time of this quantity is a measure of the fluctuation time of the light and some estimate of this can be obtained by taking advantage of the change of statistics with integration time noted above. It is not difficult to show, for example, that a measurement of the integration-time dependence of the second factorial moment of the photon-counting distribution can be used to deduce the intensity correlation function via the inversion formula

$$g^{(2)}(T) = \frac{\langle I(0)I(T)\rangle}{\langle I\rangle^2} = \frac{1}{2}\frac{d^2 n^{[2]}}{dT^2} \quad . \tag{3.9a}$$

The well-known factorization theorem for a complex Gaussian process [3.37-39]

$$g^{(2)}(T) = 1 + |g^{(1)}(T)|^2 \tag{3.9b}$$

can then be used to obtain the modulus of the first-order correlation function (3.8) in the case of Gaussian light.

Owing to the double differentiation of the data required by (3.9b) it does not represent a practicable method of spectral measurement. Indeed, the only single-interval techniques which have been used successfully in this respect were used to measure the linewidth of a specific type of optical field: Gaussian-Lorentzian light [3.40,41]. This is generated, for example, when laser light is scattered by suspensions of macromolecules or micron-sized particles undergoing Brownian motion. Such systems have in the past provided a convenient test of the theoretical predictions for photon-counting statistics. The correlation function (3.8) in this model takes the form

$$g^{(1)}(\tau) = \exp(-\Gamma\tau) \quad , \tag{3.10}$$

where $\Gamma = D\underline{k}^2$, D being the translational diffusion constant and $\underline{k}$ the scattering vector. An exact solution of the single interval statistics is usually presented as a closed form for the generating function

$$Q(s;T) = \langle (1-s)^n\rangle = \sum_{n=0}^{\infty} (1-s)^n p(n,T) \quad , \tag{3.11}$$

namely,

$$Q(s;T) = \exp(\Gamma T)\left[\cosh(yT)+\frac{1}{2}(y/\Gamma+\Gamma/y)\sinh(yT)\right]^{-1} \quad , \tag{3.12a}$$

where

$$y^2 = \Gamma^2 + 2\Gamma\langle r\rangle s \tag{3.12b}$$

and $\langle r\rangle = \langle n\rangle/T$ is the mean count rate per second. The photon-counting distributions and factorial moments are obtained from (3.12) using the formulae

$$p(n;T) = \frac{1}{n!}\left(-\frac{d}{ds}\right)^n Q(s;T) \quad s = 1 \quad . \tag{3.13a}$$

and

$$n^{[r]} = \frac{1}{\langle n\rangle^r}\left(-\frac{d}{ds}\right)^r Q(s;T) \quad s = 0 \quad . \tag{3.13b}$$

A number of experiments were carried out in the mid-sixties to verify these formulae and determine the spectral linewidth parameter Γ [3.40a].

Only for Gaussian-Lorentzian light can full analytical solution of the integrated single-interval statistics be achieved. In the case of Gaussian light with other simple spectral profiles, the lower moments can usually be calculated using (3.2) and (3.5), with the help of the factorization properties of Gaussian variables, but the photon-counting distributions are difficult to obtain. Approximate methods have been proposed [3.41-45] but the degree of agreement with exact results is more an indication of the insensitivity of integrated single-interval statistics to the spectral properties of the light than of the accuracy of the approximations. In the case of light which is not Gaussian relation (3.9b) no longer holds and the high order correlation functions contain additional information.

Because of the obvious limitation of single-interval techniques it became clear in the late 1960's that a more direct form of spectral measurement in photon-counting experiments was necessary. This involves the determination of certain twofold statistical properties of the incident light. The measurement of single-interval statistics in the absence of temporal averaging remains a useful initial statistical test in all photon-counting experiments, however, and has indeed assumed renewed importance with the current interest in non-Gaussian fields.

3.2.2 Photon-Correlation Spectroscopy

The joint probability of finding n photons in a time interval T at time t and m in a similar interval at time $t' = t + \tau$ is given by a generalization of (3.1)

$$p(n,m;T) = \int_0^\infty dU \int_0^\infty dU' \,\frac{(\eta U)^n}{n!}\,\frac{(\eta U')^m}{m!}\exp(-\eta U-\eta U')P(U,U') \quad , \tag{3.14}$$

where $P(U,U')$ is the joint distribution of the integrated intensities $U(t;T)$, $U(t';T)$. A number of statistical properties related to the twofold counting distribution (3.14) are discussed in the literature and several have been determined experimentally [3.1-3]. Inversion of the formula has never been attempted and only for the Gaussian-Lorentzian model has the generating function for the joint photon-counting distribution been calculated [3.46]. In practice, for any Gaussian optical field we seek only to measure the first-order correlation function and this suggests investigating the lowest order correlation function of the photon-counts. According to (3.14) this is directly proportional to the integrated intensity correlation function $\langle U(t;T)U(t';T)\rangle$ [and therefore to $g^{(2)}(\tau)$, (3.9a), when there is no smoothing]. In the absence of temporal integration we obtain from (3.9b)

$$\langle nm\rangle/\langle n\rangle^2 = 1 + |g^{(1)}(\tau)|^2 \quad . \tag{3.15}$$

For simple model spectra the effect of time averaging can be calculated exactly from (3.2) [3.46]. In the case of non-Gaussian fields (3.15) no longer holds, but $\langle nm\rangle$ is still of interest as it then carries information additional to that provided by the first-order spectrum (3.8). A heterodyne or coherent mixing technique may be used to obtain the latter quantity: the optical field of interest is mixed on the photocathode of the detector with light from a local oscillator source (usually the same laser) which may have a frequency off-set ω. If the local oscillator power is much greater than that of the field of interest then for a symmetric spectrum the photon-count autocorrelation function in the absence of smoothing is given approximately by

$$\langle nm\rangle/\langle n\rangle^2 = 1 + 2(\langle n_\ell\rangle/\langle n\rangle)g^{(1)}(\tau)\cos\omega\tau \tag{3.16}$$

and contains a term proportional to the desired first-order correlation function. Here, $\langle n_\ell\rangle$ is proportional to the local oscillator power and $\langle n\rangle$ is the mean count rate.

In the absence of integration effects, photon-count autocorrelation clearly goes a long way towards solving the inverse detection problem for spectral measurements. The practical difficulties encountered in constructing $\langle nm\rangle$ from the data are by no means trivial, however. Post-detection signal processing techniques used in the general area of intensity fluctuation spectroscopy may be classified according to whether they operate in the time or frequency domain and according to whether they are single channel or parallel channel methods. The wave analyser or scanning electrical filter is a single channel instrument which operates in the frequency domain. The equivalent instrument in the time domain (which is more relevant to the present problem) is the delayed coincidence count. Both these techniques make inefficient use of the signal since only one frequency or one time-delay component is taken from each signal sample analysed. This may be an important consideration if the optical signal is weak or has a long fluctuation period so that long experiments are needed to achieve a reasonable degree of statistical accuracy. It is then clearly advantageous to use parallel channel instruments – a bank of filters in the frequency domain or an autocorrelator in the time domain. Extensive discussion of the relative merits of these two types of processor are to be found in the literature [3.1,3,47] but only the second option can preserve the digital nature of the initial data and warrants further discussion in the present context.

The principles of real-time digital autocorrelators are now well known. The signal is passed continuously through a shift register. Every shift (integration) time a new bit of signal is fed into the first channel whilst the contents of the remaining channels are shuffled down a channel, the contents of the last one being discarded. During the shift time the contents of the first (incoming) channel are multiplied by the contents of each of the remaining ones and stored. Continuous cycling of the process builds up an entire autocorrelation function with resolution equal to the shift time and maximum delay equal to the shift time multiplied by the number of channels. Certain "tricks" have been invented to reduce the multiplication process to one of addition. These have been largely responsible for the rapid development of digital autocorrelation techniques in photon-counting experiments, and continue to offer processing advantages in high-speed equipment [3.48] where the electronics and/or cost might otherwise be held to limit the potential of the method. One such processing trick is based on a radar technique (originally used for "jamming" purposes) [3.49,50] known as "clipping". In the optical analogue of this technique [3.51] a clipping gate replaces the incoming signal entry into the shift register by a series of ones and zeros according to the prescription

$$\begin{aligned} n_k &= 1 \quad n > k \\ &= 0 \quad n \leq k \quad . \end{aligned} \tag{3.17}$$

Here n_k is the number of clipped counts in a shift time (integration time) T, n is the number before clipping, and k is a pre-set clip level. The quantity n_k is now used to control gates which store the delayed (unclipped) signal so as to generate the single-clipped autocorrelation function $\langle n_k m \rangle$. For Gaussian light it may be shown that in the short sample time limit

$$\langle n_k m \rangle / \langle n_k \rangle \langle n \rangle = 1 + |g^{(1)}(\tau)|^2 (1+k)/(1+\langle n \rangle) \tag{3.18}$$

so that $|g^{(1)}(\tau)|$ can be recovered as in the full correlation case and, as it turns out, with little loss of statistical accuracy [3.52-54]. Note that the τ dependence of (3.18) is identical with that of (3.15) so that single clipping introduces no distortion.

In the case of non-Gaussian optical fields (3.18) no longer holds. In this situation, the full autocorrelation function can be generated in principle by varying the clip level of a single clipping correlator according to a uniform distribution. In practice, this can be achieved approximately by ramped or saw-tooth modulation of the clipping level or by "scaling" [3.55-57]. In this latter method the true signal is correlated with a scaled version, i.e., a form which has the value unity upon the arrival of every s^{th} count but is otherwise zero. The scaling level s is set sufficiently high for the probability of registering more than one scaled count per integration time to be negligible. Techniques of this type are essential in measurements of optical scintillation which is generally not Gaussian in nature.

3.2.3 Instrumental Effects

In the preceding subsections we have described essentially two types of statistical analysis commonly employed in photon-counting experiments. Temporal integration has throughout been presented as a prime consideration and has been discussed in some detail. In this subsection we shall discuss briefly two other important instrumental effects — spatial integration and dead times.

In order to take account of spatial as well as temporal integration (3.2) must be generalized to include the effect of a finite detector area A in the neighbourhood of the point $\underline{r}$

$$U(\underline{r},t;A,T) = \int_{t-T/2}^{t+T/2} dt' \int_{A(\underline{r})} d^2r'\ I(\underline{r}',t') \quad . \tag{3.19}$$

Spatial averaging takes place when A is significant compared to the spatial structure of the intensity pattern. Deconvolution of the data in the presence of this effect is fraught with many familiar difficulties and in photon-counting experiments the traditional approach has been to ensure that spatial averaging is negligible or, if this is not feasible, to calculate the effect of averaging on the measured quantity using a reliable mathematical model. Exact analytical results for the spatial integration problem have been obtained only for the simplest statistical properties, for example the second factorial moment of p(n;T,A) and the photon-count autocorrelation function. Formulae have been derived for the case of a thermal or Gaussian source based on the field coherence function [3.34]

$$\frac{\langle V^*(\underline{r},t)V(\underline{r}',t)\rangle}{\langle |V|^2\rangle} = \frac{2J_1(k_0S|\underline{r}-\underline{r}'|/z)}{k_0S|\underline{r}-\underline{r}'|/z} \exp\left[ik_0(r^2-r'^2)/2z\right] \quad , \tag{3.20}$$

where S is the radius of the (circular) source, z is the source-detector distance and k_0 is the wave vector of the light. The first zero of the Bessel function J_1 defines a coherence length or speckle size z/k_0S over which the received light can be regarded as coherent. Using (3.20), (3.19) and (3.14) it may be shown that the second factorial moment of the photon-counting distribution for Gaussian light takes the form [3.58]

$$\langle n(n-1)\rangle/\langle n\rangle^2 = 1 + f(A) \quad , \tag{3.21}$$

where

$$f(A) = \sum_{p=0}^{\infty} \left[\frac{(2p+2)!}{[(p+1)!]^2(p+2)!}\right]^2 (-1)^p(\kappa R/2)^{2p} \quad , \tag{3.22}$$

$\kappa = k_0 S/z$, and R is the radius of the circular integration area at the detector. For the photon-count autocorrelation function a similar calculation gives [3.58]

$$\langle nm\rangle/\langle n\rangle^2 = 1 + f(A)|g^{(1)}(\tau)|^2 \quad , \tag{3.23}$$

whilst for cross-spectrally pure Gaussian-Lorentzian light temporal integration may also be included to give

$$\frac{\langle nm\rangle}{\langle n\rangle^2} = 1 + f(A)\left[\frac{\sinh(\Gamma T)}{\Gamma T}\right]^2 \exp(-2\Gamma\tau) \quad . \tag{3.24}$$

A different integration factor enters the heterodyne result (3.16) when there is spatial averaging: the second term in this equation is then multiplied by the factor [3.59,60]

$$f_D(A) = (2/\kappa R)^2\left[1-J_0^2(\kappa R)-J_1^2(\kappa R)\right] \quad , \tag{3.25}$$

where J_0 and J_1 are Bessel functions of the first kind.

Calculation of the effects of spatial integration on distributions and higher order statistical properties is difficult even for the Gaussian model. One important exact result which has been established is that for Gaussian-Lorentzian light neither spatial nor temporal integration distorts the photon-count autocorrelation function even in the presence of clipping [3.61,62]

$$\langle n_k m\rangle/\langle n_k\rangle\langle n\rangle = 1 + f(A,T,\langle n\rangle,k) \exp(-2\Gamma\tau) \quad . \tag{3.26}$$

The undistorted exponential dependence on τ allows Γ to be determined through a simple fitting procedure and has promoted the development of photon correlation as an important technique for the measurement of the diffusion constants of macromolecules in solution [3.2]. For an arbitrary spectral profile, the effects of spatial averaging are often estimated by dividing the integration area into a number of independent coherence areas over which there is essentially no averaging. For Gaussian light the generating function for the distribution of counts in each subarea is from (3.7b) and (3.11)

$$Q(s;T) = (1+\eta U_0 s/\zeta)^{-1} \quad , \tag{3.27a}$$

whilst for the entire spatially integrated distribution we have

$$Q(s;A,T) = (1+\eta U_0 s/\zeta)^{-\zeta} \quad , \tag{3.27b}$$

where ζ is the number of coherence areas per detector area and $\eta U_0 = \langle n \rangle$ is the total average count rate per sample time (assuming no time averaging). An approximate result for intermediate sized detector areas is obtained by setting $\zeta^{-1} = f(A)$ [see (3.22)]. The formula so obtained [3.63,64] is clearly asymptotically correct for large detector areas but also correctly predicts the area dependence of the second factorial moment (3.21). Note that the photon-counting distribution generated by (3.27b) is the negative binomial class

$$p(n;A,T) = \binom{n+\zeta-1}{n} \frac{(\langle n \rangle/\zeta)^n}{(1+\langle n \rangle/\zeta)^{n+\zeta}} \tag{3.28}$$

with $\zeta \geqq 1$.

In the case of non-Gaussian optical fields the effect of spatial integration on photon-counting statistics is often complicated by the presence of several scales in the intensity pattern [3.65]. A great deal of effort has been expended on this problem in the field of laser beam propagation through the atmosphere, where it is of considerable importance in connection with the question of eye safety [3.66]. The basic coherence properties of non-Gaussian light are generally not yet well understood, however, so that experiments to measure its statistical properties are usually designed to minimize the effects of spatial integration.

We shall not give a detailed account of the theory of dead-time effects here, as there is a good deal of literature on the subject which is of interest to workers in the fields of nuclear and particle physics as well as photon counting [3.67,68] (see also Sect.4.4.2). Briefly, it has been shown that the Poisson-like kernel of (3.1) is modified by the presence of a small dead time τ_D as follows [3.69]

$$\frac{(\eta U)^n}{n!} \exp(-\eta U) \rightarrow \frac{(\eta U)^n}{n!} \exp(-\eta U)\left[1+n(\eta U-n+1)\frac{\tau_D}{T} + O\left(\frac{\tau_D}{T}\right)^2\right] \quad . \tag{3.29}$$

The factorial moments of the photon-counting distribution are then given by

$$N^{[r]} = N_0^{[r]} - r(\tau_D/T)\left[(r-1)N_0^{[r]}+N_0^{[r+1]}\right] + O(\tau_D/T)^2 \quad . \tag{3.30}$$

This formula can be used to correct the measured data for small dead-time effects ($N_0^{[r]}$ is the true r^{th} factorial moment of $p(n;A,T)$ in the absence of a dead time). Useful higher order corrections for the moments have also been calculated [3.70]. Relation (3.29) can be generalized to cope with double interval statistics and the effects of small dead times on correlation functions can then be evaluated. It is found that even in the case of Gaussian light there is distortion of the time dependence of these quantities. The effect of an arbitrary dead time on photon-counting statistics is complicated and thus presents a real limitation on the speed of data acquisition and processing. In the case of simple distribution measurements (3.30) can be used to correct the data but large dead times are usually avoided where possible.

3.2.4 Noise and Statistical Accuracy

In the last part of this section we shall briefly discuss noise limitations in photon-counting experiments. We shall not be concerned with the "detection" problem encountered in optical radar systems but will consider two kinds of limitation related to the measurement of moments or correlation functions of photon-counting distributions. First we shall outline the effects of additive noise such as background radiation, stray light and dark counts, and indicate how such effects can be reduced or deconvolved from the measurements. Secondly, we shall consider the accuracy limitations presented by the statistical nature of the signal itself.

The first of the problems mentioned above is a familiar one and may be tackled in the usual way by exploiting the differences between the unwanted additive noise and the optical field of interest. For example, any background radiation in laser light scattering experiments is likely to be broadband compared to the signal so that optical filters matching the laser wavelength can be used to reduce its contribution at the detector. Any stray scattered laser light on the other hand will usually have a much finer spatial structure than the signal field. If the integration time and detector area are chosen to match the signal characteristics both types of noise will be averaged and merely add a constant intensity to the light of interest. The effect on the photon-counting statistics will then be identical with that produced by the random emission of dark counts within the detector and all three effects can be taken into account by assuming an additive constant incident noise intensity U_b. The generating function of the measured photon-counting distribution will then be the simple product

$$Q(s;A,t) = \exp(-\eta s U_b) Q_0(s;A,T) \quad , \tag{3.31}$$

where Q_0 corresponds to the light of interest and the exponential factor is just the generating function for a Poisson distribution. Significant distortion of the statistics will occur if ηU_b is comparable with the signal intensity. However, an exact deconvolution formula relating the desired moments $N_0^{[r]}$ to the measured ones $N^{[r]}$ can be derived from (3.31) using (3.13b)

$$N_0^{[r]} = \sum_{p=0}^{r} (-\langle n_b \rangle)^{r-p} N^{[p]} \binom{r}{p} \quad , \tag{3.32}$$

where $\langle n_b \rangle = \eta U_b$ is the (measurable) count rate due to unwanted additive noise. In the case of autocorrelation measurements, noise of this kind merely adds a flat background. This does not distort the time dependence but may necessitate longer experiments to achieve the same accuracy of measurement (see also Chap.4).

Even in the absence of additive noise, in any experiment of finite duration the accuracy of a measurement will be limited by the statistical nature of the photoelectron emission process. Fluctuations in the incident optical field add a further complication to this problem. Consider first an experiment to measure the rth factorial moment of the photon-counting distribution. The number of counts n_i in each of a large number M of samples of duration T separated by intervals T_p will be registered and an *estimate* of the factorial moment calculated as follows [3.71]:

$$\hat{N}^{[r]} = \frac{1}{M}\sum_{i=1}^{M} n_i(n_i-1)\ \dots\ (n_i-r+1) \quad . \tag{3.33}$$

Note that since $<\hat{N}^{[r]}> = N^{[r]}$ this is an *unbiased* estimate. In a series of identical experiments the spread of measurements will be given by the variance of $\hat{N}^{[r]}$. For periodic sampling of a stationary process this may be reduced to the form

$$\mathrm{Var}\ \hat{N}^{[r]} = \frac{1}{M}\left[\sum_{s=0}^{r} s!\binom{r}{s}^2 N^{[2r-s]} - \{N^{[r]}\}^2\right] + 2\frac{\{N^{[r]}\}^2}{M}\sum_{s=1}^{M-1}\left(1-\frac{s}{M}\right)\left(g_s^{[r]}-1\right) \quad , \tag{3.34a}$$

where

$$g_s^{[r]} = <n_p(n_p-1)\dots(n_p-r+1)n_{p+s}(n_{p+s}-1)\dots(n_{p+s}-r+1)> \quad . \tag{3.34b}$$

In the limit $<n> \ll 1$ (3.34a) is given approximately by

$$\mathrm{Var}\ \hat{N}^{[r]} = r!N^{[r]}/M \qquad <n> \ll 1 \tag{3.35}$$

whilst if the sample time is much larger than any fluctuation time of the light $g_s^{[r]} = 1$ and only the term in square brackets survives. This is also the case for coherent (laser) light when (3.34a) reduces to

$$\mathrm{Var}\ \hat{N}^{[r]} = \frac{1}{M}\sum_{s=1}^{r} s!\binom{r}{s}^2 <n>^{2r-s} \quad . \tag{3.36}$$

This result also holds for an arbitrary light field if the integration time T is much larger than the fluctuation time. For Gaussian-Lorentzian light with $\Gamma T \ll 1$ and $\Gamma NT_p \gg 1 \gg \Gamma rT_p$ we obtain the result

$$\mathrm{Var}\ \hat{N}^{[r]} = <n>^{2r}(r!)^2(MT_p\Gamma)^{-1}\sum_{s=1}^{r}\frac{1}{s}\binom{r}{s}^2 \quad . \tag{3.37}$$

Certain simple conclusions can be drawn from these results. First, the error decreases as expected when the number of samples is increased. It also increases as T_p increases for fixed M. This is because the information collected from each sample

Table 3.1. Measured and predicted moments and errors for an incoherent source [3.95]

Normalized factorial moments	Incoherent source, 10^5 samples	
	Experiment	Theory
$n^{[1]}$	1.000 ± 0.036	1.000 ± 0.038
$n^{[2]}$	2.00 ± 0.15	2.00 ± 0.16
$n^{[3]}$	6.05 ± 0.85	6.00 ± 0.85
$n^{[4]}$	24.7 ± 5.9	24.0 ± 5.7
$n^{[5]}$	130 ± 45	120 ± 49
$n^{[6]}$	805 ± 461	720 ± 510

is correlated when $\Gamma T_p < 1$ but becomes independent when $\Gamma T_p \gg 1$. Equation (3.35) is an expression of the importance of shot noise when the mean count rate per sample time is low. Equation (3.36) shows that the spread of measurements is inversely proportional to the number of coherence times per experiment time: clearly a measure of the total information in the experiment. The above formulae have in fact been found to give excellent agreement with experiment (Table 3.1) and although they have been thoroughly investigated only for coherent light and the Gaussian-Lorentzian model, the qualitative behaviour of the statistical accuracy with changing experimental parameters is not expected to be markedly different in the case of non-Gaussian statistics.

The approach described above has been extended to deal with measurements of photon-count autocorrelation functions. The results obtained have played an important role in optimizing the design and operational principles of digital correlation systems used in photon-correlation spectroscopy. The calculations involved are lengthy and tedious, whilst the results are sufficiently subtle to prohibit brief exposition. The interested reader is therefore referred to the original papers on the subject [3.47,52-54].

3.3 Scattering Theory

Whenever electromagnetic radiation encounters matter, part or all of it is scattered. This scattering may be classified as inelastic if the mean frequency of the scattered radiation is shifted from that of the incident radiation and quasi-elastic if it remains much the same. In the cases of interest here, inelastic processes (Raman, Brillouin, Compton, etc.) have much lower cross sections than quasi-elastic processes and will not be discussed further. We will also consider only media describable (actually or effectively) by a real scalar dielectric constant so that absorption can be neglected (see Chap.2). The quasi-elastic scattering process can be viewed as an excitation of the polarizability of the atoms in the medium which oscillates at the incident frequency and therefore causes reradiation or scattering. Even in the hypothetical case of a completely homogeneous medium this scattering will occur; however then, due to interference, the secondary wavelets re-sum to give a wave similar to that incident but reduced in velocity by a constant factor, the refractive index. It is convenient, therefore, to regard a homogeneous medium as not producing any scattering and to realize that what we will call scattering arises from variations (both spatial and temporal) in the refractive index of the medium.

Defined in this way quasi-elastic scattering can be subdivided into many categories. If the fluctuations in the refractive index are weak enough most of the radiation passes through the medium unscattered and the situation can be treated by the first Born (or "single-scattering") approximation as considered in Chap.2. (This leads to such familiar results as the inverse Fourier relationship between the angular envelope of the average intensity of the scattered radiation and the spatial correlation function of the refractive index fluctuations in the medium.) Of interest to us here are strong scatterers. In the extreme case radiation will enter the medium and be scattered many times through large angles. It will thus emerge having been largely "randomized" and will provide little information about the medium. In this article we consider two of the simpler special cases of strong scattering. The first is the deep random phase screen, introduced in Sect.3.3.1, where large phase shifts can be imposed on the radiation but it is assumed that the scattering medium is sufficiently localized that appreciable amplitude fluctuations do not develop within it. The second case is an extended region of medium which can produce multiple scattering. Here, however, it is assumed that the refractive index inhomogeneities are much larger in linear dimension than the light wavelength so that the scattering is strongly peaked in the forward direction. In this case it might appear to a casual observer that an incident beam of radiation is simply broadened somewhat. However closer examination would reveal a rich, fluctuating structure within the beam.

It is perhaps worth pointing out that none of the scattering systems to be considered here are in equilibrium; energy is being pumped into the medium to cause, in general, a turbulent motion. Of course, due to random molecular motions, fluctuations in density and temperature (and hence refractive index) do occur in systems in equilibrium. However, except very near critical points, these have a small spatial scale and produce only weak scattering.

3.3.1 Mechanisms and Theories for Strong Scattering

For a phase screen the obvious starting point for the theory is the Huygens-Fresnel integral which expresses the scattered field at some point beyond the screen as the integral over the perturbed emergent phase front with appropriate propagation factors [3.14,19]. With the assumption of certain statistical properties for the phase perturbations, (e.g., joint-Gaussian statistics and some particular form for their spatial correlation function) it is possible to calculate, by a mixture of analytic and numerical methods, a number of functions of the field such as the lower-order moments and correlation functions [3.19]. Although considerable progress in this direction has been made in the last few years it seems unlikely that this approach will yield, in the near future, much insight into such properties of immediate interest as the full probability distribution P(I) of the scattered intensity.

A similar situation exists regarding the extended medium problem. The small-angle or parabolic approximation allows Maxwell's equations to be used in the inhomogeneous medium to give equations obeyed by moments and correlation functions of the propagating field [3.72,73]. Again these can only be solved, frequently with difficulty, for the lower-order functions.

In this paper, therefore, we will discuss a different approach which has recently proved useful in providing insight into the behaviour of the higher-order functions, P(I) in particular. It should be emphasized that, unlike the methods outlined above, this approach is not rigorous but simply plausible. Nevertheless it is found that, in several cases, the experimental data show quantitative agreement with the predictions of this "theory" (Sect.3.5). This encourages one to look for more fundamental justifications of the approach. Work with this aim, which is far from complete, is reviewed in Sect.3.4.

This new approach, which involves the identification of "discrete" scatterers or scattering events, is perhaps illustrated best by qualitative consideration of the mechanisms by which intensity fluctuations develop in the deep phase screen problem. Figure 3.6, to be discussed further in Sect.3.5.1, shows a typical plot of σ^2, the relative variance of the intensity

$$\sigma^2 \equiv \langle I^2 \rangle / \langle I \rangle^2 - 1 \tag{3.38}$$

as a function of distance z from the screen. At $z = 0$, only phase fluctuations exist in the wave front and $\sigma^2 = 0$. From this point "rays" start to propagate normal to the local phase front; as they overlap, intensity fluctuations develop and σ^2 increases. At a certain distance z_F a maximum value of σ^2 is found. This can be identified as a mean "focussing distance" where regions of high intensity, associated with the lens-like focussing effects of the individual inhomogeneities in the scattering medium, are found in the radiation pattern (see Fig.3.7 in Sect.3.5.2). Due to the random nature of the medium this focussing is a random process and the intensity pattern at z_F is best viewed as a random network of diffraction-broadened caustics [3.5]. For $z > z_F$ the detection point can receive radiation originating from more than one independent phase front inhomogeneity. For large z ($\gg z_F$) there is a random interference between a large number of these contributions. The central limit theorem then predicts a Gaussian distribution for the field and a negative exponential distribution for P(I) (3.7a), for which $\sigma^2 = 1$ (3.7b). Clearly, then, for $z > z_F$ the scattered field can be viewed as being made up from independent contributions of variable amplitude and phase from a number of discrete scatterers. In the simplest approach these "scatterers" can be identified as individual phase front inhomogeneities. However a more detailed consideration of the multiscale nature of turbulence suggests that each of these independent "superscatterers" can itself be made up from a fluctuating number of elementary scatterers (Sect.3.4). The applicability of this discrete scatterer approach to the region $z < z_F$, where radiation is received from less than one "superscatterer", is not immediately obvious. However it has recently been suggested by several authors [3.11,74] that the addition of a coherent (non-fluctuating) component, which constitutes the whole field at the screen but tends to zero as the focussing peak is reached, can remedy this omission (Sect.3.5.2).

It is also not immediately obvious that this approach will apply to propagation in an extended medium. However in many cases it is possible to view, at least qualitatively, the effect of an extended medium in terms of an "effective deep phase screen". Thus light rays entering an extended inhomogeneous medium will pick up position-dependent phase shifts as they propagate. At first, because of the assumed large scale of the inhomogeneities the rays will not deviate significantly from their initial direction. Therefore some distance into the medium, the phase front can be strongly perturbed without the presence of significant amplitude fluctuations. On further propagation intensity fluctuations will develop as with the phase screen, the difference being that this propagation will itself be further perturbed by the inhomogeneities in the extended medium.

The rather qualitative approach of the last two paragraphs has recently been given support from a totally different point of view by DASHEN [3.75] who formulated the problem of propagation in inhomogeneous media in terms of Feynman's path integral. This not only leads to a formal similarity between phase screens and extended media but also identifies the field at a point as being the sum of contributions from a number of different ray or Fermat paths.

3.3.2 The "Discrete-Scatterer" Model [3.76-78]

The starting point of the discrete-scatterer model (also known as the "microarea" or "particle" model) is obtained by writing the complex analytic field $V(\underline{r},t)$ at detection point $\underline{r}$ and time t as

$$V(\underline{r},t) = e^{i(\underline{k}\cdot\underline{r}-\omega t)} \sum_{i=1}^{N(\underline{r},t)} a_i(\underline{r},t) \exp[-i\phi_i(\underline{r},t)] \quad , \tag{3.39}$$

where $\underline{k}$ and ω are, respectively, the wave vector and frequency of the incident radiation and unimportant constants have been set to unity. Here, $N(\underline{r},t)$ is the effective number of scatterers which can depend on both time and detection position as can the real amplitudes $\{a_i(\underline{r},t)\}$ of the contributions from the individual scatterers; the $\{\phi_i(\underline{r},t)\}$ are shifts in phase associated with the scattering events. The scattering systems of interest to us are all, in some sense, random. In general, therefore, the observed field amplitude V will fluctuate in time due to stochastic variations in some or all of the variables N, $\{a_i\}$ and $\{\phi_i\}$. Here we will assume large phase deviations, i.e.,

$$\phi_0 \equiv \left[\langle \phi_i^2 \rangle - \langle \phi_i \rangle^2 \right]^{1/2} \gg 1 \quad . \tag{3.40}$$

Thus, in the exponential factors in (3.39), we can take the $\{\phi_i\}$ to be uniformly distributed between 0 and 2π radians. We will also assume that N, $\{a_i\}$, and $\{\phi_i\}$ are all statistically independent of each other (though, for example, the $\{a_i\}$ may be statistically identical). Then, as is now well known, the sum in (3.39) can be represented graphically as a two-dimensional random walk of N steps of lengths $\{a_i\}$ in the complex V plane. This enables us to use the results, derived many years ago and discussed by PEARSON [3.80], KLUYVER [3.79], and RAYLEIGH [3.81], for the probability distribution of the resultant length of such a walk.

We first assume N and the $\{a_i\}$ to be nonfluctuating and can then apply Kluyver's equation directly to give [3.78,82,83]

$$P(I) = \frac{1}{2} \int_0^\infty u\, du\, J_0(u\sqrt{I}) \left\{ \prod_{i=1}^{N} J_0(ua_i) \right\} \quad , \tag{3.41}$$

where the intensity $I \equiv |V|^2$ and J_0 is a zeroth order Bessel function of the first kind. This result can now be averaged over fluctuations in the N and $\{a_i\}$

$$P(I) = \frac{1}{2} \int_0^\infty u\, du\, J_0(u\sqrt{I}) \left\langle \left\langle J_0(ua) \right\rangle_a^N \right\rangle_N \quad , \tag{3.42}$$

where the $\{a_i\}$ have been assumed to be statistically identical.

Equation (3.42), as it stands, is not particularly illuminating and can only be simplified further in a few special, though important, cases (see Sect.3.3.3). More direct insight comes from the moments of P(I) which are obtained from the moment generating function derived from (3.41), again with subsequent averaging over number and amplitude fluctuations

$$\langle I \rangle = \langle N \rangle \langle a^2 \rangle$$

$$\langle I^2 \rangle = 2\langle N(N-1) \rangle \langle a^2 \rangle^2 + \langle N \rangle \langle a^4 \rangle$$

$$\langle I^3 \rangle = 6\langle N(N-1)(N-2) \rangle \langle a^2 \rangle^3 + 9\langle N(N-1) \rangle \langle a^2 \rangle \langle a^4 \rangle + \langle N \rangle \langle a^6 \rangle$$

$$\begin{aligned}\langle I^4 \rangle = {} & 24\langle N(N-1)(N-2) \rangle \langle a^2 \rangle^4 + 72\langle N(N-1)(N-2) \rangle \langle a^2 \rangle^2 \langle a^4 \rangle \\ & + 18\langle N(N-1) \rangle \langle a^4 \rangle^2 + 16\langle N(N-1) \rangle \langle a^2 \rangle \langle a^6 \rangle + \langle N \rangle \langle a^8 \rangle \ldots \text{ etc.}\end{aligned} \tag{3.43}$$

In general, then the nth intensity moment depends on all factorial moments of N up to the nth and all even moments of a up to the 2nth. If we take the limit N(fixed)$\to\infty$, the first terms dominate giving $\langle I^n \rangle / \langle I \rangle^n \to n!$, the result for Gaussian light, as expected from the central limit theorem. The associated negative exponential P(I) can be derived directly from (3.42) in this limit [3.78].

Here we digress slightly to point out a subtle relationship between the amplitude factors a and the number N. We consider one of the first problems to which (3.42) was applied [3.83,84]: the scattering by identical microscopic noninteracting particles suspended in a fluid medium. The sample contains a very large number of particles N_{Tot} in a large volume v_{Tot} but only light scattered from a small volume v, defined by a focussed laser beam and suitable detection optics, is received. This volume contains a small number of particles N_p which fluctuates as particles diffuse in and out due to Brownian motion. The intensity moments for this situation can be derived by two routes. Firstly we can describe the occupation number fluctuations by taking the amplitude factors to be "counting" variables

$$a_i(t) = 1 \quad \text{if particle i is in v}$$
$$a_i(t) = 0 \quad \text{otherwise} \quad . \tag{3.44}$$

(Here we have assumed the scattering volume to be illuminated uniformly, a circumstance not always achieved easily in practice.) Clearly, for the situation described above, $\langle a^{2n} \rangle = \langle a \rangle$. In (3.43) we take $N \equiv N_{Tot}$ to be large and unfluctuating so that $\langle N(N-1) \rangle \approx N_{Tot}^2$, etc., to give

$$\langle I \rangle = N_{Tot} \langle a \rangle$$
$$\langle I^2 \rangle = 2N_{Tot}^2 \langle a \rangle^2 + N_{Tot} \langle a \rangle$$
$$\langle I^3 \rangle = 6N_{Tot}^3 \langle a \rangle^3 + 9N_{Tot}^2 \langle a \rangle^2 + N_{Tot} \langle a \rangle \quad , \text{etc.,}$$

or, defining the average number of particles in v by $\langle N_p \rangle = N_{Tot} v/v_{Tot} = N_{Tot} \langle a \rangle$, the normalized moments are given by

$$\langle I \rangle = \langle N_p \rangle$$

$$\frac{\langle I^2 \rangle}{\langle I \rangle^2} = 2 + \frac{1}{\langle N_p \rangle}$$

$$\frac{\langle I^3 \rangle}{\langle I \rangle^3} = 6 + \frac{9}{\langle N_p \rangle} + \frac{1}{\langle N_p \rangle^2}$$

$$\frac{\langle I^4 \rangle}{\langle I \rangle^4} = 24 + \frac{72}{\langle N_p \rangle} + \frac{34}{\langle N_p \rangle^2} + \frac{1}{\langle N_p \rangle^3} \quad , \text{etc.} \tag{3.45}$$

In the alternative approach we treat the number fluctuations directly, taking $N = N_p(t)$ and $\{a_i\} = \text{constant} = 1$, for simplicity. This gives

$$\langle I \rangle = \langle N_p \rangle$$

$$\langle I^2 \rangle = 2\langle N_p(N_p-1) \rangle + \langle N_p \rangle \quad , \text{ etc.} \tag{3.46}$$

We now use our a priori knowledge (which was effectively derived in the first approach) that, since the particles are not interacting and $N_{Tot}/N_p \gg 1$, $N_p(t)$ is, to a high degree of approximation, Poisson distributed (3.6b). For such a distribution the nth factorial moment is simply the nth power of the mean. Substitution of this result into (3.46) gives (3.45). Thus it is possible to treat number fluctuations either directly or through the amplitude factors $\{a_i\}$ in (3.39). We shall see another manifestation of this relationship in the next section when we consider K distributions.

3.3.3 K Distributions

K distributions were originally discovered by searching tables of integral transforms for cases in which (3.42) could be evaluated to give useful closed-form expressions [3.77]. In this approach they follow from taking N in (3.42) to be constant and the probability distributions of the amplitudes to be given by

$$p(a) = \frac{2b}{\Gamma(\beta)} \left(\frac{ba}{2}\right)^{\beta} K_{\beta-1}(ba) \quad , \quad \beta > 0 \quad , \tag{3.47}$$

where b is a constant determined by $\langle a \rangle$ and $K_{\beta-1}$ is a modified Bessel function. Substitution of (3.47) in (3.42) gives

$$P_K(I) = \frac{b^{\alpha+1} I^{\frac{1}{2}(\alpha-1)} K_{\alpha-1}(b\sqrt{I})}{2^{\alpha}\Gamma(\alpha)} \quad , \tag{3.48}$$

where

$$\alpha = N\beta \quad . \tag{3.49}$$

The moments of $P_K(I)$ can be determined directly from (3.48) [3.77,78] or via substitution of the moments of p(a) in (3.43) and are

$$\langle I \rangle = 4\alpha/b^2$$

$$\frac{\langle I^2 \rangle}{\langle I \rangle^2} = 2 + \frac{2}{\alpha}$$

$$\frac{\langle I^3 \rangle}{\langle I \rangle^3} = 6 + \frac{18}{\alpha} + \frac{12}{\alpha^2}$$

$$\frac{\langle I^4 \rangle}{\langle I \rangle^4} = 24 + \frac{144}{\alpha} + \frac{264}{\alpha^2} + \frac{144}{\alpha^3} \quad , \text{ etc.} \tag{3.50}$$

K distributions are two-parameter distributions conveniently characterized by the mean $<I>$ and the second moment $<I^2>/<I>^2$ or α. As mentioned in Sect.3.3.1 these distributions provide another example of the interplay between the number N and the amplitude factors $\{a_i\}$ in random walk problems. These enter only in the combination $\alpha = N\beta$ [β determines p(a)], so that the same K distribution can arise from a large number of wildly fluctuating scatterers (N large, β close to 0) or a smaller number with more moderate fluctuations (N smaller, β larger). From the point of view of the inverse problem, that is obtaining information about the scattering system from properties of the scattered radiation, this is not an entirely satisfactory situation. In the next section we give a derivation of K distributions as the $<N> \to \infty$ limit of a scattering system in which the number of scatterers undergoes *correlated* fluctuations. In this case the derivation becomes independent of the form chosen for p(a).

A useful feature of K distributions is that Mandel's equation (3.1) can be solved to give an expression for $P_K(n,T)$ which can be compared directly with experimental data [3.78]

$$P(n,T) = \left(\frac{b^2}{4\eta T}\right)^{\frac{\alpha}{2}} \frac{\Gamma(\alpha+n)}{\Gamma(\alpha)} \exp\left(\frac{b^2}{8\eta T}\right) W_{-\left(\frac{\alpha}{2}+n\right),\frac{1}{2}(\alpha-1)}\left[\frac{b^2}{4\eta T}\right] , \tag{3.51}$$

where $W_{K,\ell}$ is a Whittaker function.

A second useful feature is that the moments (though not the distribution function) for K-distributed noise with an added coherent component have a relatively simple form [3.11]

$$\frac{<I^r>}{<I>^r} = \frac{(r!)^2}{(1+x)^r} \sum_{m=0}^{r} \binom{m+\alpha-1}{m} \frac{(x/\alpha)^m}{[(r-m)!]^2} , \tag{3.52}$$

where x is the ratio of the mean intensity of the K-distributed noise to the mean intensity of the coherent component.

3.3.4 Correlation Functions

The first- and second-order correlation functions of the field $V(\underline{r},t)$ can be calculated from (3.39). We assume N to be fixed and the $\{a_i\}$ and $\{\phi_i\}$ to have the statistical properties outlined in Sect.3.3.1. The first-order (field) correlation function is

$$|<V^*(\underline{r}_1,t)V(\underline{r}_2,t+\tau)>| = \left|\sum_{i=1}^{N}\sum_{j=1}^{N} <a_i(\underline{r}_1,t)a_j(\underline{r}_2,t+\tau)> \times \left\langle \exp\left\{i\left[\phi_i(\underline{r}_1,t)-\phi_j(\underline{r}_2,t+\tau)\right]\right\}\right\rangle\right| , \tag{3.53}$$

where, in separating the averages over amplitude and phase, we have already used the assumed statistical independence of the $\{a_i\}$ and the $\{\phi_i\}$. Use of the statistical independence of ϕ_i and ϕ_j allows separation of the phase average for $i \neq j$ and the large phase approximation makes these individual terms average to zero. Thus only $i = j$ terms survive in (3.53) to give

$$|\langle V^*(\underline{r}_1,t)V(\underline{r}_2,t+\tau)\rangle| = N\langle a(\underline{r}_1,t)a(\underline{r}_2,t+\tau)\rangle \times |\langle \exp\{i[\phi(\underline{r}_1,t)-\phi(\underline{r}_2,t+\tau)]\}\rangle| \quad .$$

Suitable normalization gives

$$|g^{(1)}(\underline{r}_1,\underline{r}_2,\tau)| \equiv \frac{|\langle V^*(\underline{r}_1,0)V(\underline{r}_2,\tau)\rangle|}{\langle|V(\underline{r}_1)|^2\rangle^{\frac{1}{2}}\langle|V(\underline{r}_2)|^2\rangle^{\frac{1}{2}}}$$

$$= \frac{\langle a(\underline{r}_1,0)a(\underline{r}_2,\tau)\rangle}{\langle a^2(\underline{r}_1)\rangle^{\frac{1}{2}}\langle a^2(\underline{r}_2)\rangle^{\frac{1}{2}}} \times |\langle \exp\{i[\phi(\underline{r}_1,0)-\phi(\underline{r}_2,\tau)]\}\rangle| \quad . \qquad (3.54)$$

The normalized intensity correlation function is

$$g^{(2)}(\underline{r}_1,\underline{r}_2,\tau) \equiv \frac{\langle|V(\underline{r}_1,0)|^2|V(\underline{r}_2,\tau)|^2\rangle}{\langle|V(\underline{r}_1)|^2\rangle\langle|V(\underline{r}_2)|^2\rangle}$$

$$= \frac{1}{N^2\langle a^2(\underline{r}_1)\rangle\langle a^2(\underline{r}_2)\rangle} \sum_{i=1}^{N}\sum_{j=1}^{N}\sum_{k=1}^{N}\sum_{\ell=1}^{N}$$

$$\langle a_i(\underline{r}_1,0)a_j(\underline{r}_1,0)a_k(\underline{r}_2,\tau)a_\ell(\underline{r}_2,\tau)\rangle$$

$$\times \langle \exp\{i[\phi_i(\underline{r}_1,0)-\phi_j(\underline{r}_1,0)+\phi_k(\underline{r}_2,\tau)-\phi_\ell(\underline{r}_2,\tau)]\}\rangle \quad . \qquad (3.55)$$

Use of the assumed statistical properties of the phase in the last average gives rise to $N(N-1)$ terms for which $i = j \neq k = \ell$, $N(N-1)$ terms for which $i = \ell \neq k = j$ and N terms for which $i = j = k = \ell$. Thus

$$g^{(2)}(\underline{r}_1,\underline{r}_2,\tau) = \left[1 - \frac{1}{N}\right]\left[1+|g^{(1)}(\underline{r}_1,\underline{r}_2,\tau)|^2\right] + \frac{1}{N}\frac{\langle a^2(\underline{r}_1,0)a^2(\underline{r}_2,\tau)\rangle}{\langle a^2(\underline{r}_1)\rangle\langle a^2(\underline{r}_2)\rangle} \quad . \qquad (3.56)$$

Two features of (3.56) should be noted. Firstly as $N \to \infty$ it reduces to the well-known factorization property of correlation functions of Gaussian fields [3.37]. For this reason, the last term in (3.56) is frequently identified as a "non-Gaussian term". Secondly, this non-Gaussian term depends only on the amplitude factors whereas the first-order correlation function $g^{(1)}$ depends on both amplitude and phase factors.

3.4 Limit Distributions in the Random Walk Problem

In the last section we introduced K distributions as a class of functions for which the fixed-N solution of the finite random walk problem could be expressed in closed form in terms of familiar tabulated functions. It can be shown that the moments of these distributions always lie between those of the negative exponential (Gaussian) and log-normal distributions and they were originally proposed on this basis as a model for the statistics of microwave sea echo, which is non-Gaussian when the area of sea illuminated is sufficiently small [3.77]. Early attempts to fit experimental data were surprisingly successful but no particular significance was attached to this owing to the statistical spread and instrumental inaccuracies and uncertainties in the measurements [3.85]. However, K distributions have more recently been shown to provide an excellent model for the non-Gaussian statistics of light scattered by a variety of turbulent fluid systems (Sect.3.5). A large body of accurate data on these systems has now been accumulated using photon counting techniques and some underlying, unifying scattering mechanism is clearly needed to explain the experimental results. These suggest that although K distributions first arose through the notion of a limited area of illumination, containing few scatterers, there might be many-scatterer regimes (possibly associated with the multiscale nature of turbulence) where limit theorems confer the observed non-Gaussian universality on the statistics. It has been conjectured that fluctuations in the *number density* of scattering centres may play an important role in this context [3.10] and in this section, therefore, we shall review recent work on the effect of step-number fluctuations in the random walk problem.

3.4.1 The Gaussian Limit

Consider first the characteristic function corresponding to the random walk (3.39) with N fixed

$$C_N(\underline{u}) = \langle \exp(i\underline{u}\cdot\underline{V}) \rangle = \langle J_0(ua) \rangle^N = \langle J_0(u\sqrt{I}) \rangle \quad , \tag{3.57}$$

where J_0 is the zeroth order Bessel function of the first kind. In order to derive the asymptotic distribution for the case of large N it is convenient to renormalize the step length a by a factor $\sqrt{N}$ and examine the limit of the expression $\langle J_0(ua/\sqrt{N})\rangle^N$ as N tends to infinity. It is not difficult to show that *whatever the step-length distribution*

$$\lim_{N\to\infty} C_N(\underline{u}) = \exp(-u^2\langle a^2\rangle/4) \quad , \tag{3.58}$$

corresponding to the negative exponential distribution

$$p(I) = \langle I \rangle^{-1} \exp(-I/\langle I \rangle) \quad . \tag{3.59}$$

This result expresses the fact that in the large N limit the real and imaginary parts of $V(\underline{r},t)$ are zero-mean, independent Gaussian variables as predicted by the central limit theorem. A corollary of this result is that if the squares of the step-lengths $\{a_j\}$ in (3.39) are negative exponentially distributed then so is the resultant intensity I (but with scaled mean $\langle I \rangle = N\langle a^2 \rangle$) *whatever the value of N*. The distribution (3.59) is said to be *stable* with respect to the convolution (3.58). Identical results are obtained in the limit of a large *mean* number of scatterers when N varies according to a Poisson distribution, (3.65), rather than being fixed [$\langle N_p \rangle \to \infty$ in (3.45)]. This is because the normalized variance of a Poisson distribution decreases inversely as its mean so that the number fluctuations assume, relatively speaking, less and less importance as the mean increases.

We have already pointed out in Sect.3.2 that Gaussian light is frequently encountered in laser light scattering experiments. It also occurs as unwanted noise (speckle) in imaging and other optical processing systems. In the former situation only the temporal correlation properties normally provide useful information because the single interval statistics (3.59) indicate only that many scatterers are present in the illuminated region, whilst the spatial coherence properties relate only to the size of that region. In the noise modelling context, however, the low information content of the Gaussian model (requiring specification of the mean and first-order coherence functions only) together with its scaling and stability properties referred to above, are significant assets, and it is properties of this type that we seek in the development of models for non-Gaussian statistics.

3.4.2 Negative Binomial Number Fluctuations

Having established that Poisson number fluctuations lead back to Gaussian statistics in the high mean limit it is natural to investigate other familiar number distributions such as the geometric or thermal distribution (3.76). This candidate looks more interesting as its relative variance remains finite and nonzero as its mean increases. Unlike the Poisson distribution which describes a purely random collection of scatterers (such as a suspension of noninteracting Brownian particles), the geometric distribution may be used to characterize a situation in which the scatterers are correlated or bunched in space and time. The negative binomial distributions (3.28) form a more general class with this property and if (3.57) is averaged over this type of number fluctuation we obtain

$$C_{\bar{N}}(u) = [1+(\bar{N}/\alpha)(1-\langle J_0(ua/\sqrt{\bar{N}})\rangle)]^{-\alpha} \quad , \tag{3.60}$$

where we have scaled a by a factor $\sqrt{\bar{N}} = \sqrt{\langle N \rangle}$ and used the index α instead of ζ in (3.28) because in the present context this parameter is restricted only to be greater than zero. The associated limit distribution can easily be derived from (3.60)

$$\lim_{\bar{N}\to\infty} C_{\bar{N}}(u) = [1+u^2<a^2>/4\alpha]^{-\alpha} \quad ; \quad (\alpha>0) \tag{3.61}$$

corresponding to the class of K distributions (3.48) with $b = 4\alpha/<a^2>$. Whereas the Gaussian result (3.59) is a manifestation of *interference* in the coherent random walk problem, K distributions reflect the presence of both interference and step-number density fluctuations. In the incoherent case (e.g., white light scattering) there is, of course, no interference so that variations of the resultant amplitude in the limit of large average step number can only arise due to number density fluctuations. Using the distributions (3.28) we obtain [3.10,11] for this case simply a continuum analogue of the negative binomial class — the gamma distributions

$$p(I) = \alpha(\alpha I/<I>)^{\alpha-1} \exp(-\alpha I/<I>)/<I>\Gamma(\alpha) \quad . \tag{3.62}$$

As in the Gaussian case (3.59), the limit distributions (3.48,62) are obtained whatever the statistical properties of the scattering amplitudes $\{a_j\}$ appearing in (3.39). We have already noted in Sect.3.3.3 that if these amplitudes are *assumed* to be K distributed then the resultant of a fixed-N random walk is also K distributed — a property reminiscent of the corollary to the central limit theorem mentioned above. In the K-distribution case the index, rather than simply the mean, is scaled by N (K distributions are *infinitely divisible* rather than stable), but the fact that N is not introduced as an additional parameter is a very useful property in statistical modelling applications.

The above derivation of K distributions could well account for their widespread occurrence and reduces the inverse scattering problem to one of identifying the (countable) scatterers and explaining their bunching properties in terms of the properties of the scattering medium. A clue to the first point is perhaps that the data presented in the next section refer to strongly scattering systems where geometrical optics plays an important role in determining the properties of the scattered light. This suggests that the scattering centres may be specular points on the scattered wave front. Analytical investigation of the Fresnel integral formulation of the scattering problem has gone some way towards confirming this conjecture [3.86]. The second point mentioned above is much more difficult to deal with and will be discussed briefly in connection with the model proposed in Sect.3.4.3. However, since the subject of number fluctuations in the random walk problem seems to have received little attention in the literature it is perhaps worthwhile making a few more general comments at this point.

It is not difficult to construct other limit distributions which arise from step-number fluctuations in the random walk (3.39). For example, the uniform distribution $p(N) = (R+1)^{-1}$ for $N \leq R$ gives

$$\lim_{\bar{N}\to\infty} C_{\bar{N}}(u) = [1-\exp(-2u^2<a^2>)]/2u^2<a^2> \tag{3.63}$$

corresponding to the exponential-integral limit distribution

$$p(I) = (2\langle I\rangle)^{-1}E_1(I/2\langle I\rangle) \quad , \tag{3.64a}$$

where

$$E_1(x) = \int_x^\infty \frac{\exp(-y)}{y}\,dy \quad . \tag{3.64b}$$

The single parameter $\langle I\rangle$ characterizing this distribution clearly does not scale under the convolution (3.57), however, so that in a fixed-N random walk, where the $\{a_j\}$ are distributed according to (3.64), N is introduced as a second parameter. This situation can in fact always be avoided by choosing a discrete compound Poisson distribution for N [3.87], i.e.,

$$N = \sum_{r=0}^{M} n_r \quad , \tag{3.65}$$

where $\{n_r\}$ are statistically identical and independent integers and M is Poisson distributed. Averaging the characteristic function (3.57) over this type of number fluctuation distribution leads to [3.11]

$$C_{\bar{N}}(u) = \exp\left[-\bar{M}(1-\overline{\langle J_0(ua/\sqrt{\bar{N}})\rangle^n})\right] \quad , \tag{3.66}$$

where the bar indicates averaging over the fluctuations in n and $\bar{M}$ is a parameter which scales under the fixed-N convolution (3.57), i.e., $\bar{M}\to N\bar{M}$ so that as in the case of the negative exponential and K distributions the number of free parameters of the model is not increased by this procedure. Formula (3.66) reduces to the Poisson case when n is fixed and to the negative binomial result (3.60) when n is logarithmically distributed, i.e., when $p(n) = q^n/n\ln[1/(1-q)]$. It is not easy to find other simple examples, however, for which it can be evaluated and the limit distributions expressed in terms of familiar tabulated functions.

3.4.3 A Population Model

So far we have demonstrated that negative binomial fluctuations in the number of scatterers might provide a reasonable explanation of the occurrence of K distribution as the *single* interval statistics of light scattered by turbulent media. In order to develop the model further it is necessary to extend the theoretical approach to predict the field and intensity coherence functions of the scattered light. These are measurable quantities, so that the range of experimental tests of the theory will then be considerably increased. Negative binomial distributions occur widely throughout the statistics literature but perhaps assume their most important role in queuing theory and in population statistics where they occur as equilibrium distributions in the birth-death-immigration process [3.88]. We shall review the consequences of this last-mentioned model for the scattering problem under consideration here. There is, as yet, no experimental evidence either for or against this choice, but both physical and mathematical arguments support its adoption as a preliminary model. The birth-death-immigration problem is in fact closely related to the photon-distribution problem in a laser well below threshold (a single mode of thermal radiation) [3.35,36], immigration being similar to spontaneous emission. We can draw an analogy between this model and the cascade description of turbulence in which large eddies are spontaneously created (immigration), give rise to daughter eddies (birth) and finally succumb to viscous dissipation (death). Here we have implicitly connected the rather nebulous concept of eddies with our earlier interpretation of scattering centres as specular points on the scattered wave front. Note that we have shown that the statistics of the scattered light is insensitive to the detailed nature of the scatterers, being dominated by fluctuations in their number so the "size" of the various generations of eddies is immaterial, provided they introduce path differences which are comparable to or greater than the optical wavelength. From a mathematical point of view the population model is attractive because it is a first-order Markov process characterized by a single correlation function. This means that the statistics of the scattered light will be characterized by only two coherence functions, one associated with interference effects (as in the Gaussian case) and one connected with the number fluctuations. Thus the field and intensity coherence functions together with the single interval distributions completely define the statistics of the scattered field for this model.

The fluctuations of a population with birthrate λ, deathrate μ and immigration rate ν are governed by the rate equation

$$\frac{dP_N}{dt} = \mu(N+1)P_{N+1} - [(\lambda+\mu)N+\nu]P_N + [\lambda(N-1)+\nu]P_{N-1} \quad , \tag{3.67}$$

where $P_N(t)$ is the probability of finding a population of N individuals at time t. If M individuals are present initially then the transient solution for the generating function [(3.11), no integration] at time t is [3.88]

$$Q_t(s) = \{(\lambda-\mu)/[\lambda-\mu+\lambda(\theta-1)s]\}^{\nu/\lambda}\{[\lambda-\mu+(\mu\theta-\lambda)s]/[\lambda-\mu+\lambda(\theta-1)s]\}^M \quad , \tag{3.68a}$$

where

$$\theta(t) = \exp(\lambda-\mu)t \quad . \tag{3.68b}$$

An equilibrium distribution for large times exists if the deathrate μ is greater than the birthrate λ. Setting $\theta = 0$ in (3.68) gives

$$Q_\infty(s) = (1+\bar{N}s/\alpha)^{-\alpha} \quad , \tag{3.69a}$$

where

$$\bar{N} = \langle N \rangle = \nu/(\mu-\lambda) \quad \text{and} \quad \alpha = \nu/\lambda \quad . \tag{3.69b}$$

The result (3.69a) is the generating function for the negative binomial class (3.28) with $\zeta = \alpha$. Because of the Markov nature of the process all the higher order joint statistical properties can be calculated from (3.68a). The simplest of these is the bilinear moment

$$\langle N(0)N(\tau) \rangle/\bar{N}^2 = 1 + (\alpha^{-1}+\bar{N}^{-1})\theta(\tau) \quad . \tag{3.70}$$

The lowest order temporal coherence functions of the random walk representation (3.39) of a scattered optical field can be calculated exactly using these formulae. The results obtained for $g^{(1)}(\tau)$ (3.8) and $g^{(2)}(\tau)$ (3.9a) in the limit of large $\bar{N}$ are [3.11]

$$g^{(1)}(\tau) = \exp(-\mu\tau)\langle a(0)a(\tau) \rangle \langle \exp\{i[\phi(0)-\phi(\tau)]\} \rangle/\langle a^2 \rangle \tag{3.71}$$

and

$$g^{(2)}(\tau) = 1 + (1+\alpha^{-1})|g^{(1)}(\tau)|^2 + \alpha^{-1}\theta(\tau) \quad . \tag{3.72}$$

Both the lifetime of the scatterers and their cross-section fluctuations enter into (3.71) for the field correlation function, but orientational (phase) fluctuations may often dominate the time dependence of this quantity. Two types of contribution to the intensity correlation function (3.72) can be identified: a number fluctuation term $\theta(\tau)$ and an interference term proportional to $|g^{(1)}(\tau)|^2$. There is an interesting similarity of structure between this formula, which replaces the Gaussian factorization theorem (3.96), and the result obtained in the case of a finite, fixed-N random walk given in Sect.3.3.4 (3.56).

It is clear that much remains to be done, both theoretically and experimentally, if the model described in this section is to find general acceptance as a valid approach to the problem of strong scattering by turbulent media. However, the approach we have outlined shows considerable promise in this direction and certainly demonstrates the interesting *mathematical* possibilities which arise when non-Poissonian step-number fluctuations are included in the random walk problem.

3.5 Experiments

Experiments on three different scattering systems will be emphasized in this section. The first is a 'dynamic scattering' liquid crystal which, under appropriate experimental conditions, appears to behave, to good approximation, as a phase screen. Here the scattering was only studied in the far-field (Fraunhofer) region and measurements of statistics and correlation functions clearly demonstrate many of the characteristic properties of scattering in the non-Gaussian regime. The second system is a phase screen produced in the laboratory by turbulent mixing of hot and cold air. The larger spatial scale of the fluctuations in this case enabled measurements to be made in the near-field (Fresnel) region as well as the far field. This work provided what appears to be the first measurement of a complete focussing curve (discussed in Sect.3.3). Finally we discuss outdoor measurements on propagation through an extended region of turbulence. The use of accurate photon-counting methods allowed detailed observation of the changing nature of the probability distribution function P(I) as a function of propagation distance.

For all three systems K distributions were found under certain conditions. In Sect.3.5.4 we mention briefly other experiments where K distributions seem relevant.

All the experiments described in this section were performed using modern photon-counting and photon-correlation equipment whose operation was discussed in Sect.3.2. Detection apertures were chosen to be small compared to the spatial size of features in the intensity patterns; similarly electronic sampling times were small compared to characteristic fluctuation times of the scattered radiation. Thus the effects of spatial and temporal integration could be neglected. Where necessary photon-count factorial moments were corrected for any significant background contribution to the intensity (dark count and stray light) by use of (3.32). Other details of the experiments are given in the literature. Here the discussion is largely qualitative with emphasis on the physical principles illustrated.

3.5.1 Dynamic Scattering by Nematic Liquid Crystals [3.76,89-91]

A nematic liquid crystal is a material in which elongated molecules maintain orientational correlations over distances many times the molecular dimensions. In fact, due to surface effects, it is possible to have more or less complete alignment within a thin layer ($\sim 50\ \mu$m) sandwiched between two plates. Application of an electric potential (~ 10 V) across the thickness of this layer can then produce an electrically driven turbulence within the medium which causes, in addition to a bulk motion, angular perturbations of the molecules about their mean alignment direction. Associated with these perturbations are refractive index fluctuations which cause strong light scattering, the so-called dynamic scattering, (provided the confining plates are transparent). Not surprisingly for an anisotropic medium, the liquid crystal is birefringent and the scattering process quite complicated. Nevertheless, in the early days of dynamic scattering, DEUTSCH and KEATING [3.92] showed that, under certain conditions, the main effect of dynamic scattering in the liquid crystal layer was to introduce random phase retardations into a propagating light beam, i.e., to produce a phase screen. These conditions occur when incident polarization, detection polarization and alignment direction all lie in the scattering plane and the scattering angle is not too small.

In such a system the turbulence correlation length ξ (and hence that of the phase) is a few microns and the rms phase shift of order 2π. For visible light this produces a focussing distance z_F (Sect.3.3.1) itself of a few microns so that measurements near the focussing region are difficult. Accordingly experiments were confined to the far-field, Fraunhofer, region with the main aim of investigating the dependence of the properties of the scattered radiation on the magnitude of the area of phase screen illuminated. When this area A included many correlation areas, $A \gg \xi^2$, the far-field intensity pattern had the familiar 'Gaussian speckle' appearance. However, as the area was reduced, larger-scale features appeared in the pattern. These features, identified as 'single-scatterer' or non-Gaussian, contributions covered several speckles and moved slowly compared to the speckle evolution.

This system was treated theoretically by both analytic and microarea or discrete-scatterer approaches [3.76,78,90]. In keeping with the thrust of this article, we will emphasize the latter. Thus the wave front emerging from the screen was divided conceptually into independent plane facets of area $\sim\xi^2$ whose orientation fluctuates with some prescribed probability distribution. Each facet projects a lobe of radiation into the far field whose structure is determined by diffraction. When many such lobes contribute to the scattered radiation at any detection point, interference between them leads to Gaussian speckle. When few contribute, the structure of the individual lobes is evident in the intensity pattern as the larger-scale features referred to above.

Perhaps the simplest theoretical prediction stemming from this model concerns the angular dependence of the mean intensity and the excess or non-Gaussian part of the second moment of P(I). Clearly $\langle I \rangle$ should be proportional to the mean number $\langle N_F \rangle$ of facets pointing in the observation direction. On the other hand the excess second moment should be determined by fluctuations in this number. The theories outlined in Sect.3.3, together with the assumed independence of the facets, predict

$$\frac{\langle I^2 \rangle}{\langle I \rangle^2} - 2 \propto \frac{1}{\langle N_F \rangle} \quad . \tag{3.73}$$

Thus the excess second moment and the reciprocal scattered intensity should show the same angular dependence. Figure 3.1 shows this to be so to a reasonable degree of accuracy.

The second prediction of theory which can easily be tested in this system is the dependence of the second moment on the area A of the liquid crystal illuminated (at a fixed detection angle). From (3.73) the excess second moment is proportional to

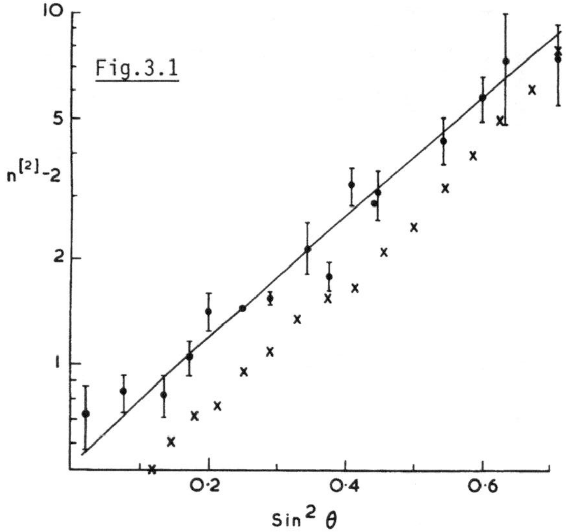

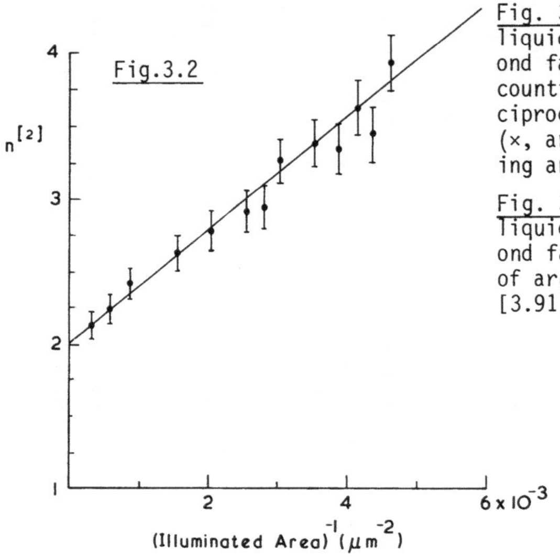

Fig. 3.1. Dynamic scattering by a liquid crystal: dependence of second factorial moment of photon-counting distribution (●) and reciprocal mean scattered intensity (×, arbitrary units) on scattering angle θ [3.78]

Fig. 3.2. Dynamic scattering by a liquid crystal: dependence of second factorial moment on reciprocal of area of sample illuminated [3.91]

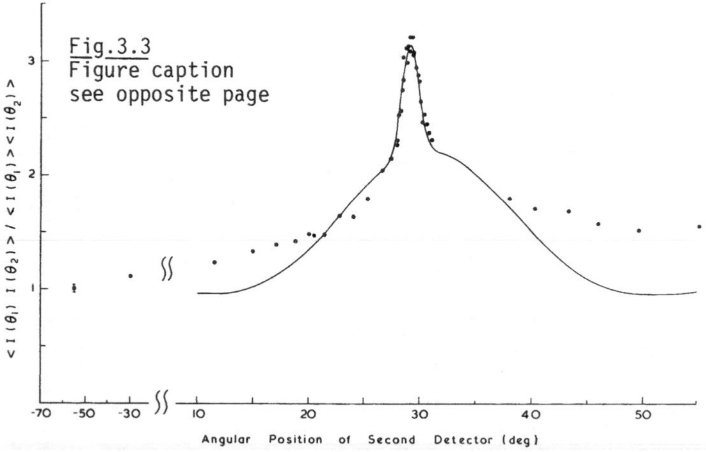

Fig.3.3
Figure caption see opposite page

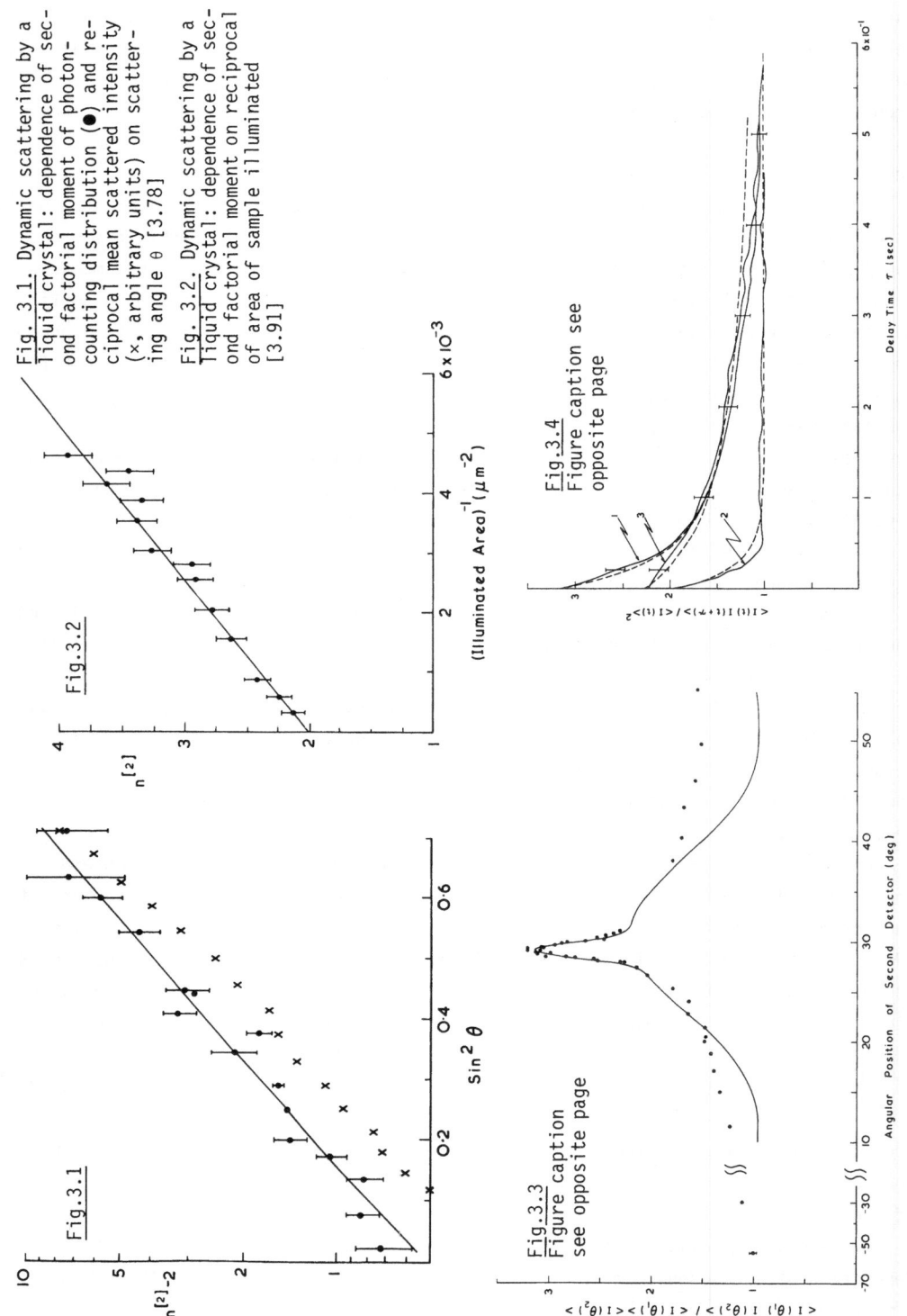

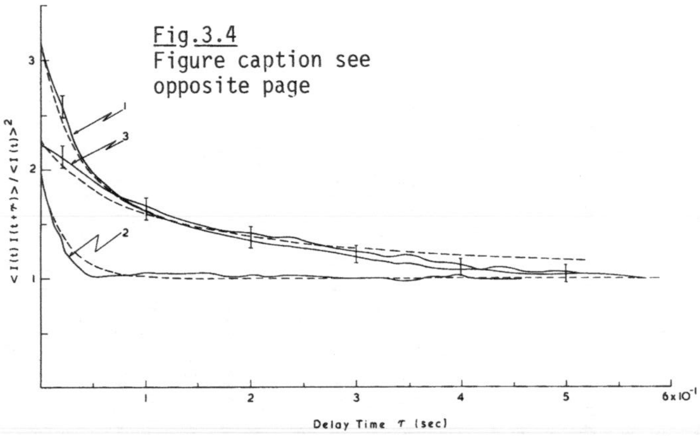

Fig.3.4
Figure caption see opposite page

$\langle N_F \rangle^{-1}$ and hence A^{-1}. The data (Fig.3.2) show this dependence though it must be mentioned that, in this figure, a correction has been applied to A to account for some spreading of the laser beam within the sample.

Measurements were also made of spatial and temporal correlations. Figure 3.3 shows the cross-correlation at zero time between a detector placed at $\theta_1 = 29.1^\circ$ and a second detector whose angular position was varied in the range $-55^\circ < \theta_2 < 55^\circ$. These data clearly show two spatial scales. The small-scale feature can be identified with the Gaussian speckle size and is determined purely by geometrical considerations [3.58,78]. The larger-scale feature is the non-Gaussian contribution, the last term in (3.56). Its structure is determined by the diffraction-broadened lobe scattered by a *single* facet.

The temporal autocorrelation function can also be separated into a rapidly decaying Gaussian component and a non-Gaussian component which decays more slowly. This separation, not immediately obvious in Fig.3.4, can be checked by spatio-temporal correlation measurements with detectors separated by more than a single speckle dimension but by much less than the non-Gaussian spatial coherence length [3.91]. From (3.56,54) it is seen that the temporal decay of the Gaussian term in the autocorrelation function depends on the product of phase and amplitude terms. For a deep phase screen, the phase term decays more rapidly with characteristic time equal to the time taken for the contributions from two different (independent) facets to change their relative phase difference by about 2π radians. By contrast the non-Gaussian term depends only on the amplitude scattered by a single facet (3.56) and its characteristic decay time is roughly the time taken for the radiation lobe to sweep across the detector.

Finally we discuss the higher-order, single-interval statistics, starting with a brief digression on data presentation. As mentioned in Sect.3.2, one measures the photon-counting distribution $p(n,T)$ and, in general, requires to compare this with theoretical expressions for the intensity distribution $P(I)$. It is frequently not possible to compute $p(n,T)$ for a given $P(I)$; (K distributions are a notable exception to this rule). Usually, therefore, we exploit the identity of the normalized factorial moments $n^{[r]}$ of $p(n,T)$ (3.5) and the normalized moments $\langle I^r \rangle / \langle I \rangle^r$ of $P(I)$, i.e., measured $n^{[r]}$ are compared with calculated $\langle I^r \rangle / \langle I \rangle^r$. The theoretical

---

◄ Fig. 3.3. Dynamic scattering by a liquid crystal, spatial coherence: cross-correlation between intensities detected at two different scattering angles. Detector 1 was fixed at $\theta_1 = 29.1^\circ$ and position of detector 2 was varied [3.91]

Fig. 3.4. Dynamic scattering by a liquid crystal, temporal coherence: *Curve 1*, autocorrelation function for detector 1 at $\theta_1 = 29.1^\circ$ with small illuminated region; *Curve 2*, autocorrelation function for detector 1 at $\theta_1 = 29.1^\circ$ with larger illuminated region (Gaussian term only); *Curve 3*, *cross*-correlation function for detector 1 at $\theta_1 = 29.1^\circ$ and detector 2 at 27.5° (non-Gaussian term only, since detectors are separated by more than one "speckle dimension"). Note that the time-dependent part of the full autocorrelation (curve 1) is equal to the sum of the time-dependent parts of the Gaussian (curve 2) and non-Gaussian (curve 3) contributions [3.91]

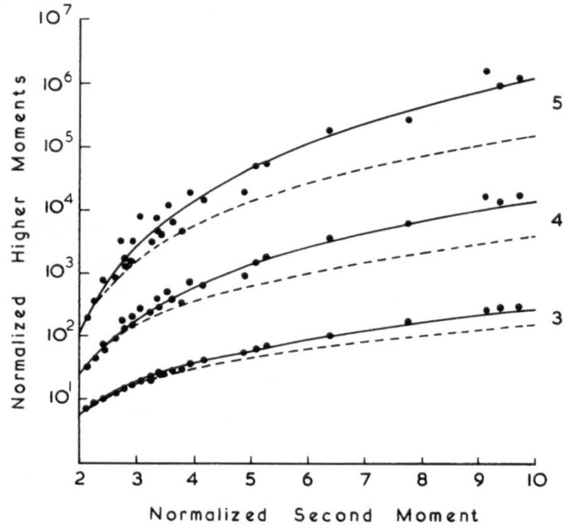

Fig. 3.5. Dynamic scattering by a liquid crystal, single-interval statistics: higher normalized moments $n^{[r]}$ ($r=3,4,5$) as a function of second normalized moment $n^{[2]}$ for various scattering angles and illuminated areas [3.10]

distributions $P(I)$ of interest are frequently characterized by two parameters. The first, related to the mean, is removed by normalization (effectively working in terms of a reduced intensity $I/\langle I\rangle$). The second parameter can always be written as a function of the second moment $\langle I^2\rangle/\langle I\rangle^2$. Comparison between experiment and theory can then be made by plotting higher normalized moments, both theoretical and experimental, as a function of the second moment.

Figure 3.5 shows such a plot for the liquid crystal data obtained in a range of experimental configurations including different scattering angles and illumination areas A. Initially these data were compared with predictions of the facet model, outlined above, shown as dotted lines in the figure. There are clear differences between experiment and theory. These experiments predated K distributions so that immediate comparison was not possible. However, after the wide-ranging applicability of K distributions became apparent, such a comparison was made. It was then found that agreement between experiment and K distributions (the solid lines in Fig.3.5) was remarkably good.

3.5.2 Hot-Air Phase Screen [3.5,24]

While the experiments on the liquid crystal described above demonstrated many features of non-Gaussian scattering in the Fraunhofer regime it did not seem a promising system with which to investigate the Fresnel regime. Because of both the importance of Fresnel region phase screen effects to many natural phenomena and the large extant body of related but untested theory we decided to construct a system on which such experiments could be performed in a controlled laboratory environment. We found that the plume of hot air rising above a suitably constructed electric heater would, at a height of about 15 cm above the heater, entrain the cooler surrounding air to form a region of fairly isotropic turbulent mixing. The phase screen produced in this way provided an rms phase shift $\sim\pi$ radians, a phase correlation length ξ of a few mm, and a focussing distance z_F of a few meters.

Measurements of the photon-counting distribution were made on axis (i.e., at zero scattering angle) as a function of distance z from the phase screen. The distance-dependence of the second moment, the 'focussing curve', is shown in Fig.3.6 where a peak due to focussing effects, is clearly evident. The failure of the second moment to reach the 'Gaussian speckle' value of 2 at large distance is probably a residual far-field non-Gaussian effect since the cross-sectional area of the laser beam in the turbulence contained a finite number of coherence areas ξ^2. The solid line in Fig.3.6 shows an attempt to fit the data by a model which assumed the spatial correlation function of the phase to be Gaussian, i.e., to have a single dominant length scale. The data have a wider peak than that predicted by this theory indicating, consistent with the expected structure of the turbulence, a wider range of scale sizes.

Figure 3.7 shows photographs of the intensity pattern as a function of propagation distance z. The exposure time was 0.5 ms, short enough to effectively 'freeze' motions in the pattern. These patterns are consistent with the qualitative description given in Sect.3.3.1. For $z \ll z_F$, the intensity modulations are relatively weak whereas in the focussing region, $z \approx z_F$, a rich, deeply modulated structure is seen. The centre of the far-field pattern, $z \gg z_F$, has a typical, random, Gaussian speckle form.

The higher factorial moments of p(n,T) were compared with the moments of various two-parameter distributions for P(I). For distances within the focussing region, $z < z_F$, no distribution was found to fit over a wide range of z though, possibly for-

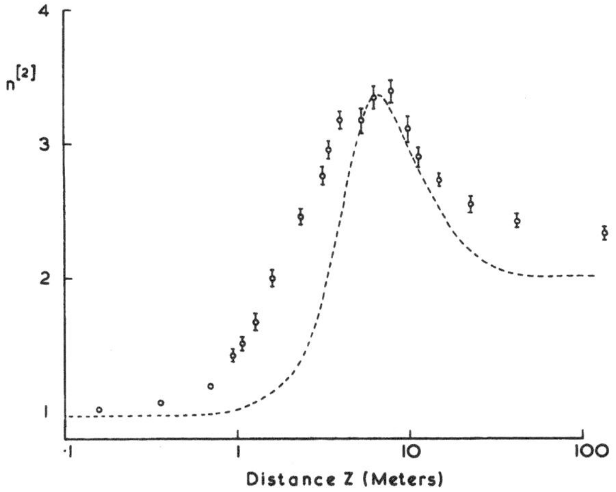

Fig. 3.6. Focussing curve for phase screen: dependence of normalized second moment on distance from screen with detector on axis (θ=0). Note that scintillation index, (3.38), is given by $\sigma^2 = n^{[2]} - 1$ [3.5]

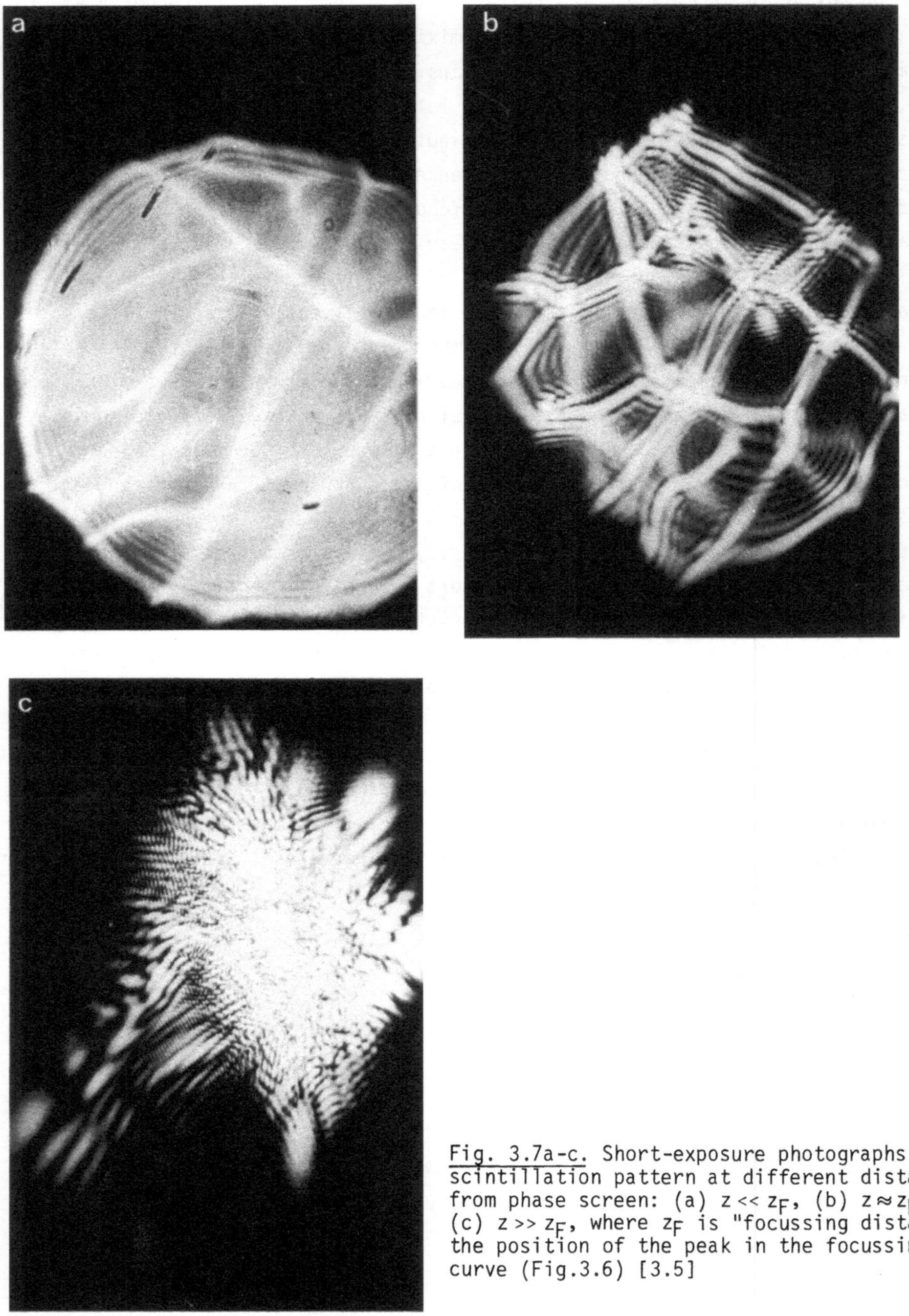

Fig. 3.7a-c. Short-exposure photographs of scintillation pattern at different distances from phase screen: (a) $z \ll z_F$, (b) $z \approx z_F$, (c) $z \gg z_F$, where z_F is "focussing distance", the position of the peak in the focussing curve (Fig.3.6) [3.5]

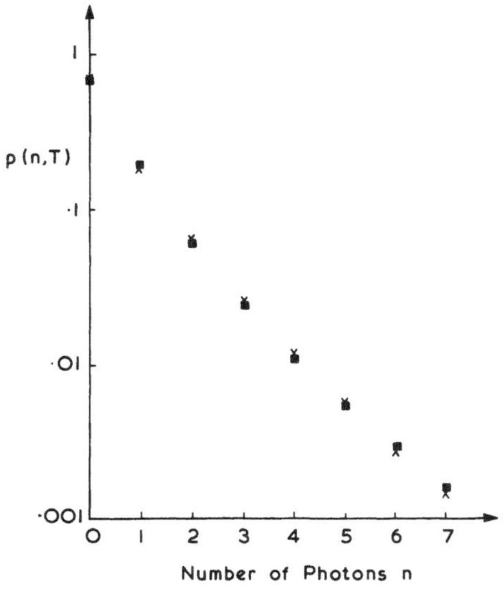

Fig. 3.8. Probability density p(n,T) of photon number fluctuations for phase screen at $z > z_F$: ■, experimental data; ×, theoretical values with same first and second moments (3.51) [3.5]

tuitously, the log-normal distribution (see Sect.3.5.3) gave a good description of the data over a limited range of z. For $z > z_F$, K distributions were found to fit, to within experimental error. It is in this region, where the detector can receive radiation from several independent coherence areas in the turbulence, that one would expect K distributions to be useful (see Sects.3.3,4). This experiment was the first to provide indications of the quantitative applicability of K distributions. It was subsequently found that the three-parameter K+ coherent component distributions, discussed in Sect.3.3.3 gave a good description of the data over the whole range of z [3.11]. The parameter x^{-1}, the ratio of coherent component to K-distributed noise varied from very large at $z = 0$ to about 0.1 for $z > z_F$. Since the two-parameter distributions also fitted well at large ranges we conclude that the data are not accurate enough to distinguish between $x^{-1} = 0.1$ and $x^{-1} = 0$.

Figure 3.8 shows a direct comparison of the experimental photon-counting distribution p(n,T) with the theoretical expression for K-distributed fluctuations (3.51).

Measurements were also made of spatio-temporal correlations. Except in the far field, $z \gg z_F$, there was not a marked separation of spatial and temporal scales for the Gaussian and non-Gaussian contributions. Correlation functions measured with two detectors separated in the direction of mean flow in the turbulent region showed peaks at delay times determined by the transit of features in the intensity pattern from one detector to the other.

3.5.3 Extended Atmospheric Turbulence [3.93]

The first prediction for the probability distribution P(I) of the light propagating in an extended random medium came from the so-called Rytov approximation [3.22,94] which leads to a log-normal distribution for which

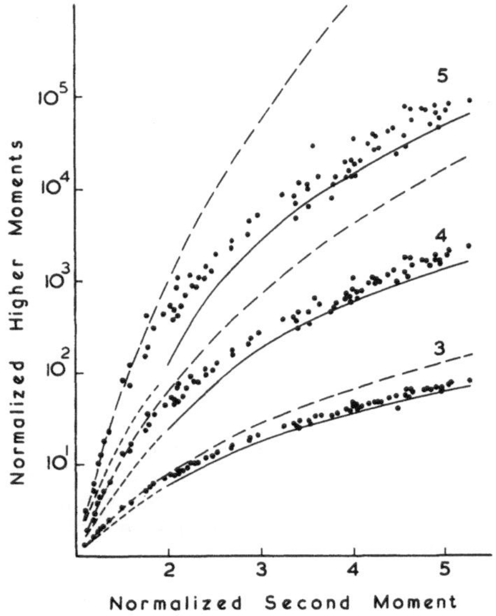

Fig. 3.9. Propagation through extended turbulent medium, single-interval statistics: higher normalized moments $n^{[r]}$ (r=3,4,5) as a function of second normalized moment $n^{[2]}$. Abscissa is roughly proportional to range z, varying from 0 at $n^{[2]}=1$ to ~700 m at $n^{[2]}=5$ [3.93]

$$\frac{\langle I^r \rangle}{\langle I \rangle^r} = \left[\frac{\langle I^2 \rangle}{\langle I \rangle^2}\right]^{r(r-1)/2} \quad . \tag{3.74}$$

This approximation is clearly closely related to the Born approximation and, as such, is expected to apply only for small intensity fluctuation $\sigma^2 \ll 1$, and therefore, small propagation distances. Nevertheless early experiments seemed to indicate a wider range of applicability for the Rytov approximation and, for a while, there was a widespread acceptance that the log-normal distribution was valid under all conditions of atmospheric propagation. More recently this view has been questioned and the proper limitations of the Rytov approximation re-emphasized [3.22]. At the same time there has been growing acceptance that, at very long range, where the detector 'sees' many independent scattering events, P(I) should be a negative exponential (3.7a) by virtue of the central limit theorem [3.22].

The current theoretical situation thus leaves a considerable gap at intermediate ranges where the structure of P(I) is not understood. Furthermore this intermediate range is important in the operation of many electrooptical systems such as communication links.

Figure 3.9 shows data obtained by photon-counting methods for atmospheric propagation of visible laser light over paths up to 700 m above an aircraft runway in reasonably warm weather [3.93]. For $\sigma^2 \lesssim 0.75$ the data are well described by the log-normal distribution. In view of the comments made above this large range of validity for a theory based on a first-order approximation remains a surprise. Nevertheless, as the propagation distance was increased, σ^2 increased and deviations from the log-normal distribution became apparent. At large values of σ^2, found at z = 500-700 m, the data are tending towards the K distributions though the experimental moments are still slightly high. Data have not yet been obtained for $z > 700$ m with photon-counting equipment. By analogy with the phase screen we expect a peak in σ^2 as a function of z, followed by K-distributed fluctuations with decreasing σ^2 tending to a negative exponential as $z \to \infty$. This remains to be established experimentally.

Finally we note that the maximum values of σ^2 found, ~4, are considerably higher than those found with the laboratory-generated phase screen (Sect.3.5.2) indicating an intensity pattern with even greater modulations than that shown in Fig.3.7b.

3.5.4 Other Experiments

As mentioned in Sect.3.4 K distributions were first introduced as a model for the statistics of microwave sea echo. Because of the low energy of a microwave photon, experiments in this area are, by nature, analog and we will not discuss them further here.

An area in which photon-counting methods have recently been applied is the scintillation or twinkling of starlight [3.4,6], which is generally thought to be due largely to turbulence in the tropopause, a layer of the atmosphere at a height of about 10 km. An early experiment on the star Sirius, when low in the sky, gave a second moment of about 2.9 with higher moments slightly above those for a K distribution. Comparison with data for the laboratory-generated phase screen (Sect.3.5.2) would then suggest that the detection point was just on the near-field side of the focussing peak.

Unpublished measurements have also been made in our laboratory by Dr. G. Parry on a phase screen formed by turbulent mixing of hot and cold water. The results were qualitatively similar to those found with the hot air systems.

3.6 Concluding Remarks

We have emphasized on several occasions in this chapter the advantages of using photon-counting techniques to establish the statistical and correlation properties of scattered optical fields. Our assertions have been illustrated by a discussion of the application of these techniques to the measurement of optical scintillation which has led directly to new theoretical models for the scattering process. The research in this area has now reached an interesting stage in that a connection appears to exist between the observation of K distributions for scattered radiation and the turbulent nature of the scattering systems themselves. However, this connection has not yet been established conclusively and future research must be directed towards this end. Several aspects of the model discussed in Sect.3.4 have not yet been tested experimentally. Comparisons between coherent and white light scattering from the same systems have not been made, nor have the theoretical predictions for the intensity correlation function been investigated. The concept of correlated scattering centres (Sect.3.4) should only apply when the centres are separated by distances less than or comparable to the outer scale of turbulence. Thus one future direction for experiments should be the study of single-interval statistics and factorization of correlation functions when the linear dimension of the illuminated region is less than this scale. In all such statistical work at optical frequencies, photon-counting techniques will continue to be the most appropriate method, both from the noise and accuracy viewpoints, for the collection and processing of scattering data.

Acknowledgement. We are grateful to Dr. G. Parry, our collaborator over the past three years, for many valuable discussions.

References

3.1 H.Z. Cummins, E.R. Pike (eds.): *Photon Correlation and Light Beating Spectroscopy* (Plenum Press, New York 1974)
3.2 E.R. Pike, E. Jakeman: Adv. Quantum Electro. *2*, 1 (1974)
3.3 B. Saleh: *Photoelectron Statistics*, Springer Series in Optical Sciences, Vol.6 (Springer, Berlin, Heidelberg, New York 1978)
3.4 E. Jakeman, E.R. Pike, P.N. Pusey: Nature *263*, 215 (1976)
3.5 G. Parry, P.N. Pusey, E. Jakeman, J.G. McWhirter: Opt. Commun. *22*, 195 (1977)
3.6 E. Jakeman, G. Parry, E.R. Pike, P.N. Pusey: Contemp. Phys. *19*, 127 (1978)
3.7 C.J. Oliver: "Correlation Techniques", in Ref.[3.1], pp.151-236
3.8 E. Jakeman, C.J. Oliver, E.R. Pike: Adv. Phys. *24*, 349 (1975)
3.9 C.J. Oliver: Infrared Phys. *18*, 303 (1978)
3.10 E. Jakeman, P.N. Pusey: Phys. Rev. Lett. *40*, 546 (1978)
3.11 E. Jakeman: J. Phys. A *13*, 31-48 (1980)
3.12 A. Zardecki: "Statistical Features of Phase Screens from Scattering Data", in *Inverse Source Problems in Optics*, ed. by H.P. Baltes, Topics in Current Physics, Vol.9 (Springer, Berlin, Heidelberg, New York 1978) pp.155-189
3.13 H.G. Booker, J.A. Ratcliffe, D.H. Shinn: Phil. Trans. Roy. Soc. (London) *242*, 579 (1950)
3.14 R.P. Mercier: Proc. Camb. Phil. Soc. *58*, 382 (1962)
3.15 E.E. Salpeter: Astrophys. J. *147*, 433 (1967)
3.16 R. Buckley: Austral. J. Phys. *24*, 351, 373 (1971)
3.17 V.I. Shishov: Izv. Vuz. Radiofuz. *14*, 85 (1971)
3.18 L.S. Taylor, C.J. Infosino: J. Opt. Soc. Am. *65*, 78 (1975)
3.19 E. Jakeman, J.G. McWhirter: J. Phys. A *9*, 1599 (1977)
3.20 J.W. Strohbehn: Proc. IEEE *56*, 1301 (1968)
3.21 R.S. Lawrence, J.W. Strohbehn: Proc. IEEE *58*, 1523 (1970)
3.22 R.L. Fante: Proc. IEEE *63*, 1669 (1975)
3.23 A.M. Prokhorov, F.V. Bunkin, K.S. Gochelashvily, V.I. Shishov: Proc. IEEE *63*, 790 (1975)
3.24 G. Parry, P.N. Pusey, E. Jakeman, J.G. McWhirter: "The Statistical and Correlation Properties of Light Scattered by a Random Phase Screen", in *Coherence and Quantum Optics IV*, ed. by L. Mandel, E. Wolf (Plenum Press, New York 1978) pp.351-361
3.25 R.J. Glauber: "Optical Coherence and Photon Statistics", in *Quantum Optics and Electronics*, ed. by C. DeWitt, A. Blandin, C. Cohen-Tannoudji (Gordon and Breach, New York 1965) pp.65-185

3.26 N. Tornau, A. Bach: Opt. Commun. *11*, 46 (1974)
3.27 D. Stoler: Phys. Rev. Lett. *33*, 1397 (1974)
3.28 H.D. Simaan, R. Loudon: J. Phys. A *8*, 539 (1975)
3.29 H.J. Carmichael, D.F. Walls: J. Phys. B *9*, L43 (1976)
3.30 H.J. Kimble, M. Dagenais, L. Mandel: Phys. Rev. Lett. *39*, 691 (1977)
3.31 L. Mandel: Proc. Phys. Soc. *74*, 233 (1959)
3.32 E. Wolf, C.L. Mehta: Phys. Rev. Lett. *13*, 705 (1964)
3.33 S.O. Rice: Bell Syst. Tech. J. *23*, 282 (1944)
3.34 L. Mandel, E. Wolf: Rev. Mod. Phys. *37*, 231 (1965)
3.35 R. Loudon: *The Quantum Theory of Light* (Clarendon Press, Oxford 1973) Chap.10
3.36 A. Schell, R. Barakat: J. Phys. A *6*, 826 (1973)
3.37 A.J.F. Siegert: MIT Rad. Lab. Rpt. No.465 (1943)
3.38 D. Middleton: Quart. Appl. Math. *5*, 455 (1948)
3.39 I.S. Reed: IRE Trans. Inf. Theory *IT-8*, 194 (1962)
3.40a E. Jakeman, C.J. Oliver, E.R. Pike: J. Phys. A *1*, 406 (1968)
3.40b E.E. Serrallach, M. Zulauf: J. Appl. Math. Phys. *26*, 669 (1975)
3.41 G. Bedard, J.C. Chang, L. Mandel: Phys. Rev. *160*, 1496 (1967)
3.42 G. Lachs: J. Appl. Phys. *42*, 602 (1971)
3.43 S.K. Srinivasan, S. Sukavanam: J. Phys. A *5*, 682 (1972)
3.44 B.E.A. Saleh: J. Appl. Phys. *46*, 943 (1975)
3.45 S.K. Srinivasan, S. Sukavanam: Phys. Lett. *60A*, 287 (1977)
3.46 E. Jakeman: J. Phys. A *3*, 201 (1970)
3.47 V. Degiorgio, J.B. Lastovka: Phys. Rev. A *4*, 2033 (1971)
3.48 R.G.W. Brown, R.D. Callan, W. Jenkins, I. Miller, C.J. Oliver, E.R. Pike, D.J. Watson, M. Wood: In "Photon Correlation Techniques in Fluid Mechanics", Proc. 2nd Intern. Conf., Stockholm, Sweden, June 14-16, 1978, Phys. Scr. *19*, 365 (1979)
3.49 J.H. Van Vleck: Harvard Univ. Rad. Res. Lab. Rpt. No.51 (1943)
3.50 J.H. Van Vleck, D. Middleton: Proc. IEEE *54*, 2 (1966)
3.51 E. Jakeman, E.R. Pike: J. Phys. A *2*, 411 (1969)
3.52 E. Jakeman, E.R. Pike, S. Swain: J. Phys. A *3*, L55 (1970)
3.53 E. Jakeman, E.R. Pike, S. Swain: J. Phys. A *4*, 517 (1971)
3.54 A.J. Hughes, E. Jakeman, C.J. Oliver, E.R. Pike: J. Phys. A *6*, 1327 (1973)
3.55 P.N. Pusey, W.I. Goldburg: Phys. Rev. A *3*, 766 (1971)
3.56 D.W. Schaefer, B.J. Berne: Phys. Rev. Lett. *28*, 475 (1972)
3.57 E. Jakeman, C.J. Oliver, E.R. Pike, P.N. Pusey: J. Phys. A *5*, L93 (1972)
3.58 E. Jakeman, C.J. Oliver, E.R. Pike: J. Phys. A *3*, L45 (1970)
3.59 M. Born, E. Wolf: *Principles of Optics* (Pergamon, London, New York 1959) p.397
3.60 E. Jakeman: In Ref.[3.1], pp.75-149
3.61 D.E. Koppel: J. Appl. Phys. *42*, 3216 (1971)
3.62 S.H. Chen, P. Tartaglia: Opt. Commun. *6*, 119 (1972)
3.63 G. Lachs: Phys. Rev. *138*, B1012 (1965)
3.64 J. Peřina: Phys. Lett. *24A*, 333 (1967)
3.65 E. Jakeman, J.G. McWhirter, P.N. Pusey: J. Opt. Soc. Am. *66*, 1175 (1976)
3.66 A.S. Clarke: RARDE Tech. Rpt. 7/77 (1977)
3.67 F.A. Johnson, R. Jones, T.P. McLean, E.R. Pike: Phys. Rev. Lett. *16*, 589 (1966)
3.68 G. Bedard: Proc. Phys. Soc. *90*, 131 (1967)
3.69 I. Delotto, P.F. Manfredi, P. Principio: Energia Nucl. (Milan) *11*, 557 (1964)
3.70 R.F. Chang, V. Korenman, C.O. Alley, R.W. Detenbeck: Phys. Rev. *178*, 612 (1969)
3.71 E. Jakeman, E.R. Pike: J. Phys. A *1*, 690 (1968)
3.72 Y.N. Barabanenkov, Y.A. Kravstov, S.M. Rytov, V.I. Tatarskii: Usp. Fiz. Nauk. *102*, 1 (1970)
3.73 B.J. Uscinski: *The Elements of Wave Propagation in Random Media* (McGraw-Hill, New York 1977)
3.74 B.J. Uscinski: J. Atmos. Terrest. Phys. *40*, 1257 (1978)
3.75 R. Dashen: Stanford Res. Inst. Tech. Rpt. JSR-76-1 (1977)
3.76 E. Jakeman, P.N. Pusey: J. Phys. A *6*, L88 (1973)
3.77 E. Jakeman, P.N. Pusey: IEEE Trans. AP-*24*, 806 (1976)
3.78 P.N. Pusey: "Statistical Properties of Scattered Radiation", in *Photon Correlation Spectroscopy and Velocimetry*, ed. by H.Z. Cummins, E.R. Pike (Plenum Press, New York 1977) pp.45-141
3.79 J.C. Kluyver: Proc. Roy. Acad. Sci. Amsterdam *8*, 341 (1905)

3.80 K. Pearson: "A Mathematical Theory of Random Migration", in *Draper's Company Research Memoirs Biometric Series III*, No.15 (1906)
3.81 Lord Rayleigh: Philos. Mag. *6*, 321 (1919)
3.82 D.W. Schaefer, P.N. Pusey: Phys. Rev. Lett. *29*, 843 (1972)
3.83 P.N. Pusey, D.W. Schaefer, D.E. Koppel: J. Phys. A *7*, 530 (1974)
3.84 S.H. Chen, P. Tartaglia, P.N. Pusey: J. Phys. A*6*, 490 (1973)
3.85 E. Jakeman, P.N. Pusey: *Radar 77* (IEE, London 1977) p.105
3.86 B.J. Hoenders, E. Jakeman, H.P. Baltes, B. Steinle: Opt. Acta *26*, 1307-1319 (1979)
3.87 W. Feller: *An Introduction to Probability Theory and its Applications*, Vol.1 (Wiley, New York 1968) p.285
3.88 M.S. Bartlett: *An Introduction to Stochastic Processes* (Cambridge University Press, Cambridge 1966) Chap.3
3.89 E. Jakeman, P.N. Pusey: Phys. Lett. *44A*, 456 (1973)
3.90 E. Jakeman, P.N. Pusey: J. Phys. A *8*, 369 (1975)
3.91 P.N. Pusey, E. Jakeman: J. Phys. A *8*, 392 (1975)
3.92 C. Deutsch, P.N. Keating: J. Appl. Phys. *40*, 4049 (1969)
3.93 G. Parry, P.N. Pusey: J. Opt. Soc. Am. *69*, 796 (1979)
3.94 V.I. Tatarskii: *Wave Propagation in a Turbulent Medium* (Dover, New York 1967)
3.95 E. Jakeman, C.J. Oliver, E.R. Pike: J. Phys. A *1*, 497 (1968)

4. Microscopic Models of Photodetection

A. Selloni

With 3 Figures

In this chapter we study the problem of determining statistical properties of stochastic radiation fields by means of photoelectric counting detectors. This problem is an important step in the retrieval of information on stochastic scattering systems considered in the preceding chapter. We start with the definition of the detection problem and a general comparison of "ideal" and "real" photodetection (Sect.4.1). In Sect.4.2, we present an overview of models describing the ideal detection process and discuss formal procedures of inversion for counting probabilities. In Sect.4.3 we report the recent progress in understanding the quantum process of photodetection in terms of microscopic models, namely the open-system detection scheme. In Sect.4.4 we review selected results on disturbing effects such as noise, dead time and partial coherence. Using the open-system approach we develop a first-principle theory of the intrinsic thermal noise in photodetection and show how the statistical properties of the incoming radiation are affected by the detector temperature (Sect.4.5). Pertinent techniques from statistical physics and quantum optics are summarized in Sects.4.6 and 4.7.

4.1 Photoelectron and Photon Statistics

In principle, the aim of photoelectric counting experiments is to determine the statistical state of the incoming radiation field and the pertinent derived quantities such as the moments of the photon statistics or time and space correlations and their spectra. In practice, photodetection experiments yield nothing but some statistical properties of the resulting fluctuating current of photoelectrons, which require careful analysis. Usually there is no simple one-to-one correspondence between photoelectrons and incoming photons: the number of photoelectrons is not identical with the number of photons and the observed statistics of the photoelectrons is not simply that of the radiation field. The problem outlined in this section is how the statistical properties of the radiation field are transferred to the detector signal in the presence of noise and other disturbing effects, and how they can be recovered from the resulting photoelectric signal.

4.1.1 Definition of the Problem

There is clearly no need for a general review-type chapter on photodetection. Several books devoted wholly or in part to photoelectric counting are available. We mention the books by KLAUDER and SUDARSHAN [4.1], PEŘINA [4.2], LOUDON [4.3], CROSSIGNANI et al. [4.4], SALEH [4.5], and BARAKAT and BLAKE [4.6]. There exist also a number of pertinent review articles, for example those by PIKE [4.7], ARECCHI and DEGIORGIO [4.8], BERTOLOTTI [4.9], and JAKEMAN [4.10]. We also make reference to a review by MEHTA [4.11] for the semiclassical viewpoint and to one by HELSTROM [4.12] for the link with decision theory. For the technical aspects of the various types of detectors we refer to the review by MELCHIOR [4.13] and to the recent books by KEYES [4.14] and KINGSTON [4.15]. It is not the goal of this chapter to provide an exhaustive bibliography of photodetection. Referencing several hundred pertinent papers, the above books and reviews are an excellent guide to the original articles.

In view of the existing review literature, we feel entitled to restrict our considerations as follows. We shall focus our interest on the photon statistics (moments, quasi-probability distribution) of the radiation field rather than studying the time-space correlations or spectral properties. Hence we shall, in general, assume a single-mode field characterized by the quasi-probability $P(\alpha)$ with α denoting the stochastic complex amplitude and a pointlike detector model. Our objective is the relation between the statistics of the resulting photoelectric detector signal as described by, e.g., the probability $p(n,T)$ of observing n events in a time interval of length T and the statistical state $P(\alpha)$ of the incoming field. We shall also pursue the likewise important problem of measuring the power of the incoming field [4.14,15]. The main goal of this chapter is to derive the relation between photon statistics $P(\alpha)$ and photoelectron statistics $p(n,T)$ from *microscopic models of the detection process* including an adequate description of the interaction of radiation with photosensitive matter (Sect.4.3). We thus are mainly interested in the *quantum theory of photodetection* as initiated by GLAUBER [4.16] and KELLEY and KLEINER [4.17] and continued by MOLLOW [4.18], GLAUBER [4.19], and SCULLY and LAMB [4.20] and, more recently, by SELLONI et al. [4.21-23]. We are less interested in semiclassical theory as reviewed by, e.g., MEHTA [4.11] and SENITZKY [4.24] and mathematical statistical problems as covered by SALEH [4.5]. As for disturbing effects, we shall emphasize the statistical aspects (e.g., thermal noise, Sect.4.5), though we shall also discuss a few selected space-time effects in Sect.4.4.

4.1.2 Ideal and Real Detection

A source sends radiation through an optical system which may include stochastic scatterers or media affecting the photon statistics as studied in Chap.3. The resulting radiation field interacts with the photosensitive material of the detector and produces photoelectrons, which are amplified and processed electronically to give an output signal. Ideally, the electronic output signal should directly exhibit the statistical features of the radiation field impinging on the detector, but in practice many causes of inaccuracies are involved in the transfer of the field statistics to the resulting output signal [4.5-10,13,25-33].

As the first group of phenomena which disturb the "ideal" detection process we mention background radiation, dark current, and the various types of noise, such as shot noise, gain fluctuation, generation-recombination noise (e.g., of thermal origin), Johnson noise, and 1/f noise. Under the heading of "time and space effects" we comprise a second group of phenomena involved in photodetection, namely dead time, rise time, and time resolution of the detector (or the electronics), the time correlation of the detector response, the counting time window T as compared with the coherence time τ_c (inverse of the bandwidth $\Delta\omega$) of the incident radiation, and the effective detector area A as compared with the coherence area A_c of the radiation field. Related time and space effects are the frequency dependence and threshold of the detector response and the dependence of sensitivity of the detecting surface on the angle of incidence of the radiation and on the position on the surface. Moreover, we have to be aware of the nonideal quantum efficiency of the detection process and of the ultimate lack of statistical accuracy due to the finite number of samples. Finally, instability of the source (as far as they are not of physical interest and the aim of the statistical study), gain fluctuations in the electronic amplifier, threshold fluctuations in the discriminator, detector afterpulses, possible nonlinearities in the detector response [4.34,35] also affect the accuracy of the detection experiment.

Model theories describing the transfer of statistics in an idealized detection process are reported in Sects.4.2 and 4.3. In view of the above list of disturbing effects, a definition of "ideal detection" can be attempted in terms of the following requirements:

- zero background, dark current, and noise
- zero dead time, rise time, and response correlation time
- "mono-mode detection": counting time window small compared with coherence time, viz., $T \ll \tau_c$, frequency-independent response
- "pointlike detector": detector area small compared with coherence area, viz., $A \ll A_c$, angle- and position-independent response
- infinite number of samples
- stable source of radiation
- zero intensity threshold, linear intensity response
- no losses (black detector surface)
- 100% quantum efficiency.

We observe that nonideal quantum efficiency can be accepted, if one is not interested in the absolute intensity, by normalizing the moments of the statistics with respect to the first moment [4.8].

In principle, there are two ways of eliminating disturbing effects in photodetection, namely by refining the experimental procedure and technique or by evaluating the experimental data in terms of correction formulae resulting from theories that account for these effects. Both procedures require theoretical knowledge of the detection process.

4.2 Models for Ideal Detection - a Review

Theories of photodetection have been developed on various levels of sophistication: macroscopic theory describing the mathematical statistical aspect only and using a phenomenological efficiency without studying the interaction of the radiation with the photosensitive material, perturbation theory of the interaction without accounting for the field attenuation, and closed-system theory describing the field attenuation properly, but not the relaxation of the detector levels. These theories are reviewed in the present section, whereas the open-system theory (with relaxation) is presented in Sect.4.3.

4.2.1 Mandel's Formula

Photocounting is treated as a classical problem in probability theory on two basic assumptions [4.11,36]:

I) the probability $W_1(t)\Delta t$ of registering a photoelectric count in the small time interval Δt is proportional to Δt and to the instantaneous intensity $I(t)$ of the light falling on the detector

$$W_1(t)\Delta t = \xi I(t)\Delta t \tag{4.1}$$

with ξ denoting the quantum efficiency;

II) the different counting events are statistically independent.

Consider a finite time interval $(t_0,\ t_0+T)$ and let $p(n;\ t_0,\ t_0+T)$ be the probability of registering n photocounts in the interval. We divide $(t_0,\ t_0+T)$ in N small subintervals of length $\Delta T = T/N$ so that $t_k = t_0 + k\Delta T$ $(k = 1,\dots,N)$ denotes the end of the kth subinterval. For any given realization of the stochastic process $I(t)$, we obtain $p(n;\ t_0,\ t_0+T)$ by letting $\Delta T \to 0$ in the product of the probabilities of n "successes" in n arbitrary subintervals $t_{k_1}, \dots, t_{k_n}$ and (N-n) "failures" in the remaining (N-n) subintervals, viz.,

$$\begin{aligned} p(n;\ t_0,\ t_0+T) &= \lim_{\Delta T\to 0} \sum_{k_1=1}^{N} \dots \sum_{k_n=1}^{N} \xi^n I(t_{k_1}) \dots I(t_{k_n})(\Delta T)^n \\ &\quad \times \prod_{k=1}^{N} [1 - \xi I(t_k)\Delta T] / \prod_{j=1}^{n} [1 - \xi I(t_{k_j})\Delta T] \\ &= \frac{[\Omega(t_0,t_0+T)]^n}{n!} \exp[-\Omega(t_0,t_0+T)] \end{aligned} \tag{4.2}$$

where

$$\Omega(t_0, t_0 + T) = \xi \int_{t_0}^{t_0+T} I(t')dt' \qquad (4.3)$$

is proportional to the light intensity falling on the detector during the counting interval. The measured photocounting distribution is obtained by taking the ensemble average of the right-hand side of (4.2) over the probability density $W(\Omega)$ of $\Omega(t_0,t_0 + T)$, viz.,

$$p(n; t_0,t_0 + T) = \int_0^\infty d\Omega W(\Omega) \frac{\Omega^n}{n!} e^{-\Omega} \quad . \qquad (4.4)$$

Equation (4.4) is independent of t_0 in the case of stationary light beams. Experimentally the counting distribution of stationary light is obtained not from an ensemble of identical systems, but from periodic sampling of a single system (see Sect.4.4.3). This means that the measured counting distribution is actually the average of (4.2) over the initial time t_0. This time average is assumed equivalent to the ensemble average by the usual ergodic hypothesis for stationary light sources [4.3,16,36]. In the case of transient light, however, the system must be prepared in the same initial state at each measurement (see, e.g. [4.37,38]).

Equation (4.4) describes photocounting as a compound Poisson process with weight $W(\Omega)$. This result indicates that two sources of randomness are present: the randomness inherent in the incident light and that of the photoelectric events. For example the variance of the number n of photoelectrons is given by

$$\langle(\Delta n)^2\rangle = \langle(\Delta\Omega)^2\rangle + \langle n\rangle \quad , \qquad (4.5)$$

which shows explicitely the contribution of both the light fluctuations and the fluctuations in the photoelectric emissions (quantum or shot noise). Recently the possibility of a quantum/shot noise-free measurement was discussed by AOKI [4.62].

The above purely statistical results have to be implemented by physical models. For instance (4.1) can be justified by first-order perturbation theory in the framework of a semiclassical approach [4.39].

4.2.2 Perturbation Approach

Consider a detector made up of a large number N of identical atoms in the ground state. The detector is shielded from the stationary radiation field under investigation by a shutter, which controls the length of time for which light falls on the detector. At time $t = 0$ the shutter is opened for a length of time T. We ask for the probability $p(n,T)$ that n atoms undergo a photoelectric transition during the time interval $(0,T)$.

The calculation can be divided in two parts. Probabilistic arguments are first used to show that the generating function

$$Q(s) = \sum_{n=0}^{N} (1-s)^n p(n,T) \tag{4.6}$$

can be written in the form [4.16]

$$Q(s) = \sum_{m=0}^{N} (-s)^m \sum_{\substack{m\text{-fold}\\ \text{combinations}}} p^{(m)}_{j_1 \dots j_m} , \tag{4.7}$$

where $p^{(m)}_{j_1 \dots j_m}$ is the probability that each atom of the set $(j_1 \dots j_m)$ absorbs a photon, regardless of what all the other atoms do. The proof of (4.7) involves treating the total number of counts in (0,T) as a random variable N_A, which is the sum of N independent random variables, one for each atom of the detector

$$N_A = \sum_{j=1}^{N} zj , \tag{4.8}$$

where z_j can take only the two values 1 (the jth atom absorbs a photon) and 0 (the jth atom does not absorb a photon).

Second, lowest-order perturbation theory with the dipolar atom-field interaction yields $p^{(m)}$. In the limit $N \to \infty$ one finds

$$Q(s) = \sum_{m=0}^{\infty} \frac{(-s)^m}{m!} \xi^m \int_0^T dt_1 \dots \int_0^T dt_m \int_V dr_1^3 \dots \int_V dr_m^3$$
$$G^{(m)}(r_1 t_1, \dots, r_m t_m, r_m t_m, \dots, r_1 t_1) \quad , \tag{4.9}$$

where V denotes the volume and ξ the quantum efficiency of the detector, and $G^{(m)}$ stands for the mth order correlation function of the (linearly polarized) electric field (see Sect.4.6). The efficiency depends on the matrix elements of the photoelectric process, on the density of atoms in the detector and on the probability that an electron ejected by photoabsorption is really counted [4.16]. This last factor approximately accounts for effects such as the finite escape length of the electrons. It is assumed here that the range of frequencies to which the detector responds is much broader than the field bandwidth.

Equation (4.9) can be expressed in the form

$$Q(s) = \langle \hat{N} e^{-s\Omega_{op}} \rangle \, , \tag{4.10}$$

where $\hat{N}$ is the normal ordering operator, $\langle \dots \rangle$ denotes the average over the state ρ of the field, and

$$\Omega_{op} = \xi \int_V d^3 r \int_0^T dt E^{(-)}(rt) E^{(+)}(rt) \tag{4.11}$$

with $E^{(-)}$, $E^{(+)}$ denoting the electric field operators. In addition, time ordering must be taken into account when the interaction of the radiation field with its sources is considered [4.17,40,41]. The pertinent ordering operator becomes

$$\langle T_N \Omega^n_{op} \rangle \sim \langle E^{(-)}(t_1) \dots E^{(-)}(t_n) E^{(+)}(t_n) \dots E^{(+)}(t_1) \rangle$$
$$\text{with } t_1 \leqq t_2 \leqq \dots \leqq t_n \, . \tag{4.12}$$

Using (4.6,10) we calculate the counting probabilities

$$p(n,T) = \langle \hat{N} \frac{\Omega^n_{op}}{n!} \exp(-\Omega_{op}) \rangle \quad . \tag{4.13}$$

Equation (4.13) is usually evaluated in terms of the P representation of the field density matrix leading to

$$p(n,T) = \int P(\{\alpha_k\}) \frac{\Omega^n(\{\alpha_k\})}{n!} e^{-\Omega(\{\alpha_k\})} \prod_k d^2\alpha_k , \qquad (4.14)$$

where $|\alpha_k\rangle$ is an eigenstate of the kth cavity mode annihilation operator in the expansion of $E^{(+)}(r,t)$, and $\Omega(\{\alpha_k\})$ is the expectation value of the operator (4.11) in a multimode coherent state.

An equivalent form of p(n, T) is

$$p(n, T) = \int W(\Omega) \frac{\Omega^n}{n!} e^{-\Omega} d\Omega , \qquad (4.15)$$

where

$$W(\Omega') = \int P(\{\alpha_k\})\delta[\Omega' - \Omega(\{\alpha_k\})] \prod_k d^2\alpha_k \quad . \qquad (4.16)$$

Equation (4.15) is formally identical to the classical result (4.4). Normal ordering is however, essential to derive (4.15), i.e., the compound Poisson form of the counting distribution is now strictly related to the quantized nature of the radiation field. It is well known that, unlike the classical case, $P(\{\alpha_k\})$ is not necessarily positive definite. In addition, the similarity between the counting distribution and the photon distribution has been pointed out by a number of authors [4.3,7,20]. Contrasting opinions about the necessity of a full quantum mechanical description of photocounting are summarized in an exchange between MANDEL and WOLF [4.42], on the semiclassical, and JAKEMAN and PIKE [4.43], on the quantum side. The semiclassical approach has been recently rediscussed by SENITZKY [4.24].

The perturbation-theoretical approach has been applied to discuss a variety of effects such as electron correlations in the detector [4.44-46], time resolution on the scale of optical oscillation [4.47], two-photon-absorption detection [4.48,49], and frequency dependence of the detector response [4.50].

A slightly different derivation of (4.15) was given by KELLEY and KLEINER [4.17] (see also [4.41]). Instead of identifying the probability of n photocounts in (0,T) with the probability that n excited atoms are present at time T, they consider each photoionization event as an observation on the system. This picture applies to photon detection at low light levels, when single electron pulses are observed (see Sect.4.4.1). After each measurement the detector is assumed to return immediately to the initial state. This assumption prevents possible saturation effects, whereas lowest-order perturbation theory for high count number is inappropriate in the framework of GLAUBER's theory [4.16]: if a large fraction of excited atoms is present, emission back into the field would occur. To avoid this kind of effects a mechanism must be supplied by the measuring apparatus which withdraws the excited electrons from the system [4.7].

The theory of photon detection discussed in this section is essentially a generalization to many atoms of a perturbation theory calculation of transition probabilities for a single atom. Meaningless results may appear when the number of detecting atoms becomes very large [4.18] or when the time becomes very large [4.51]. For example, when a coherent light beam with mean photon number $\bar{n}$ passes successively through a large number M of identical (transparent) detectors with quantum efficiency ξ, perturbation theory predicts that the average number of photoelectrons is $\xi\bar{n}M$. For M sufficiently large this number becomes much greater than $\bar{n}$, which is an unacceptable result as long as we assume that no gain mechanisms are present inside the detectors [4.18]. Another meaningless result is that the probability for a single atom to be ionized by a monomode light grows linearly with time instead of being bounded by unity [4.51]. These and other similar examples suggest that field attenuation should be taken into account. This implies the inclusion of multiple transitions which are neglected by conventional perturbation theory.

4.2.3 Field Attenuation

Two equivalent theories of field attenuation in photodetection were formulated by MOLLOW [4.18] and SCULLY and LAMB [4.20]. Following [4.20], the mechanism of the field attenuation can be clarified in terms of the following detection scheme.

I) Having established the steady-state photon distribution, remove all the atomic sources from the cavity. The radiation density matrix is thus time-independent.

II) Insert into the cavity a group of photosensitive atoms as detector.

III) After a time T remove the detector and count the number of excited atoms.

IV) Repeat the procedure a large number of times, every time starting from the same initial conditions, in order to determine the probability p(m,T) of finding m excited atoms at time T.

We take the detector to consist of N noninteracting but distinguishable atoms each of which has a ground state and a continuum of excited states. The atoms are placed in the cavity at t = 0 in their ground state, while the monomode radiation field is initially described by the diagonal density matrix $\rho_{n,n}(0)$. The atoms are coupled to the field via dipolar interaction.

In order to calculate the probability p(m,T) we first establish the equation of motion describing the field-detector evolution. To this end, we use the expansion

$$p(m, T) = \sum_{n} P_{n,m}(T) , \tag{4.17}$$

where $P_{n,m}$ is the probability of finding the field in the state $|n\rangle$ with m photoelectrons ejected. The equation of motion for $P_{n,m}(T)$ is obtained by reduced density matrix techniques (see Sect.4.7), where the role of the reservoirs is played by the detector atoms. For $N \gg m$, $P_{n,m}(T)$ is found to obey the following differential equation

$$\dot{P}_{n,m} = -2\kappa_0 n P_{n,m} + 2\kappa_0(n+1)P_{n+1,m-1} \quad , \tag{4.18}$$

where κ_0 is the reservoir-induced attenuation which depends on the dipolar coupling constant and on the density of final atomic states. The form of (4.18) is typical of a birth-and-death process [4.52]: the field-detector system evolution is due to transitions by which one photon is absorbed and simultaneously one photoelectron is produced, with probability proportional to the number of photons present before the absorption. Subject to the condition that initially $n_0 = n + m$ photons are present with all the atoms in the ground state, the solution to (4.18) is

$$P_m(n_0,T) = \binom{n_0}{m}\eta_0^m(1-\eta_0)^{n_0-m}\rho_{n_0,n_0}(0) \quad , \tag{4.19}$$

where $P_m(n_0,T) \equiv P_{n_0-m,m}(T)$ and

$$\eta_0 = 1 - \exp(-2\kappa_0 T) \quad . \tag{4.20}$$

Here $P_m(n_0, T)$ is a Bernoulli's distribution for m successful events (counts) and $n_0 - m$ failures, each event having a probability η_0. The counting probability as given by (4.17) is obtained summing over all possible states of the field

$$p(m,T) = \sum_{n_0 \geq m}\binom{n_0}{m}\eta_0^m(1-\eta_0)^{n_0-m}\rho_{n_0,n_0}(0) \quad . \tag{4.21}$$

In order to show the relation of the above expression to the perturbation theory result (4.14), we write $\rho(0)$ in the $P(\alpha)$ representation

$$\rho_{nn} = \int d^2\alpha P(\alpha)\frac{|\alpha|^{2n}}{n!}\exp(-|\alpha|^2) \quad , \tag{4.22}$$

so that (4.21) becomes [4.18,20,51]

$$p(m,T) = \int d^2\alpha P(\alpha)\frac{\eta_0^m|\alpha|^{2m}}{m!}\exp(-\eta_0|\alpha|^2) \quad . \tag{4.23}$$

Contrary to (4.14) p(m, T) as given by (4.23) is valid at all times. In the limit $T \ll \kappa_0^{-1}$, $\eta_0 \sim 2\kappa_0 T$, (4.23) reduces to the perturbation theory result for a monomode field and a detector efficiency $\xi = 2\kappa_0$. In the limit $T \gg \kappa_0^{-1}$, p(m, T) becomes the photon distribution (4.22) at time T = 0, which means that all the photons initially present are absorbed. In particular if the field is initially in an n-photon state, p(m, T) vanishes when $m > n$, while it is equal to the probability that (n-m) photons are present at time T when $m \leq n$.

As pointed out by SCULLY and LAMB [4.20], the detection scheme I) - IV) where the combination of field and detector is left undisturbed until time T does not correspond to the usual experimental situation. Here, the measuring apparatus continuously withdraws the excited electrons from the detector, counts them, replenishes new electrons in the ground state and thus provides also the mechanism for the reservoirlike behavior of the detector atoms. An improved description of the detection process should therefore include the measuring system as an essential component of the detector model, together with the field and the photosensitive atoms. This feature characterizes the open-system model discussed in Sect.4.3.

4.2.4 Inversion Problem

In the previous sections we considered the "direct problem" of determining the photocounting statistics from a given initial field state. Let us now turn to the corresponding "inverse problem" of calculating the initial state of the field from the counting distribution (4.15) or (4.23).

The following formal procedure of inversion was proposed by WOLF and MEHTA [4.53]. Referring for example to (4.23) we define the function

$$F(x) = \sum_{n=0}^{\infty} (ix/\eta_0)^n p(n, T) \quad . \tag{4.24}$$

The initial field distribution $P(\alpha)$ is then recovered as

$$\int_0^{2\pi} d\varphi \; P(\alpha) = \pi^{-1} \exp(\eta_0|\alpha|^2) \int dx F(x) \exp(-i|\alpha|^2 x) \; , \tag{4.25}$$

where φ denotes the phase angle of α. We notice that we can retrieve only the phase average of $P(\alpha)$ since only field intensity correlations are involved in (4.23). When (4.15) holds, a procedure similar to (4.24,25) yields $W(\Omega)$, which is related to the statistical state of the field $P(\{\alpha_k\})$ by (4.16). When the counting time T can be adjusted so that it is much shorter than the field coherence time τ_c, $W(\Omega)$ reduces to the probability distribution of the light intensity.

The method (4.24,25) is only useful when a closed analytic expression for p(n, T) is known [4.11]. In practice p(n, T) is experimentally determined only for a finite number of small values of n. However the expansion (4.24) cannot be truncated even if p(n, T) is small for larger values of n, since the inverse Fourier transform in (4.25) would then always yield a singular function. In such cases approximation methods should be used. For example, an approximate formula for $W(\Omega)$ may be obtained by using the expansion [4.54]

$$W(\Omega) = \sum_{m=0}^{\infty} \left[\sum_{k=0}^{m} (-1)^k \binom{m}{k} p(k,T) \right] L_m(\Omega) \; , \tag{4.26}$$

where L_m is the Laguerre polynomial of order m. The problem then is how fast the summations in this formula converge and which is the error due to the inaccuracies in the experimental values of p(n, T).

Other approximation methods are based on the equality between the factorial moments $\langle n^{(k)}\rangle$ of the counting distribution and the corresponding moments of $W(\Omega)$, i.e.,

$$\langle n^{(k)}\rangle = \int d\Omega\, \Omega^k W(\Omega) \quad . \tag{4.27}$$

From the knowledge of the first few moments, one can determine some general features and approximate expression for $W(\Omega)$ [4.54]. When $T \ll \tau_c$ the factorial moment $\langle n^{(k)}\rangle$ reduces to the kth moment of the intensity distribution. The problem of obtaining the phase information is discussed by ARECCHI and DEGIORGIO [4.8]. We also refer to the recent review of the phase retrieval problem due to FERWERDA [4.55].

The onefold photoelectron statistics does not give information on the time evolution of the stochastic field. This information is provided by multifold photoelectron statistics. Consider for example a monomode field as in (4.23). The joint probability $p(m_1t_1;\, m_2t_2)$ of registering m_1 counts in the interval $(t_1,\, t_1+T)$ and m_2 counts in the interval $(t_2,\, t_2+T)$ is related to the joint field probability $W_2(\alpha_1t_1,\, \alpha_2t_2)$ (see Sect.4.6.3) by [4.56]

$$p(m_1t_1,\, m_2t_2) = \int d^2\alpha_1 d^2\alpha_2 W_2(\alpha_1t_1,\, \alpha_2t_2)$$

$$\frac{\eta_0^{m_1}|\alpha_1|^{2m_1}}{m_1!}\, \frac{\eta_0^{m_2}|\alpha_2|^{2m_2}}{m_2!}\, \exp(-\eta_0|\alpha_1|^2)\, \exp(-\eta_0|\alpha_2|^2) \quad . \tag{4.28}$$

This equation can be formally inverted by a procedure similar to (4.24,25). While W_2 suffices to characterize the statistical behavior of a Markov field, higher-order joint photocount distributions are needed for non-Markovian fields. The direct problem of multifold statistics is treated in the books by SALEH [4.5] and BARAKAT and BLAKE [4.6] and references therein.

The temporal behavior of the field can be also characterized by its correlation functions. In practical cases, relevant information **is** obtained by measuring only the lowest order correlation functions, i.e., the field and the intensity correlation functions [4.9,10,26,27]. The first- and second-order correlation functions are for instance sufficient to determine if the field is fully coherent or not, provided one knows that $P(\alpha)$ is a positive definite distribution [4.57]. Similarly, if we expect that the field has a particular statistics (e.g., a Gaussian one) it can be sufficient to verify that the second-order correlation function satisfies the appropriate relations. In the case of Gaussian fields, moreover, the second-order correlation function completely characterizes the statistical behavior of the field.

4.3 Open-System Detection Scheme

Open-system detector schemes include mechanisms for withdrawing the excited electrons from the detecting system and repopulating the lower states. In these detection schemes the various ad hoc assumptions made in the closed-system theories presented in Sect.4.2 can be dispensed with. The idea of open-system detection theory was proposed by PIKE [4.7]; in his own words: "To describe an absorption detector, which is the more usual kind, we withdraw the electron from contact with the heatbath, populate the lower state only, and then watch the development of its reduced density matrix ρ as a function of time, under the action of the field. To be quite accurate, we should not allow the upper state to become appreciably populated, since emission would then take place back into the field and it would be partly an emission detector. We can therefore imagine the upper state to be greater than some ionisation energy, and that a mechanism is supplied which will withdraw the excited electron from the system. This process will constitute the 'counting' of a photon". This program was carried through only recently [4.21,22]. The results for a zero-temperature heatbath are reported here.

Our model calculation shows the following features. We consider a pointlike model detector that consists of many identical two-level atoms. We describe the atoms and the field by the DICKE Hamiltonian [4.101]. We introduce an additional coupling of the atoms to a reservoir with rapidly decaying internal correlation. We derive the dynamics of the detector atoms and the field by using the reduced-density-matrix formalism for the successive elimination of reservoir and atomic, or field, variables. We assume that the atom-field coupling is strong compared with the atom-bath coupling and consider times that are large compared with the atomic relaxation time. We establish the relations between the time evolution of the atomic and the field moments of any order. We finally obtain the photocounting rate in terms of the initial field state. Our results are formally analogous to those of [4.18,20,51], with the quantum efficiency, however, depending not only on the atom-field, but also on the atom-bath, coupling parameter.

Our detector model is presented in Sect.4.3.1. In Sect.4.3.2 we establish the connection between atomic and field moments. The corresponding photocounting probability is discussed in Sect.4.3.3.

4.3.1 Detector Model

We take the photodetector to consist of N-independent equivalent, but distinguishable, two-level atoms with level spacing ε. The atoms interact with a single-mode field of frequency ω in a cavity of unit volume. Atoms, field, and interaction are described by the DICKE Hamiltonian [4.101]

$$H_D = H_F + H_A + \lambda H_{AF} = \omega a^+ a + \varepsilon S_3 + \lambda(S^+ a + a^+ S^-) \ , \tag{4.29}$$

where a^+, a denote the field operators and where

$$S_3 = \sum_{j=1}^{N} s_{3j} \quad \text{and} \quad S^{\pm} = \sum_{j=1}^{N} s_j^{\pm}$$

denote the collective, and s_{3j} and $s_j^{\pm}$ the individual, atomic variables. The coupling constant is $\lambda = \omega^{\frac{1}{2}} g_{12}$ with g_{12} denoting the dipole transition matrix element of the atoms ($\hbar \equiv 1$). An interaction of the atoms with a reservoir is required in order to provide a stationary distribution of the atoms in the initial state and a mechanism for withdrawing excited atoms and allowing the definition of a photoelectric current. We therefore introduce a bath described by the Hamiltonian H_R and its interaction with the atoms described by V_{AR}. The full detector model is thus given by

$$H = H_D + H_R + V_{AR} \quad . \tag{4.30}$$

The terms H_R and V_{RA} have not to be specified in detail, since the reservoir variables are eliminated by standard methods (see Sect.4.7.1). The reservoir is characterized by the condition that internal correlations decay in a time τ small compared with any other characteristic time involved in the dynamics, e.g., $\tau << \lambda^{-1}$. We thus obtain the equation of motion of the reduced density matrix $\rho_D(t)$ of the Dicke system whose atoms are affected by the interaction with the reservoirs [4.88,97,98]

$$\dot{\rho}_D(t) = -i[H_D, \rho_D(t)] + \mathcal{D}_A \rho_D(t) \ , \quad \mathcal{D}_A = \sum_{j=1}^{N} \mathcal{D}_j \tag{4.31}$$

with the operators $\mathcal{D}_j$ obeying

$$\begin{aligned} \mathcal{D}_j \rho_D = 1/2\{ & \gamma_1([s_j^-, s_j^+ \rho_D] + [\rho_D s_j^-, s_j^+]) \\ & + \gamma_2([s_j^+, s_j^- \rho_D] + [\rho_D s_j^+, s_j^-])\} \ , \end{aligned} \tag{4.32}$$

where γ_1 and γ_2 are inverse atomic relaxation times determined by the atom-reservoir coupling strengths and the reservoir correlation function.

The detector model described by (4.29,30) offers a number of straightforward extension which could be easily incorporated in our calculations, e.g,, multimode fields, various kinds of atoms, position dependence, loss mechanism for the field. When the detector temperature is zero and the number of excited atoms is a small fraction of the total number N, the atomic system can be described as well by the oscillator approximation [4.18,19,21].

$\hbar = h/2\pi$ (normalized Planck's constant)

4.3.2 Relation Between Atomic and Field Dynamics

Our aim is the retrieval of field statistics from observed detector behavior. Hence we have to find the relation between the statistical state of the field and that of the atomic system representing the detector. This is done in terms of relations between field moments $\langle a^{+\nu}a^{\nu}\rangle$ and atomic moments. The latter moments are defined as expectation values of the powers of the operator N_2 defined as

$$N_2 = S_3 + N/2 = \sum_{j=1}^{N} s_j^+ s_j^- \quad . \tag{4.33}$$

Field Dynamics

From (4.31) we first derive the equation of motion for the reduced density matrix $\rho_F(t)$ of the field, assuming the atomic system to be initially in the ground state $|0\rangle_A = |S = N/2, S_3 = -N/2\rangle$. The reduction is achieved by the projector $P_A = |0\rangle_A\langle 0|\mathrm{Tr}_A$ leading to

$$\rho_F(t) = \mathrm{Tr}_A P_A \rho_D(t) \quad . \tag{4.34}$$

We assume that the atomic relaxation due to the bath is rapid compared with the characteristic time of the atom-field interaction, i.e.,

$$\gamma^{-1} \ll \lambda^{-1} N^{-1/2} \quad , \tag{4.35}$$

where $\gamma \equiv (\gamma_1 + \gamma_2)/2$. Thus we are entitled to use the Born approximation and the Markov limit

$$t \gg \gamma^{-1} \quad . \tag{4.36}$$

In the above limits, $\rho_F(t)$ satisfies the equation

$$\dot{\rho}_F(t) = -i(\omega + \Delta)[a^+a, \rho_F(t)] - \kappa_0[1 + (\varepsilon - \omega)^2/\gamma^2]^{-1}\{[a^+, a\rho_F(t)] + [\rho_F(t)a^+, a]\} \tag{4.37}$$

with the resonance absorption constant

$$\kappa_0 = \lambda^2 N/\gamma \tag{4.38}$$

obeying $\kappa_0 \ll \gamma$, and with the shift $\Delta \equiv \kappa_0\gamma^{-1}(\omega - \varepsilon)[1 + (\omega - \varepsilon)^2/\gamma^2]^{-1}$. Using the P representation for $\rho_F(t)$, (4.37) can be transformed into a first-order partial differential equation for the quasi-probability $P_F(\alpha,t)$. In the case of resonance this equation reads

$$\frac{\partial}{\partial t} P_F(\alpha,t) = \kappa_0\{\frac{\partial}{\partial\alpha}[\alpha P_F(\alpha,t)] + \frac{\partial}{\partial\alpha^*}[\alpha^* P_F(\alpha,t)]\} \quad . \tag{4.39}$$

From (4.39) we derive the equation of motion of the νth field moment in the resonant case, viz.,

$$\frac{d}{dt} \langle a^{+\nu} a^{\nu} \rangle_t = -2\nu\kappa_0 \langle a^{+\nu} a^{\nu} \rangle_t \quad . \tag{4.40}$$

The nonresonant case is described by the same equation, but with κ_0 replaced by $\kappa_0[1 + (\omega - \varepsilon)^2/\gamma^2]^{-1}$. Equation (4.40) explicitly shows the attenuation of the field due to the absorption process. We observe that the absorption constant κ_0 does not only depend on the atom-field coupling constant λ, but also on the atom-bath relaxation constant γ. We recall that no restriction other than the natural time scales $N^{-\frac{1}{2}} \lambda^{-1} \gg \gamma^{-1} \gg \tau$ has been adopted to derive (4.39,40).

Dynamics of the Atomic Moments

The factorial moments $N_2^{(\nu)}$ of N_2 are evaluated with respect to the reduced density operator ρ_A of the atomic system. We obtain ρ_A from the Dicke operator ρ_D obeying (4.31) by taking the partial trace with respect to field variables. Being interested in the deviations from the initial ground state we consider

$$\begin{aligned} \rho_A(t) - \rho_A(0) &\equiv Tr_F\{\rho_D(t) - \rho_D(0)\} \\ &= Tr_F\{(1 - P_A)\rho_D(t)\} \quad . \end{aligned} \tag{4.41}$$

Because of (4.34), (4.41) offers relations between atomic and field expectation values. We calculate

$$\langle N_2 \rangle_t = Tr_A\{S^+S^-[\rho_A(t) - \rho_A(0)]\} \tag{4.42}$$

evaluating (4.41) at resonance $\omega = \varepsilon$ up to the second order in λ for $t \gg \gamma^{-1}$, and find the linear relation

$$\langle N_2 \rangle_t = \zeta_0 \langle a^+ a \rangle_t \tag{4.43}$$

with the proportionality constant

$$\zeta_0 \equiv \lambda^2 N/\gamma^2 = \kappa_0/\gamma \tag{4.44}$$

obeying $\zeta_0 \ll 1$. This property of ζ_0 justifies the second-order approximation in the evaluation of $\langle N_2 \rangle_t$. Out of resonance, ζ_0 has to be multiplied by $[1 + (\omega - \varepsilon)^2/\gamma^2]^{-1}$. Thus the photocurrent is found to be proportional to the field intensity at any time $t \gg \gamma^{-1}$. The calculation of higher-order factorial moments is reported in [4.22]. The final result reads

$$\begin{aligned} \langle N_2^{(\nu)} \rangle_t &\equiv \langle N_2(N_2 - \mathbb{1}) \ldots (N_2 - \nu + \mathbb{1}) \rangle_t \\ &= \zeta_0^{\nu} \langle a^{+\nu} a^{\nu} \rangle_t \quad . \end{aligned} \tag{4.45}$$

This relation between atom and field statistics is basic to the calculation of the photocounting probability.

4.3.3 Photocounting Probability

In this section we discuss the experimentally accessible counting rate. Combining (4.40) and (4.45) for the case $\nu = 1$ leads to the rate equation

$$\frac{d}{dt} \langle a^+ a \rangle_t = -2\gamma \langle N_2 \rangle_t \ , \tag{4.46}$$

which shows that the time variation of the number of photons inside the cavity is equal to the number of excited atoms (photoelectrons) which leave the cavity in the time interval γ^{-1}. Since γ^{-1} is very small, $\gamma^{-1} \ll \kappa_0^{-1}$, the atoms follow the field adiabatically. We evaluate (4.46) by using the coherent state representation for the field. The average number $N_A(T)$ of photocounts in the time interval (0,T) is given by

$$\begin{aligned} N_A(T) \equiv 2\gamma \int_0^T \langle N_2 \rangle_t dt &= \int d^2\alpha P(\alpha) |\alpha|^2 [1 - \exp(-2\kappa_0 T)] \\ &= n_F(0) - n_F(T) \end{aligned} \tag{4.47}$$

with $n_F(t)$ denoting the average number of photons at time t and $P(\alpha) \equiv P_F(\alpha, 0)$. Thus the number of photocounts (excited atoms) in the time interval (0,T) is equal to the number of photons which are absorbed in the same time interval. This result is not unexpected since our model does not include mechanisms for field losses. Similarly, the νth factorial moment of the photoelectron number is given by

$$\begin{aligned} N_A^{(\nu)}(T) \equiv (2\gamma)^\nu \int_0^T dt_1 \ .. \int_0^T dt_\nu \\ \langle N_2(t_1)[N_2(t_2) - \mathbb{1}] \ \dots \ [N_2(t_\nu) - \nu + \mathbb{1}] \rangle \ . \end{aligned} \tag{4.48}$$

As we are dealing with a Markovian stochastic process, all joint probability distributions are determined by the solution of (4.39). We thus obtain

$$N_A^{(\nu)}(T) = \int d^2\alpha P(\alpha) |\alpha|^{2\nu} [1 - \exp(-2\kappa_0 T)]^\nu \ . \tag{4.49}$$

The photocounting probability is calculated by using the standard generating function. This leads to

$$p(n,T) = \int d^2\alpha P(\alpha) \exp(-\eta_0 |\alpha|^2)(\eta_0 |\alpha|^2)^n / n! \tag{4.50}$$

with

$$\eta_0 \equiv 1 - \exp(-2\kappa_0 T) \ . \tag{4.51}$$

Hence the probability p(n,T) of observing n excited atoms in the time interval (0,T) equals the probability that n photons are absorbed in the same time interval.

The form of (4.50) is identical to (4.23). However, the physical meaning of the efficiency factor η_0 in (4.50) is different. This constant depends not only on the atom-field interaction (coupling constant λ), but also on the atomic relaxation due to the coupling to the reservoirs (relaxation constant γ), whereas the corresponding parameter in the previous results allows only for atom-field interaction. The constant η_0 represents a condensed description of the experimental setup (responsible for, e.g., the extraction of photoelectrons from the detector). Under the simplification that the measuring apparatus can be described by a single parameter, (4.50) shows how this parameter affects the observed counting rate. The essential result of the present approach is (4.45) between atomic and field moments. This relation is valid for times large compared with the atomic relaxation time γ^{-1}, where typically $\gamma^{-1} \ll 10^{-8}$ s. We therefore can reconstruct the initial field statistics from counting rates measured at practically any time. Our model settles the usual difficulties of detection theory such as radiative decay, saturation, and stimulated emission, and incorporates previous results on this subject.

4.4 Disturbing Effects

Distortions of the photoelectron statistics can be due to a variety of effects, as listed in Sect.4.1.2. Some of these effects are now discussed in more detail. Section 4.4.1 is an overview of some features of noise in real detectors (see also [4.102]). Dead time effects are considered in Sect.4.4.2. Section 4.4.3 deals with coherence and sampling effects, i.e., the effects due to the finite counting interval, the finite detector area and the finite duration of an experimental run.

4.4.1 Dark Currents and Noise

Internal noise in photon detectors was extensively discussed by MELCHIOR [4.13] and KRUSE [4.25][1] with particular reference to the problem of detecting very small signals in the detector output. The present discussion is mainly concerned with phenomena affecting counting experiments. After a general overview on various types of detectors and related noise problems, we focus our attention on some effects relevant for photomultipliers.

Photodetectors

Three basic types of photodetectors are used: photoemissive detectors, photoconductive detectors and solid-state photodiodes (in which we include avalanche photo-

---

[1] See also Physics Today, Nov. 1977.

diodes). In photoemissive detectors electrons emitted from the surface of the photocathode are collected by an anode; very often the photocathode is followed by some structure providing gain, such as the dynodes in photomultipliers. In photoconductive detectors (semiconductors) charge carriers are excited between valence and conduction bands (intrinsic detectors) or between impurity states and one of the bands (extrinsic detectors), giving rise to a change in the conductivity of the semiconductor. In photodiodes an internal potential barrier with a built-in electric field separates a photoexcited electron-hole pair (e.g., at a p-n junction).

Silicon photodiodes are the most widely used detectors for the near-ultraviolet, visible and near-infrared part of the spectrum. A wide variety of photoconductors and photodiodes is available for the infrared and far-infrared region. For photocounting applications, photomultipliers are used below 0.9 μm wave length. Much effort is now spent in studying avalanche photodiodes which seem to be the most suitable detectors for lightwave communication[2].

Noise in Photoconductive Detectors

Johnson noise - due to random motion of charge carriers - is present in all resistive materials. It occurs in the absence of electrical bias as a fluctuating current or voltage proportional to the square root of the temperature (Nyquist's theorem) [4.58]. It can set the limit to the minimum detectable signal in detectors with wide bandwidth and no internal current gain.

The most important type of noise in photoconductors is usually that due to fluctuations in the rates of thermal generation and recombination of charge carriers, thereby giving rise to a fluctuation in the average carrier concentration. Fluctuations in the generation rate may be also due to background radiation. Recombination is essentially due to impurities and defects, acting either as electron-hole recombination centers, or as trapping centers for one type of carriers only. The magnitude of generation-recombination (g-r) noise depends on the detailed features of the semiconductors, e.g., extrinsic or intrinsic generation, densities of carriers, lifetimes, mobilities [4.13,25]. A statistical description of g-r noise can be given in terms of a rate equation for the probability distribution of the fluctuating number of carriers [4.40,52].

At lower modulation frequencies, smaller than ~100 Hz, 1/f noise can be important. In general, 1/f noise appears to be associated with the presence of potential barriers at the contacts, interior, or surface of the semiconductor. The 1/f noise mechanism was recently discussed by PUTTERMAN [4.59] and VOSS [4.60].

---

[2] H. Melchior: Physics Today 1977.

Noise in Photomultipliers

Shot noise due to dark currents is usually the most important source of noise in photomultipliers (PMT). When weak light is involved it can set the limit to the minimum detectable signal. Dark currents in PMT are mainly due to thermionic emission from the photocathode. An estimate of the dark current per unit area can be obtained from the Richardson equation, which indicates a strong dependence on the temperature and on the cathode work function. Through cooling, dark currents can be reduced by several orders of magnitude down to values where only a few electrons are emitted per second. Dark pulses can also be produced by ions of gases adsorbed on the envelope of the tube, by the thermionic emission from the dynodes or by cosmic rays. It can usually be assumed that noise due to PMT dark currents is uncorrelated with the optical field under investigation (the signal), i.e., the dark pulses are produced randomly (see also Sect.3.2.4). In this case the resulting photocount distribution p(n) is the convolution of the signal distribution $p_S(n)$ and the noise distribution $p_N(n)$ [4.8,61]. The corresponding average is $\langle n\rangle = \langle n\rangle_S + \langle n\rangle_N$, while the variance is

$$\langle(\Delta n)^2\rangle = \langle(\Delta n)^2\rangle_S + \langle(\Delta n)^2\rangle_N \quad . \tag{4.52}$$

If the dark pulses are not emitted at random (e.g., positive ions afterpulses), correlations are introduced in the output signal [4.8,26-28].

Another source of spurious pulses in a PMT is background radiation (background-induced current). In this case the superposition of the background field and the input signal must be considered, that is the convolution property of the probability distributions is transferred to the $P(\alpha)$ density functions. By assuming that the two fields are at the same frequency and at least one of them is isotropic in phase [4.16] the measured first moment is $\langle n\rangle = \langle n\rangle_S + \langle n\rangle_N$, while the variance is given by

$$\langle(\Delta n)^2\rangle = \langle(\Delta n)^2\rangle_S + \langle(\Delta n)^2\rangle_N + 2\langle n\rangle_S\langle n\rangle_N \quad . \tag{4.53}$$

Thus the interference between signal and noise gives rise to an additional broadening with respect to the uncorrelated case (4.52). In practice the difference between this case and the uncorrelated one is important only when the background field is almost at the same frequency as the signal. When the frequency difference between background and signal is larger than the inverse of the counting time T, noise and signal counts become uncorrelated.

PMT Statistics

In photomultipliers single photoelectrons emitted from the photocathode are multiplied by a cascaded secondary emission process to produce pulses of charge at the anode. At high light levels, these pulses overlap, giving rise to a continuous anode current which is a measure of the incident light intensity. At lower light levels, however, the greatest amount of information is obtained when single-electron pulses (SER) are detected. The average duration of the SER is the PMT resolution time τ_R. It is due to the transit time dispersion of the secondary electrons and is typically of the order of a few nanoseconds. The resolution time determines the upper frequency limit of the time-dependent effects that can be observed with a single photomultiplier. The area of the SER corresponds to the number of charges provided by the gain structure. Gains as high as 10^8 are quite typical. Both the area and the shape of the pulse fluctuate, the main reasons being the gain fluctuations and the spread in the electron transit times. A complete statistical characterization of the PMT requires the knowledge of the correlation functions of the random function of time describing the SER. When, however, the correlation time τ_c of intensity fluctuations in the optical field is much larger than τ_R, one is interested only in the gain (SER area) statistics [4.8,28,63,64]. The effects of the PMT statistics are of importance when analog processing of the PMT outputs is used. In photocounting experiments the analog method can be used to reduce dead time losses. It can be used also in intensity-correlation spectroscopy when $\tau_c \sim \tau_R$ [4.8]. In this case knowledge of the PMT response correlation function is required [4.65]. More usually, however, digital processing of the PMT output is used, since it removes to a first approximation the effects of gain instability which are marked in the analog method. In addition, photocounting correlations, such as those caused by positive ion after-pulsing, can be made negligible by suitable selection of the discriminator dead time [4.28].

4.4.2 Dead Time Effects

Dead time in a photon-counting experiment is essentially due to the pulse-height discriminator following the detector. The dead time is usually of the order of 10 ns, its lower limit being set by the detector response time. It determines the maximum count rate the detector can handle.

We consider a "nonparalizable dead time counter", i.e., a counter that does not record pulses during a time interval of fixed duration t_D after recording a given pulse. For coherent constant light the probability $p(n,T,t_D)$ of registering n counts in the time interval T is given by [4.32,33]

$$p(n,T,t_D) = \sum_{k=0}^{n} \lambda^k [T - nt_D]^k \, e^{-\lambda(T - nt_D)}/k! - \sum_{k=0}^{n-1} \lambda^k [T - (n-1)t_D]^k \, e^{-\lambda(T - (n-1)t_D)}/k! \; , \qquad (4.54)$$

where λ is the photoelectron counting rate and $n \leq T/t_D$. By the definition of dead time, count numbers greater than T/t_D are forbidden. In the limit $t_D/T \ll 1$, (4.54) becomes [4.66] (see also Sect.3.2.3)

$$p(n,T,t_D) = \frac{(\lambda T)^n}{n!} e^{-\lambda T}[1 + \frac{t_D}{T}(\lambda T - n + 1)] \quad (4.55)$$

showing explicitly that dead time effects increase with (t_D/T), with the photoelectron rate and with the count number. The mean and the variance of the distribution (4.54) are given by

$$\langle n \rangle = \bar{n}_0(1 + \bar{n}_d)^{-1} + 1/2\ \bar{n}_d^2(1 + \bar{n}_d)^{-2} \quad (4.56)$$

and

$$\langle (\Delta n)^2 \rangle = \bar{n}_0(1 + \bar{n}_d)^{-3}\ , \quad (4.57)$$

where $\bar{n}_0 = \lambda T$ is the average number of counts in the absence of dead time and $\bar{n}_d = \bar{n}_0(t_D/T)$ is the average number of counts in a dead time. As expected, the mean number of observed pulses is reduced by dead time effects. Also the variance is reduced, i.e., dead time gives rise to an "antibunching" of the resulting photoelectrons.

When λ is a fluctuating function of time with coherence time $\tau_c \gg T$, the counting probabilities are obtained by averaging the right-hand side of (4.54) with respect to the probability density of λ. The result is [4.32,33]

$$p(n,T,t_D) = \sum_{k=0}^{n} p_0(k, T - nt_D) - \sum_{k=0}^{n-1} p_0[k, T -(n-1)t_D]\ , \quad (4.58)$$

where p_0 denotes the probability distribution in the absence of dead time effects.

Equation (4.58) may be used to predict the results of experiments with light sources of known properties. From an experimental point of view, however, the problem is how to derive information on the undisturbed distribution when the dead-time-modified distribution and/or its moments are measured. Dead time effects on counting distributions were experimentally observed by JOHNSON et al. [4.67] and CHANG et al. [4.68]. JOHNSON et al. [4.67] studied the counting distribution produced by a tungsten lamp, in the limit $T \gg \tau_c$. The experimental results were fitted by a Poisson distribution corrected to the first order in t_D/T. CHANG et al. [4.68] studied laser radiation in the threshold region. Dead time corrections in the form of a power series in t_D/T were applied to the measured factorial moments of the counting distribution. In practice this is the most convenient method for the analysis of experimental results [4.61], in particular at low count rates when a few correcting terms are usually sufficient.

More recently some attention has been given to the problem of dead time corrections in the case of arbitrary values of the ratio T/τ_c and variable dead time [4.69,70].

4.4.3 Coherence and Sampling Effects

Time Effects

In an ideal detector there is no lower limit to the counting time interval T. In a real detector, however, T cannot be chosen arbitrarily small, its lower limit being set by the system dead time: the value of the ratio T/τ_c is of the order of, or greater than, unity when $\tau_c \leqq 10^{-8}$ s. This value determines the number of independent samples of the field which are lumped together during a measurement. The specific form of the counting distribution for different statistics and arbitrary T/τ_c is discussed in detail in the books by SALEH [4.5] and BARAKAT and BLAKE [4.6] and many references given therein.

Let us now neglect any "disturbing effect" other than the finite sampling time. In this case the counting probabilities are given by (4.15,16), where (for simplicity we set $\xi \equiv 1$)

$$\Omega = \int_0^T dt|E(t)|^2 \tag{4.59}$$

for a pointlike detector. In order to determine the probability density $W(\Omega)$ we use a Karhunen-Loève [4.5,6] expansion in the time interval (0,T). It is then necessary to find the eigenvalues of the integral equation

$$\int_0^T \langle E^*(t)E(t')\rangle\Phi_k(t')dt' = \nu_k\Phi_k(t) \quad , \tag{4.60}$$

where the set $\{\Phi_k(t)\}$ is complete and orthonormal in (0,T). The eigenvalues ν_k allow a straightforward calculation of the moment generating function

$$Q_\Omega(s) \equiv \langle e^{-s\Omega}\rangle \quad . \tag{4.61}$$

The probability density $W(\Omega)$ is obtained as the inverse Laplace transform of $Q_\Omega(s)$.

For example, in the case of a linearly polarized thermal field, $Q_\Omega(s)$ reads

$$Q_\Omega(s) = \prod_k (1 + s\nu_k)^{-1} \quad . \tag{4.62}$$

When $T \ll \tau_c$, then (4.60) has only one eigenvalue $\nu_1 = \langle\Omega\rangle$. Thus $W(\Omega)$ reduces to the exponential distribution, $W(\Omega) = (\langle\Omega\rangle)^{-1}\exp(-\Omega/\langle\Omega\rangle)$, and $p(n)$ is the Bose-Einstein distribution, $p(n) = \bar{n}^n/(1+\bar{n})^{n+1}$ with $\bar{n} = \langle\Omega\rangle$ denoting the average number of photoelectrons in the counting interval. When $T \gg \tau_c$, a very large number $N_t = T/\tau_c$ of approximately equal eigenvalues $\nu = \langle\Omega\rangle/N_t$ exists. In this case we obtain

$$Q_\Omega(s) \sim (1 + s\langle\Omega\rangle/N_t)^{-N_t} \tag{4.63}$$

and

$$W(\Omega) = [N_t^{N_t}\Omega^{N_t-1}/\Gamma(N_t)\langle\Omega\rangle^{N_t}]\exp(-N_t\Omega/\langle\Omega\rangle) \quad . \tag{4.64}$$

This expression for $W(\Omega)$ is known as a chi-square distribution of order $2N_t$, which corresponds to the sum of N_t-independent exponentially distributed modes. In the limit $N_t \to \infty$, (4.62) reduces to

$$Q_\Omega(s) \sim e^{-s<\Omega>} , \tag{4.65}$$

which is recognized as the generating function of a coherent field of intensity $<\Omega>$. This result corresponds to the fact that the long integration time washes out the intensity fluctuations. The intermediate region $T \sim \tau_c$ cannot be discussed in a general way, since the solutions depend on the field spectrum. Analytical solutions exist for a few cases, such as a Lorentzian spectrum [4.71]. In most instances, however, the eigenvalues of (4.60) must be calculated numerically.

Experimental counting distributions are usually analyzed in terms of factorial moments or cumulants. For Gaussian light the mth normalized cumulant $K^{(m)}(T)$ of the integrated intensity Ω is given by [4.36]

$$K^{(m)}(T) = B_m K^{(m)}(0) , \tag{4.66}$$

where $K^{(m)}(0)$ is the normalized cumulant in the limit of zero counting time and

$$B_1 = 1$$

$$B_m = T^{-m} \int_0^T dt_1 \ldots \int_0^T dt_m g(t_1 - t_2) g(t_2 - t_3) \ldots g(t_m - t_1) \quad . \tag{4.67}$$

Here $g(t)$ is the normalized first-order correlation function of the field. Equation (4.66) can be used to estimate the effect of the finite sampling time when the field spectrum is known. For example, in the case of a Lorentzian spectrum with halfwidth Γ, the normalized cumulants deviate from their zero-counting time limits by 3,5, and 7% for $m = 2,3$, and 4, respectively, for ΓT as small as 0.1 [4.68]. The spectrum halfwidth in this case can be extracted from the normalized second factorial moment of the counting distribution [4.72].

Spatial Effects

The relevant parameter in this case is the ratio A/A_c, where A is the effective detector area and A_c is the coherence area of the optical field. The ratio A/A_c is the number of independent samples of the field covered by the detector. The total intensity falling on the detector during the time interval $(0,T)$ is

$$\Omega = \int_A d^3r \int_0^T dt |E(r,t)|^2 . \tag{4.68}$$

In order to determine the probability density $W(\Omega)$, we have to find the eigenvalues of the integral equation

$$\int_A d^3r' \int_0^T dt' <E^*(rt)E(r't')> \Phi_{jk}(r't') = \nu_{jk} \Phi_{jk}(rt) , \tag{4.69}$$

where the set $\{\Phi_{jk}\}$ is complete and orthonormal in (0,T) and A. This problem is greatly simplified in the case of cross-spectrally pure fields, i.e., when the spatial and temporal dependences of the correlation function are factorized. The same factorization holds then for the eigenfunctions and the eigenvalues, i.e., $\Phi_{jk}(rt) = \lambda_j(r)\zeta_k(t)$ and $\nu_{jk} = \mu_j\lambda_k$.

For instance, in the case of a polarized thermal field the generating function becomes

$$Q_\Omega(s) = \prod_j \prod_k (1 + s\mu_j\lambda_k)^{-1} \quad . \tag{4.70}$$

When $A \ll A_c$ one spatial mode only is important and we are back to the case discussed previously. When $A \gg A_c$ we have to take into account $N_s = A/A_c$ spatial modes, together with the temporal modes. When $T \ll \tau_c$ the photocounting distribution is the Bose-Einstein distribution in the limit of small detector area, but gradually approaches a Poisson distribution with increasing A. This effect is due to the averaging out of uncorrelated intensity fluctuations in different coherence areas. In most counting experiments the effective detector area is reduced so as to avoid finite aperture corrections [4.28]. When necessary these corrections are applied to the factorial moments or to the factorial cumulants of the counting distribution.

Sampling Effects

Experimentally photocounting distributions are obtained by taking periodically repeated samples of the number n of photoelectrons emitted in a time T by a photoelectric detector. The measured p(n,T) is then given by

$$p(n,T) = \frac{1}{M}\sum_{p=1}^{M} \delta_{n_p n} \quad , \tag{4.71}$$

where n_p is the number of counts in the pth sample and M is the total number of samples. An estimate of the statistical accuracy of (4.71) is given by the variance of the measured factorial moments $\mathcal{N}^{(r)}$, viz.

$$\mathrm{Var}\,\mathcal{N}^{(r)} = \frac{1}{M^2}\sum_{p,q=1}^{M} [\langle n_p(n_p - 1) \ldots (n_p - r + 1)n_q(n_q - 1) \ldots (n_q - r + 1)\rangle - \{N^{(r)}\}^2] \quad , \tag{4.72}$$

where $N^{(r)} = \langle\mathcal{N}^{(r)}\rangle$ correspond to $M \to \infty$ in (4.71), as shown by JAKEMAN and PIKE [4.29]. The terms with p different from q in (4.72) describe correlations among different samples. These terms vanish when the period $\tau(\tau > T)$ of the sampling is greater than any coherence time of the field. Otherwise, correlations between samples increase the statistical errors.

As a general rule at given constant total measuring time $M\tau$, the optimum separation between successive samples is of the order of τ_c, i.e., no further reduction of statistical errors is achieved by increasing M. When M is fixed, however, the error decreases as τ increases, since correlations between samples are reduced. When the samples are uncorrelated it is simpler to consider the ordinary moments.

Denoting the measured and true k^{th} moments by $\mathcal{M}_k$ and M_k, respectively we have

$$\mathrm{Var}\ \mathcal{M}_k = (M_{2k} - M_k^2)/M \quad . \tag{4.73}$$

This equation was applied by ARECCHI et al. [4.73] to calculate the minimum value of detectable perturbations superposed to a coherent field.

Other Counting Experiments

A detailed analysis of these and other statistical errors in photon-correlation spectroscopy was given by DEGIORGIO and LASTOVKA [4.30] and HUGHES et al. [4.31] for Gaussian-Lorentzian sources. In principle, each of the above effects gives rise to an inverse problem, namely the problem of retrieving the pure signal from the data measured in the presence of the disturbing effects. In general, the corresponding mathematical inversion procedure has not been investigated hitherto. The only exception seems to be the effect of the coherence time τ_c, where BARAKAT and BLAKE [4.6] established the formal inversion procedure for determining the spectral profile from the measured photoelectron correlation function. The same authors treated the problem of incomplete spatial coherence in correlation spectroscopy, with particular reference to the work of KELLY [4.74]. The effect of the gate time in the measurement of the time interval distribution of Gaussian-Lorentzian light was recently treated by SONODA et al. [4.75].

4.5 Temperature Effects in Photodetection

In this section we drop the restriction of zero temperature of the heatbath of Sect. 4.3 and allow the photosensitive atoms to be coupled to reservoirs at finite temperature. This generalization results in a first-principle theory of thermal noise in photodetection, which is particularly relevant for the power and fluctuation measurement of infrared radiation. The detector temperature is an effective parameter which measures the number of excited atoms in the initial state of the detector. It can thus describe not only thermal effects, but also any other source of atomic excitation. The initial number of excited atoms corresponds to the "intrinsic dark current" of the detector. The fluctuation of this number constitutes the "intrinsic noise". Since zero detector temperature cannot be achieved experimentally, it is of interest to know how this intrinsic noise interferes with the detection process.

The main objective of our model calculation is to establish the relations between atomic and field moments, which determine the transfer of the statistical state of the incoming field to the detector signal. The derivation of the relationship in question is technically involved if we use the master equation technique of Sect.4.3. In this section we rather follow a different approach based on Langevin equations. We apply the Langevin approach to the monomode pointlike DICKE model

[4.101] with atoms undergoing relaxation due to interaction with external reservoirs. The resulting equations are formally identical to those describing a monomode laser without field losses [4.79]. We stress, however, the different meaning of the atomic reservoir, which in our model describes the measuring system rather than the atomic pumping and damping mechanism.

In Sect.4.5.1 we derive the relations between atomic and field moments. These relations are basic to the derivation of the photocounting probability presented in Sect.4.5.2. Specific applications of the counting probability are discussed in Sect.4.5.3.

4.5.1 Langevin Equations of Motion

We consider the monomode pointlike Dicke model at resonance with the atoms coupled to external reservoirs at finite temperature. The reservoirs are responsible for the atomic relaxation and for the stochastic forces occurring in the Langevin equations, viz.,

$$\dot{a} = -i\lambda \sum_i s_i^- \equiv -i\lambda S^- \tag{4.74}$$

$$\dot{s}_i^- = -\gamma s_i^- + 2i\lambda s_{3i} a + f_i^- \tag{4.75}$$

$$\dot{s}_{3i} = -2\gamma(s_{3i} - 1/2\ \sigma_0) + i\lambda(a^+ s_i^- - s_i^+ a) + f_{3i} \tag{4.76}$$

with $a^\pm$ denoting e.m. field operators, $s_i^\pm$ and s_{3i} atomic operators referring to atom i and λ the dipolar coupling constant. By $f_i^\pm$, f_{3i} we denote the stochastic forces on the ith atom. The description of a dead time effect, namely the deadtime related to the extraction of excited atoms, is included in the atomic relaxation constant γ. We make the usual hypotheses, namely: I) the forces on different atoms are uncorrelated; II) the forces on a single atom are Gaussian δ-correlated random operators with zero average and correlation functions obeying the fluctuation-dissipation theorem [4.40,79,99,100]

$$\begin{aligned}
&\langle f_i^\pm(t) f_i^\pm(t')\rangle_R = 0\\
&\langle f_i^\pm(t) f_i^\mp(t')\rangle_R = \gamma(1 \pm \sigma_0)\delta(t - t')\\
&\langle f_{3i}(t) f_{3i}(t')\rangle_R = \gamma(1 - \sigma_0^2)\delta(t - t') \quad .
\end{aligned} \tag{4.77}$$

Here $\langle \dots \rangle_R$ denotes the average over the reservoir state and $\sigma_0 \equiv -\tanh(\beta\varepsilon/2)$ is the stationary value of the population difference operator $\langle s_{3i}\rangle_R$, ε being the energy spacing between the atomic levels and β the inverse temperature. In contrast to laser theory no field losses are assumed here.

The Field Equation

By a well-known procedure [4.79], namely the linearization of the atomic Langevin equation (4.76) around the stationary solution, from (4.74-76) we can derive a closed equation of motion for the field variables. This equation exhibits the damping and the fluctuations of the field produced by the coupling to the atoms, viz.,

$$\dot{a}(t) = -\kappa a(t) + g(t) \quad . \tag{4.78}$$

Here $g(t)$ is the induced fluctuating force and κ is the temperature dependent attenuation constant

$$\kappa = \kappa_0(-\sigma_0) \equiv \lambda^2\gamma^{-1}N(-\sigma_0) \quad . \tag{4.79}$$

We remark that at any finite temperature κ is smaller than its corresponding zero temperature value κ_0, i.e., the atomic absorption decreases by increasing the fraction of excited atoms. Equation (4.78) is obtained introducing the same approximations as in Sect. 4.3, namely $\lambda N^{\frac{1}{2}} \ll \gamma$ and $t \gg \gamma^{-1}$. The equation of motion for the moments of the field reads

$$\frac{d}{dt}\langle a^{+m}a^m\rangle_t = -2\kappa m\langle a^{+m}a^m\rangle_t + 2\kappa m^2\bar{n}\langle a^{+m-1}a^{m-1}\rangle_t \ , \tag{4.80}$$

where $\bar{n} = [\exp(\beta\varepsilon) - 1]^{-1}$ is the thermal equilibrium expectation value for a^+a and $\langle X\rangle_t \equiv \langle X(t)\rangle$ denotes the average over the states of the total system (atoms, e.m. field, and reservoirs). For zero temperature (4.80) reduces to the equation of motion (4.40) in Sect. 4.3. We consider the solution to (4.80) for two different sets of initial conditions.

I) *No field initially present inside the detector.* Introducing Glauber's coherent state representation, the initial state of the field is characterized by

$$\langle a^{+m}a^m\rangle_{t=0} = \int d^2\alpha|\alpha|^{2m}P(\alpha) \ , \tag{4.81}$$

where $P(\alpha)$ is the statistical distribution of the incoming field at time $t=0$. Subject to the initial condition (4.81), the solution to (4.80) takes the form

$$\langle a^{+m}a^m\rangle_t = m!\bar{n}_t^m \int d^2\alpha P(\alpha)L_m(-|\alpha_t|^2/\bar{n}_t) \ , \tag{4.82}$$

where L_m denotes the Laguerre polynomial of order m, $\bar{n}_t = \bar{n}[1-\exp(-2\kappa t)]$ and $|\alpha_t|^2 = |\alpha|^2\exp(-2\kappa t)$. For $t \gg \kappa^{-1}$, (4.82) reduces to the thermal equilibrium value $m!\bar{n}^m$. Thus κ^{-1} measures the thermalization time for the field.

II) *A resonant thermal field with average photon number $\bar{n}$ is initially present inside the detector.* The statistical field distribution at time $t=0$ is in this case given by the convolution of the incoming field distribution, $P_{th}(\alpha) = (\pi\bar{n})^{-1}\exp(-|\alpha|^2/\bar{n})$ (see Sect.4.4.1). This leads to

$$\langle a^{+m}a^m\rangle_{t=0} = \int d^2\alpha d^2\alpha'|\alpha|^{2m}P(\alpha-\alpha')P_{th}(\alpha') \quad . \tag{4.83}$$

The solution to (4.80) takes now the form

$$\langle a^{+m}a^m\rangle_t = m!\bar{n}^m \int d^2\alpha P(\alpha)L_m(-|\alpha_t|^2/\bar{n}) \quad . \tag{4.84}$$

Connection Between Atomic and Field Dynamics

In order to relate the photon and the photoelectron statistics, we consider the moments of the atomic variables (see Sect.4.3.2). For the first moment we obtain [4.23]

$$\langle N_2\rangle_t - N_{2,eq} = \zeta(\langle a^+a\rangle_t - \bar{n}) \ , \tag{4.85}$$

where N_2 denotes the excited-atom-number operator (4.33), $N_{2,eq} = N(1+\sigma_0)/2 = N(-\bar{n}\sigma_0)$ is the number density of excited atoms at thermal equilibrium, and

$$\zeta = \zeta_0(-\sigma_0) = \kappa\gamma^{-1} \ . \tag{4.86}$$

In (4.86) memory effects on the time scale γ^{-1} are neglected, in agreement with the condition $t \gg \gamma^{-1}$. In terms of the coherent state representation for the field variables, the higher-order factorial moments are as follows.

I) *Initial condition (4.81):*

$$\langle N_2^{(\nu)}\rangle_t = \nu![N!/(N-\nu)!N^\nu](N_{2,eq} - \zeta\bar{n}\,e^{-2\kappa t})^\nu \cdot\int d^2\alpha P(\alpha)L_\nu[-\zeta|\alpha_t|^2/(N_{2,eq} - \zeta\bar{n}\,e^{-2\kappa t})] \ . \tag{4.87}$$

II) *Initial condition (4.83):*

$$\langle N_2^{(\nu)}\rangle_t = \nu![N!/(N-\nu)!N^\nu]N_{2,eq}^\nu \cdot\int d^2\alpha P(\alpha)L_\nu(-\zeta|\alpha_t|^2/N_{2,eq}) \ . \tag{4.88}$$

4.5.2 Photocounting Probability

We calculate the photocounting probability from the factorial moments of the number of photocounts in the time interval (0,T), as defined by (4.48) in Sect.4.3. By straightforward integration of (4.87) or (4.88) we obtain $N_A^{(\nu)}$ in terms of the initial field distribution $P(\alpha)$, viz.,

$$N_A^{(\nu)}(T) = \nu!u^\nu \int d^2\alpha P(\alpha)L_\nu(-\eta|\alpha|^2/u) \ , \quad \eta = 1 - \exp(-2\kappa T) \ . \tag{4.89}$$

The quantity u reads as follows.

I) *Initial condition (4.81):*

$$u = 2\gamma TN_{2,eq} - \eta\bar{n} \ . \tag{4.90}$$

II) *Initial condition (4.83):*

$$u = 2\gamma TN_{2,eq} \ . \tag{4.91}$$

We observe that (4.90) is always positive because of $\gamma \gg \kappa_0$. The average number of photocounts ($\nu = 1$) reads

$$N_A(T) = u + \int d^2\alpha P(\alpha)\eta|\alpha|^2 \quad . \tag{4.92}$$

Thus u can be interpreted as the average number of photocounts due to the thermal noise (dark current). Equation (4.91) as well as the first term of (4.90) represent the stationary flow of excitation from the atomic system to the reservoir (external circuit), while the second term of (4.90) is due to the thermal emission which is responsible for the buildup of the equilibrium photon distribution inside the cavity.

In order to connect the result (4.89) with the usual formula for the moments, we consider the integrated moments of

$$\hat{N}_2(\alpha) = N_2(\alpha) - N_2(0) \quad ,$$

where by $N_2(\alpha)$ we stress the dependence of the number operator N_2 on the field amplitude. Now $\hat{N}_2(\alpha)$ is the occupation number operator biased with respect to zero incoming field, i.e., without noise. Proceeding as before we obtain the well-known expression

$$\hat{N}_A^{(\nu)}(T) = \int d^2\alpha P(\alpha)\eta^\nu|\alpha|^{2\nu} \quad , \tag{4.93}$$

but with temperature-dependent efficiency factor η. We remark that, due to $\kappa < \kappa_0$, the efficiency η is smaller than its corresponding zero temperature value η_0 (see Fig.4.1). In particular for small attenuation, $2\kappa_0 T \ll 1$, and low temperature, $\bar{n} \ll 1$, the ratio $\hat{R}_\nu = (\eta/\eta_0)^\nu$ between (4.93) and the zero temperature moments (4.49) becomes

$$\hat{R}_\nu \sim 1 - 2\nu\bar{n} \quad . \tag{4.94}$$

We also remark that $\hat{N}_A^{(\nu)}$ is the leading term of (4.89) in the limit of large signal to noise ratio, $(\eta|\alpha|^2/u) \gg 1$.

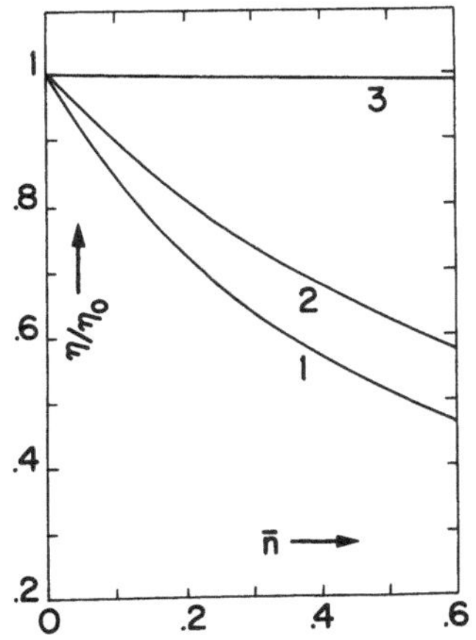

Fig. 4.1. Ratio (η/η_0) vs the average thermal photon number $\bar{n}$. (Curve 1) $2\kappa_0 T = 0.1$; (Curve 2) $2\kappa_0 T = 1$; (Curve 3) $2\kappa_0 T = 10$

By the standard procedure, from (4.89) we construct the generating function, which finally leads to the photocounting probability

$$p(n,T) = \int d^2\alpha P(\alpha)[u^n/(1+u)^{n+1}]$$

$$L_n\left(-\eta|\alpha|^2/u(1+u)\right) \exp[-\eta|\alpha|^2/(1+u)] \quad . \tag{4.95}$$

In the limit of zero temperature ($u \to 0$), (4.95) reduces to (4.51) in Sect.4.3. For zero incoming field intensity ($\alpha = 0$), (4.95) gives the photocounting probability of the dark current, viz.,

$$p(n,T) = u^n/(1+u)^{n+1} \quad ,$$

which corresponds to the statistics of a thermal field with effective photon number u. In the case of an incoming coherent field, (4.95) becomes formally identical to the ideal (zero temperature) counting distribution for the superposition of a coherent and a Gaussian field [4.5,8] [see also (4.125)]. Hence the finite temperature acts as a source of an intrinsic Gaussian field, which is coherently superimposed to the incoming field. Of course our results become meaningless in the limit of high temperature. In this case the atomic system is initially in a highly excited state, so that a linear theory is invalid [4.79].

4.5.3 Applications

We now consider some statistical distributions for the incoming field. For simplicity we restrict our discussion to the factorial moments of the counting distribution. We compare the cases of zero and nonzero detector temperature in terms of the ratio R_ν between the finite temperature moments, (4.89), and the zero temperature ones, (4.49). We take the dark current u as given by (4.91). The difference between (4.90) and (4.91) is in practice negligible, since $2\gamma TN_{2,eq} = \gamma TN(-\bar{n}\sigma_0)$, with N denoting the density of photosensitive atoms, is very large in comparison to $\eta\bar{n}$.

I) *Coherent radiation.* For a coherent incident field of intensity $|\alpha_0|^2$ we obtain

$$R_\nu = \nu!(u/\eta_0|\alpha_0|^2)^\nu L_\nu(-\eta|\alpha_0|^2/u) \quad . \tag{4.96}$$

For low temperature, $\bar{n} \ll 1$, large incoming intensity, $\eta|\alpha_0|^2 \gg 1$, and small attenuation, $2\kappa_0 T \ll 1$, (4.96) reduces to

$$R_\nu \sim 1 - 2\nu\bar{n} + \nu^2\gamma N\bar{n}/\kappa_0|\alpha_0|^2 \quad . \tag{4.97}$$

II) *Chaotic radiation.* We continue with a Gaussian field of intensity N, leading to

$$R_\nu = (\eta/\eta_0)^\nu(1 + u/\eta N)^\nu \quad . \tag{4.98}$$

For $\bar{n} \ll 1$, $\eta N \gg u$ and $2\kappa_0 T \ll 1$, we find

$$R_\nu \sim 1 - 2\nu\bar{n} + \nu\gamma N\bar{n}/\kappa_0 N \quad . \tag{4.99}$$

III) *Superposition of coherent and chaotic radiation.* We finally consider the superposition of a coherent field of intensity $|\alpha_0|^2$ with a Gaussian field of intensity N. For this distribution we obtain

$$R_\nu = (\eta/\eta_0)^\nu(1 + u/\eta N)^\nu L_\nu[-\eta|\alpha_0|^2/(u + \eta N)]/L_\nu(-|\alpha_0|^2/N) \quad . \tag{4.100}$$

For $\bar{n} \ll 1$, $\eta|\alpha_0|^2 \gg \eta N \gg u$ and $2\kappa_0 T \ll 1$, (4.100) reduces to (4.97).

Numerical Examples and Discussion

In Fig.4.2 we show the temperature effect on the first moment of the counting distribution for an incoming field with average photon density $N = 10^8$ (we remark that the first moment does not depend on the incident field statistics). To this end we plot the ratio R_1 as a function of $(\beta\varepsilon)^{-1}$ for fixed values $2\gamma T = 10(2\kappa_0 T) = 1$ of the atomic relaxation and field attenuation parameters and various values N of the density of photosensitive atoms. A similar plot of R_2 is shown in Fig.4.3 for Gaussian and coherent radiation with the same photon density $|\alpha_0|^2 = N = 10^8$.

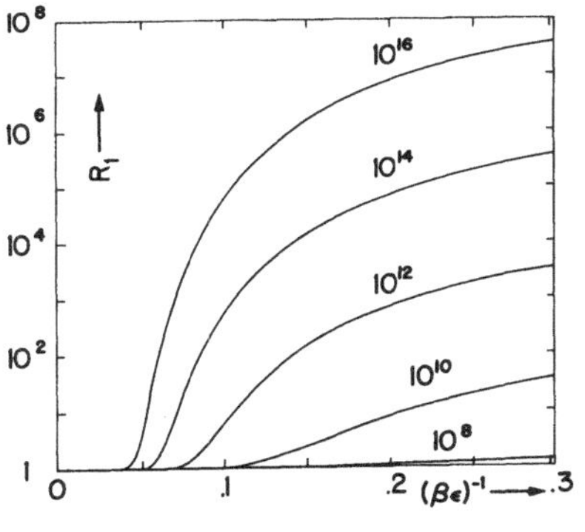

Fig. 4.2. Relative first moment vs $(\beta\varepsilon)^{-1}$ where $2\gamma T = 10(2\kappa_0 T) = 1$ and the average photon density is 10^8. The numbers labelling the different curves denote the density N of the photosensitive atoms

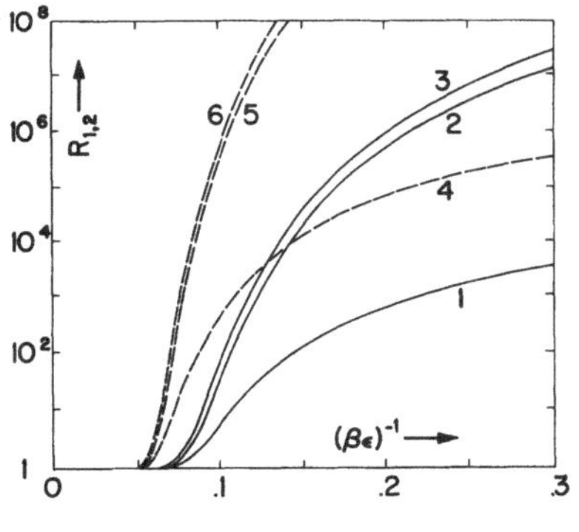

Fig. 4.3. Relative first and second moment R_1, R_2 vs $(\beta\varepsilon)^{-1}$ where $2\gamma T = 10(2\kappa_0 T) = 1$ and the average photon density is 10^8. (Curve 1) first moment; (Curve 2) second moment for a Gaussian field; (Curve 3) second moment for a coherent field; these 3 curves are for $N = 10^{12}$. Curves 4,5, and 6 are the same as 1,2, and 3 but with $N = 10^{14}$

Figures 4.2,3 show that for any fixed value (N/N) there is a "threshold temperature" $(\beta\varepsilon)^{-1}_{th}$ below which R_ν does not appreciably deviate from unity and above which the dark current contribution is rapidly increasing with increasing temperature. The value of $(\beta\varepsilon)^{-1}$ for which the deviation $(R_1 - 1)$ is of order, say, 10^{-2} can be evaluated from (4.99) or (4.97). Neglecting the small term $(-2\nu\bar{n})$ we find $(\beta\varepsilon)^{-1}$ = 0.062, 0.048, 0.039 and 0.033 for (N/N) = 10^4, 10^6, 10^8 and 10^{10}, respectively.

The value $(\beta\varepsilon)^{-1} = 0.033$ corresponds for instance to a temperature of about 44 K for a CO_2 laser beam (wavelength 10.6 μm) and to a temperature of about 380 K at 1 eV energy. The temperature 44 K at 10.6 μm compares very well with the operating temperature or with the maximum temperature for BLIP (Background Limited Infrared Photodetectors) of infrared detectors in the range 10-12 μm [4.25].

From the above results we infer that the temperature dependence of the measured moments is largely dominated by the dark current contribution for any apparently reasonable value of the ratio $(N/\mathcal{N})$. The temperature dependence of the efficiency factor η, however, may become important if one measures the noiseless moments (4.93), i.e., subtracts the detector intrinsic noise. As previously remarked, $\hat{N}_A^{(\nu)}$ is smaller than the corresponding zero temperature moment (4.50) because of the reduction of the atomic absorption; for $\bar{n} \ll 1$ and $2\kappa_0 T \ll 1$, (4.94) shows that $(\hat{R}_\nu - 1)$ is proportional to $\bar{n}$ and to the order ν of the moment. For a CO_2 laser $R_1 - 1 \sim 2\%$ when the detector is at room temperature.

To conclude we recall that two parameters are essential in our analysis of temperature effects in photodetection. These are the dark current u and the temperature-dependent efficiency factor η. The first is essential in the determination of the detector operating temperature, for instance in relation to the detection of weak signals. The second can become important if one is able to measure the noiseless moments $\hat{N}_A^{(\nu)}$ or if $(N/\mathcal{N}) \lesssim 1$. We also recall that our detector temperature is an effective parameter measuring the fraction of excited atoms in the detector. Thus all the above considerations are not restricted to thermal effects only, but apply also to situations, where an equilibrium population of excited states is maintained by any other excitation mechanism.

4.6 Summary of Statistical Methods

In this section we collect some standard definitions and results on random variables (Sect.4.6.1), stochastic processes (Sect.4.6.2) and statistical properties of the radiation field (Sect.4.6.3). The arguments of Sect.4.6.1,2 are treated in many textbooks, e.g., FELLER [4.52], STRATONOVICH [4.76], and KENDALL [4.77]. Some pertinent references are given in Sect.4.6.3.

4.6.1 Random Variables

The characteristic function and the moments of the random variable X are defined by

$$\theta(\lambda) \equiv \langle e^{i\lambda X}\rangle = \int dx\, e^{i\lambda x}\, W(x) \tag{4.101}$$

and

$$M_k \equiv \langle X^k\rangle = \int dx\, x^k\, W(x) \quad , \tag{4.102}$$

where W(x) denotes the probability density. When X is nonnegative, it is possible to define $\tilde{Q}(\lambda) = \theta(i\lambda)$ which is called the (moment) generating function. The moments are related to the characteristic function by

$$M_k = \left(-i \frac{d}{d\lambda}\right)^k \theta(\lambda)|_{\lambda=0} \quad . \tag{4.103}$$

These equations show that the knowledge of *all* the moments M_k allows the determination of W(x) by the Fourier inversion of $\theta(\lambda)$. In practical cases, however, only a finite number of moments, $M_k \pm \Delta_k$, is known, Δ_k being some experimental error (see Sect.4.2.4). By taking the logarithm of $\theta(\lambda)$, we can define the cumulants K_m

$$\ln \theta(\lambda) = \sum_{n=0}^{\infty} \frac{(i\lambda)^n}{n!} K_n \quad . \tag{4.104}$$

For example, the relation between the first three cumulants and moments is

$$K_1 = M_1$$
$$K_2 = M_2 - M_1^2$$
$$K_3 = M_3 - 3M_1M_2 + 2M_1^3 \quad . \tag{4.105}$$

In the case of a discrete random variable N which takes nonnegative integer values only, it is also useful to define the factorial moments

$$\langle N^{(k)}\rangle = \sum_n n(n-1) \ldots (n-k+1)p(n) \quad . \tag{4.106}$$

The corresponding generating function is given by

$$Q(s) = \sum_n (1-s)^n p(n) = \sum_k \frac{(-s)^k}{k!} \langle N^{(k)}\rangle \quad . \tag{4.107}$$

When the factorial moments are known, (4.107) can be used to determine the probability distribution.

Examples

I) For the Gaussian (normal) distribution

$$W(x) = (2\pi)^{-\frac{1}{2}}\sigma^{-1} \exp[-(x-\mu)^2/2\sigma^2] \tag{4.108}$$

one has

$$M_1 = \mu \ , \quad K_2 = \sigma^2 \ , \quad K_m = 0 \quad \text{for } m > 2 \quad . \tag{4.109}$$

II) The exponential distribution

$$W(y) = \mu^{-1} \exp(-y/\mu) \tag{4.110}$$

is characterized by

$$Q(s) \equiv \langle e^{-sy}\rangle = (1 + s\mu)^{-1} \tag{4.111}$$

and

$$M_n = n!\mu^n \ , \quad K_n = (n-1)!\mu^n \ . \tag{4.112}$$

The exponential distribution describes the intensity of polarized thermal light at any given space-time point.

III) The Poisson distribution

$$p(n) = \mu^n \, e^{-\mu}/n! \tag{4.113}$$

is characterized by

$$Q(s) \equiv \langle (1-s)^N \rangle = e^{-s\mu} \tag{4.114}$$

with

$$\langle N \rangle = \mu \ , \quad \langle (\Delta N)^2 \rangle = \mu \ , \quad \langle N^{(k)} \rangle = \mu^k (k > 2) \ , \tag{4.115}$$

where $\langle (\Delta N)^2 \rangle$ denotes the variance.

IV) For the geometric or Bose-Einstein distribution

$$p(n) = \frac{\mu^n}{(1+\mu)^{n+1}} \tag{4.116}$$

one finds

$$Q(s) \equiv \langle (1-s)^N \rangle = (1+s\mu)^{-1} \tag{4.117}$$

with

$$\langle N \rangle = \mu \ , \quad \langle (\Delta N)^2 \rangle = \mu^2 + \mu \ , \quad \langle N^{(k)} \rangle = k!\mu^k (k > 2) \ . \tag{4.118}$$

Equation (4.116) describes, for instance, the photon distribution of single-mode thermal light.

V) The "ideal detector" counting distribution of a monomode field is

$$p(n) = \int d^2\alpha P(\alpha) \, \frac{\eta^n |\alpha|^{2n}}{n!} \ . \tag{4.119}$$

The generating function (4.107) and the corresponding factorial moments are given by

$$Q_N(s) \equiv \langle (1-s)^N \rangle = \int d^2\alpha \; e^{-s\eta|\alpha|^2} P(\alpha) \tag{4.120}$$

and

$$\langle N^{(k)} \rangle = \int d^2\alpha \eta^k |\alpha|^{2k} P(\alpha) \ . \tag{4.121}$$

We now apply the above results to some specific field models.

a) For a coherent field, $P(\alpha) = \delta^{(2)}(\alpha - \alpha_0)$, p(n) is the Poisson distribution (4.113), with $\mu = \eta|\alpha_0|^2$. The variance contains only the shot noise term.

b) For a thermal field,

$$P(\alpha) = \frac{1}{\pi\bar{n}} \exp(-|\alpha|^2/\bar{n}) \ , \tag{4.122}$$

p(n) is the Bose-Einstein distribution (4.116) with $\mu = \eta\bar{n}$.

c) The coherent superposition of a Gaussian field with average photon number $\bar{n}$ and a coherent field of amplitude α_0 is described by the distribution

$$P(\alpha) = \frac{1}{\pi \bar{n}} \exp(-|\alpha - \alpha_0|^2/\bar{n}) \quad . \tag{4.123}$$

The corresponding generating function (4.107) is given by

$$Q_N(s) = (1 + sN_b)^{-1} \exp[-Ss/(1 + sN_b)] \; , \tag{4.124}$$

where $S = \eta|\alpha_0|^2$ is the average number of photocounts due to the coherent field and $N_b = \eta\bar{n}$ is the mean number of photocounts due to the Gaussian field. The counting distribution is

$$p(n) = \frac{N_b^n}{(1+N_b)^{n+1}} \exp[-S/(1 + N_b)]L_n\Big(-S/N_b(1 + N_b)\Big) \; , \tag{4.125}$$

where L_n is the Laguerre polynomial of order n. The mean and the variance associated with this distribution are [see (4.53)]

$$\langle N_b \rangle = S + N_b \quad \text{and} \quad \langle (\Delta N)^2 \rangle = S + N_b(1 + N_b) + 2SN_b \quad .$$

4.6.2 Stochastic Processes

A stochastic process is a random function X(t) of an independent variable t, say, the time. We may think that X(t) is obtained from an infinite sequence of random variables $\{X_k\}$ when we let the indices become a continuous variable. For any fixed t, X(t) is a random variable which may take the value x(t) with probability W(xt). Similarly, for any fixed set $(t_1, \ldots, t_n)$, $X(t_1) \ldots X(t_n)$ constitute a discrete family of random variables described by the joint probability density $W_n(x_1t_1 \ldots, x_nt_n)$. Full information on the stochastic process is given by the hierarchy of probability distributions $W_n(x_1t_1, \ldots, x_nt_n)$ for all n = 1,2, ... , or equivalently by the correlation functions $\langle X(t_1) \ldots X(t_n) \rangle$ for all sets $(t_1, \ldots, t_n)$, including repeated indices.

If t is restricted to a finite interval (0,T), we can alternatively describe X(t) by the joint probability distribution, $P(\{c_k\})$, of the random coefficients $\{c_k\}$ in the expansion

$$X(t) = \sum_k c_k \Phi_k(t) \; , \tag{4.126}$$

where the set of deterministic functions $\Phi_k(t)$ is complete and orthogonal in (0,T). The basis $\{\Phi_k(t)\}$ is usually chosen in such a way that the coefficients c_k are statistically independent (Karhunen-Loève expansion). This condition requires that $\Phi_k(t)$ satisfies the eigenvalue equation [4.5]

$$\int_0^T \langle X(t)X(t') \rangle \Phi_k(t')dt' = \mu_k \Phi_k(t) \quad . \tag{4.127}$$

The eigenvalues μ_k are then the variances of the coefficients c_k. The expansion

(4.126,127) is used to evaluate the effect of the sampling time T on the counting statistics (see Sect. 4.4.3).

The stochastic process $X(t)$ is said to be *stationary* if the correlation functions do not depend on the choice of the origin on the time axis. In particular $M_2(t_1,t_2) \equiv \langle X(t_1)X(t_2)\rangle$ depends on $(t_1 - t_2)$ only.

It can be shown [4.40] that $M_2(t_1,t_2)$ is completely determined by the average spectral properties of the process (Wiener-Khintchine theorem): $M_2(t_1,t_2)$ is the Fourier transform of the spectral density

$$S(\omega) = \langle | \int X(t)\, e^{i\omega t} dt|^2 \rangle \quad . \tag{4.128}$$

For example, if $X(t)$ is the fluctuating amplitude of a light beam, this theorem states the relation between the average spectral energy, as measured by a square law detector plus spectrum analyzer, and the correlation function of the field, as measured by a correlator.

There are two examples of random processes which are particularly relevant in practical applications. The first is the Gaussian (or normal) process which describes the statistical fluctuations of macroscopic systems at thermal equilibrium. For such processes all cumulants other than the first two vanish. The statistics of the process are then completely determined by the correlation function $M_2(t_1,t_2)$ for all t_1 and t_2, viz.,

$$\begin{aligned} &M_{2k+1}(t_1, \ldots, t_{2k+1}) = 0 \ , \\ &M_{2k}(t_1, \ldots, t_{2k}) = \sum_P M_2(t_1,t_{j_1})M_2(t_2,t_{j_2}) \ldots M_2(t_{2k},t_{j_{2k}}) \ , \end{aligned} \tag{4.129}$$

where P denotes the permutations of the indices $(j_1, \ldots, j_{2k})$ and $M_1(t)$ is assumed to be zero. The second example is the Markov process, such as Brownian motion or, more generally, the evolution of macroscopic variables which change slowly compared to the motion of their subsystems. The Markov process is defined by [3]

$$W_n(x_1t_1, \ldots, x_nt_n) = P(x_nt_n|x_{n-1}t_{n-1}) \ldots P(x_2t_2|x_1t_1)W(x_1t_1)$$

$$\text{with} \quad t_n \geqq t_{n-1} \geqq \ldots \geqq t_1 \quad . \tag{4.130}$$

Equation (4.130) is equivalent to [4.40,78]

$$\frac{\partial W(xt)}{\partial t} = \sum_{s=1}^{\infty} \frac{1}{s!} \left(- \frac{\partial}{\partial x}\right)^s D_s(xt)W(xt) \ , \tag{4.131}$$

where

$$D_s(xt) = \lim_{\tau \to 0} \tau^{-1} \left\langle \left(X(t + \tau) - X(t)\right)^s \right\rangle \Big|_{X(t)=x} \quad . \tag{4.132}$$

Apart from cases such as instabilities (phase transitions), it is generally sufficient to terminate the expansion (4.131) after the first two derivatives. The resulting equation is known as the Fokker-Planck equation

$$\frac{\partial}{\partial t} W(xt) = - \frac{\partial}{\partial x} [D_1(xt)W(xt)] + \frac{1}{2} \frac{\partial^2}{\partial x^2} [D_2(xt)W(xt)] \quad . \tag{4.133}$$

---

[3] $P(x_kt_k|x_{k-1}t_{k-1})$ denotes the conditional probability.

The Green's function of (4.133) is the transition probability $P(xt|x_0t_0)$, which by definition satisfies the condition

$$P(xt_0|x_0t_0) = \delta(x - x_0) \quad .$$

The coefficients $D_1(xt)$ and $D_2(xt)$ are called the drift and the diffusion coefficient, respectively, and do not depend on t if the process is stationary. They are often calculated from the Langevin equations of motion for the physical system under consideration (see Sect.4.7). An important example (e.g., Brownian motion of non-interacting particles) of (4.133) is when $D_1(xt) = -\Gamma x$ and $D_2 \equiv Q$; the solution is then a Gaussian process [4.40,78].

4.6.3 The Statistical Description of the Radiation Field

A complete characterization of the e.m. field[4] is provided by the set of correlation functions $G^{(n)}$ for any value of n

$$G^{(n)}(r_1t_1, \ldots, r_{2n}t_{2n}) = \mathrm{Tr}\{\rho E^{(-)}(r_1t_1) \ldots E^{(-)}(r_nt_n) \\ E^{(+)}(r_{n+1}t_{n+1}) \ldots E^{(+)}(r_{2n}t_{2n})\} \tag{4.134}$$

Here ρ is the density operator of the field and $E^{(+)}$ $(E^{(-)})$ is the positive (negative) frequency field operator, i.e.,

$$E(rt) = i \sum_k (\hbar\omega_k/2V)^{\frac{1}{2}}(a_k\, e^{ikr-i\omega_k t} - a_k^+\, e^{-ikr+i\omega_k t}) \;, \tag{4.135}$$

where $a_k(a_k^+)$ is the annihilation (creation) operator for the kth cavity mode and V is the volume of the cavity. The operators a_k^+, a_k in the Heisenberg picture are time independent if the field is free, while they do depend on time if the field interacts with charges and currents inside V.

The correlation functions (4.124) can be studied using the diagonal, coherent state representation of ρ [4.16]. Quantum averages become then formally identical to classical averages, provided normal and time ordering of the field operators is taken into account (time ordering is required in the interacting case only [4.40, 41,79]). In the case of free fields, (4.134) is usually expressed in the form

$$G^{(n)}(x_1, \ldots, x_{2n}) = \int E^*(x_1\{\alpha_k\}) \ldots E(x_{2n}\{\alpha_k\})P(\{\alpha_k\}) \prod d^2\alpha_k \;, \tag{4.136}$$

where $x \equiv (rt)$, E^*,E are the expectation values of $E^{(-)}$, $E^{(+)}$ in coherent states and $P(\{\alpha_k\})$ is the corresponding quasi-probability distribution. The time dependence of $G^{(n)}$ given by (4.136) is all contained in the exponential factors $\exp(\pm i\omega_k t)$. In the case of interacting fields, the general rules to evaluate $G^{(n)}$ are given by HAKEN [4.79] and LAX [4.40]. We also refer to [4.1-3,11,16,19,80-86] for a discussion of the coherence properties of optical fields.

---

[4] For simplicity we dispense with the vector character of the field.

4.7 The Statistical Description of Open Systems

In this section we give a short account of the techniques used in Sects. 4.3,5. In Sect.4.7.1 we describe the elimination of reservoir variables in the Schrödinger picture, which leads to an equation of motion for the reduced density matrix. In Sect.4.7.2 we derive the equations of motion of the detector model considered in Sect.4.5.

4.7.1 Equation of Motion of the Reduced Density Matrix

Two systems R and S defined by Hamiltonians H_R, H_S are coupled by an interaction Hamiltonian V and form a system R + S defined by

$$H = H_0 + gV \ , \quad H_0 = H_R + H_S \ , \tag{4.137}$$

where g denotes the coupling constant. Let the density matrix $\rho(t)$ describe the time evolution of the total system

$$\rho(t) = \mathcal{U}_t\rho(0) \equiv \exp(\mathcal{L}t)\rho(0) \ , \quad \mathcal{L}\rho(t) = -i[H,\rho(t)] \ , \tag{4.138}$$

where $\mathcal{U}_t$ is the time evolution operator (or propagator) generated by the Liouville operator $\mathcal{L}$ and $\rho(0)$ is the initial state of the system. The statistical state of the system S "dressed" by the interaction with R is described by the reduced density matrix

$$\rho_S(t) = \mathrm{Tr}_R\rho(t) \equiv \mathrm{Tr}_R\mathcal{U}_t\rho(0) \ , \tag{4.139}$$

where Tr_R denotes the partial trace on the subspace corresponding to H_R.

We establish the equation of motion for $\rho_S(t)$ on the following assumptions [4.87].

I) The interaction is switched at t = 0 so that at this time S and R are independent, viz.,

$$\rho(0) = \rho_R(0) \otimes \rho_S(0) \quad .$$

II) The initial state of R is stationary with respect to free bath motion, i.e.,

$$[\rho_R(0) \ , \ H_R] = 0 \quad .$$

III) The expectation value of the coupling term V in the state $\rho_R(0)$ is zero,

$$\mathrm{Tr}_R V\rho_R(0) = 0 \quad .$$

The above assumptions are strictly valid for a reservoir at thermal equilibrium only. Results valid for more general types of reservoirs are given by HAAKE [4.88]. Let P be the projector operator($P^2 = P$) defined by [4.89,90]

$$P\rho(t) = \rho_R(0) \otimes \mathrm{Tr}_R\rho(t) \quad . \tag{4.140}$$

The reduced density matrix (4.139) can be expressed in the form

$$\rho_S(t) = \mathrm{Tr}_R P\rho(t) \equiv \mathrm{Tr}_R P U_t P\rho(0) \ , \tag{4.141}$$

which shows that the time dependence of $\rho_S(t)$ is completely determined by the restriction PU_tP. This operator can be evaluated from the Dyson equation for U_t

$$U_t = U_t^g + g \int_0^t U_{t-s}^g (PV + VP) U_s ds \ , \tag{4.142}$$

where

$$U_t^g = \exp(L_g t)$$

$$L_g = L_0 + g(I - P)V(I - P) \ , \tag{4.143}$$

L_0 and V being the Liouville operators corresponding to H_0 and V, respectively, and I being the identity. Equation (4.141) then takes the form

$$\rho_S(t) = \mathrm{Tr}_R U_t^0 \rho_R(0)$$

$$+ g^2 \mathrm{Tr}_R \int_0^t ds \int_0^s dv U_{t-s}^0 K_g(s - v) \rho_R(0) \otimes \rho_S(v) \ , \tag{4.144}$$

where U_t^0 is the free propagator

$$U_t^0 = \exp(L_0 t)$$

and

$$K_g(t) = PVU_t^g VP \quad . \tag{4.145}$$

Differentiating (4.144) with respect to t, we finally obtain

$$\frac{d}{dt} \rho_S(t) = -i[H_S, \rho_S(t)] + g^2 \mathrm{Tr}_R \int_0^t ds K_g(t - s) \rho_R(0) \otimes \rho_S(0) \quad . \tag{4.146}$$

This equation of motion cannot be solved as it stands. In practical applications a perturbation expansion in g is used for $K_g(t)$. In addition it is often sufficient to treat the R-S interaction to order g^2 (Born approximation), so that $K_g(t)$ reduces to

$$K_g(t) \sim K_0(t) = \rho_R(0) \otimes \mathrm{Tr}_R\{V, U_t^0[V, \rho_R(0) \otimes \mathrm{Tr}_R]\} \quad . \tag{4.147}$$

Assuming $V = V_R \otimes V_S$, each term in (4.147) is a reservoir correlation function multiplying an operator of the system S. Therefore, if the characteristic correlation times of the reservoirs are short (with respect to the free-evolution time scale of S) the dynamics (4.146) becomes Markovian

$$\frac{d}{dt}\rho_S(t) = -i[H_S, \rho_S(t)] + g^2 \mathrm{Tr}_R \int_0^\infty ds K_0(s)\rho_R(0) \otimes \rho_S(t) \quad . \tag{4.148}$$

Both the approximations (4.147) and (4.148) become exact results in the case of a "large" (thermodynamic limit) reservoir at thermal equilibrium in the Van Hove (or weak coupling) limit $g \to 0$, $t \to \infty$, $g^2 t$ = const [4.91,92]. These conditions hold for instance in the case of a material system coupled to blackbody radiation or in the case of a radiation field coupled to a large assembly of atoms at thermal equilibrium.

Equation (4.148) can be used to calculate the expectation value of any observable O_S of the system S from the equation of motion

$$\frac{d}{dt}\langle O_S\rangle_t = -i\mathrm{Tr}_R[H_S, \rho_S(t)]O_S + g^2 \mathrm{Tr}_S O_S \mathrm{Tr}_R \int_0^\infty ds K_0(s)\rho_R(0) \otimes \rho_S(t) \quad . \tag{4.149}$$

We can also solve (4.148) directly for the density matrix in a given representation. When S is the radiation field and the P representation is used for ρ_S, (4.148) becomes usually a Fokker-Planck equation such as (4.143). In any case the effect of the second term in (4.148) is to introduce damping as well as fluctuations in the motion of the system S.

The equations of motion (4.31,37,39) in Sect. 4.3 can be derived by a straightforward application of the above techniques. Other examples are discussed in a number of articles and books, e.g., [4.78,79,88,90,93-96].

4.7.2 Langevin Equations

By elimination of the reservoir variables the Heisenberg equations of motion of the coupled system R + S reduce to Langevin equations of motion for the variables of the system [4.40,78,79,84,94-96,99,100]. We discuss the elimination procedure in a specific example, i.e., the detector model in Sect.4.5.

Consider a two-level atom in which the transition between the levels is coupled to a heatbath. The free atom is described by the Hamiltonian

$$H_A = \varepsilon s^+ s^- \quad . \tag{4.150}$$

As a model for the reservoir we take a system of free fermions

$$H_R = \sum_k \omega_k \Gamma_k^+ \Gamma_k \quad . \tag{4.151}$$

The reservoir is in a thermal equilibrium state, i.e.,

$$\langle \Gamma_k^+ \Gamma_{k'} \rangle_R = \delta_{kk'} n_k \quad , \tag{4.152}$$

where $\langle \dots \rangle_R$ denotes the average over the reservoir state and $n_k = [\exp(\beta\omega_k) + 1]^{-1}$, β being the inverse temperature. The operators $\Gamma_k^\pm$ anticommute with $s^\pm$. We take the atom-bath interaction of the form

$$V = \sum_k (g_k s^+ \Gamma_k + g_k^* \Gamma_k^+ s^-) \quad . \tag{4.153}$$

In the interaction representation, the Heisenberg equations of motion for the coupled atom-reservoir system read

$$\dot{\Gamma}_k = -ig_k^* s^- \, e^{+i(\omega_k-\varepsilon)t} \tag{4.154}$$

$$\dot{s}^- = -i \sum_k g_k \Gamma_k \, e^{-i(\omega_k-\varepsilon)t} \tag{4.155}$$

$$\dot{s}_3 = -i \sum_k g_k s^+ \Gamma_k \, e^{-i(\omega_k-\varepsilon)t} + \text{h.c.} \tag{4.156}$$

We eliminate the reservoir operators Γ_k in (4.155) by substituting the formal integral of (4.154)

$$\dot{s}^- = -i \sum_k g_k \Gamma_k(0) \, e^{-i(\omega_k-\varepsilon)t} - \sum_k |g_k|^2 \int_0^t e^{-i(\omega_k-\varepsilon)(t-t')} s^-(t')dt' \quad . \tag{4.157}$$

We next assume the Markov property [4.79]

$$\sum_k |g_k|^2 \, e^{-i(\omega_k-\varepsilon)(t-t')} = 2\gamma\delta(t-t') \ , \tag{4.158}$$

where $2\gamma = \pi\mathcal{D}(\varepsilon)|g(\varepsilon)|^2$, with $\mathcal{D}(\omega)$ denoting the reservoir density of states. Remark that the imaginary part has been omitted in (4.158).

From (4.157,158) we obtain

$$\dot{s}^-(t) = -\gamma s^-(t) - i \sum_k g_k \Gamma_k(0) \, e^{-i(\omega_k-\varepsilon)t} \equiv -\gamma s^-(t) + f^-(t) \quad . \tag{4.159}$$

Here the second term, which depends on the bath variables only, can be interpreted as a fluctuating force. Using (4.152,158) we obtain the correlation functions between the stochastic forces $f^\pm(t)$(f^+ denotes the force acting on s^+), viz.,

$$\langle f^\pm(t) f^\pm(t') \rangle_R = 0 \ ,$$

$$\langle f^+(t) f^-(t') \rangle_R = \gamma(1 + \sigma_0)\delta(t - t') \ ,$$

$$\langle f^-(t) f^+(t') \rangle_R = \gamma(1 - \sigma_0)\delta(t - t') \ ,$$

where $\sigma_0 \equiv -\tanh(\beta\varepsilon/2)$. The Langevin equation of motion for s_3 is calculated by a similar procedure, leading to

$$\dot{s}_3 = -2\gamma(s_3 - 1/2 \ \sigma_0) + f_3(t) \quad . \tag{4.160}$$

The stochastic force $f_3(t)$ is given by

$$f_3(t) = -i \sum_k g_k s^+(t) \Gamma_k(0) \, e^{-i(\omega_k-\varepsilon)t}$$
$$+ i \sum_k g_k^* \Gamma_k^+(0) s^-(t) \, e^{i(\omega_k-\varepsilon)t} - \gamma(1 + \sigma_0) \ ,$$

where the constant term $\gamma(1+\sigma_0)$ has been subtracted in order to have $\langle f_3(t)\rangle_R = 0$. The calculation of the correlation function $\langle f_3(t)f_3(t')\rangle_R$ is quite involved, since $f_3(t)$ depends on the atomic variables. A much more convenient procedure is then to evaluate the correlation functions by the fluctuation-dissipation theorem (or generalized Einstein relations), which relates the diffusion coefficients to the average motion of system operators including reservoir effects. The general rules to carry out this procedure are given by LAX [4.40] and HAKEN [4.79]. The connection between Langevin equations and Fokker-Planck equations is well known in the literature. A good discussion can be found in [4.40,76,79].

Acknowledgments

I am indebted to H.P. Baltes, A. Quattropani, and P. Schwendimann for constant interest and collaboration. I thank R. Loudon, R. Fivaz and J. Geist for careful reading and criticism of the manuscript and J.-F. Moser of Landis & Gyr Zug AG for his interest and encouragement. Particular thanks go to Miss Ch. Roethlisberger who typed the several drafts. I appreciate support from Landis & Gyr Zug AG.

References

4.1 J.R. Klauder, E.C.G. Sudarshan: *Fundamentals of Quantum Optics* (Benjamin, New York 1968)
4.2 J. Peřina: *Coherence of Light* (Van Nostrand Reinhold, London 1972)
4.3 R. Loudon: *Quantum Theory of Light* (Oxford University Press, Oxford 1973)
4.4 B. Crossignani, P. DiPorto, M. Bertolotti: *Statistical Properties of Scattered Light* (Academic Press, New York 1975)
4.5 B. Saleh: *Photoelectron Statistics. With Application to Spectroscopy and Optical Communications*, Springer Series in Optical Sciences, Vol. 6 (Springer, Berlin, Heidelberg, New York 1978)
4.6 R. Barakat, J. Blake: *Theory of Photoelectron Counting Statistics: an Essay*, Physics Reports *60*, 225-340 (1980)
4.7 E.R. Pike: "Photon Statistics", in *Quantum Optics*, ed. by S.M. Kay, A. Maitland (Academic Press, New York 1970) pp. 127-176
4.8 F.T. Arecchi, V. Degiorgio: "Statistical Properties of Optical Fields", in *Laser Handbook*, Vol.1, ed. by F.T. Arecchi, E.O. Schulz-Dubois (North-Holland, Amsterdam 1972) pp.191-264
4.9 M. Bertolotti: "Photon Statistics", in *Photon Correlation and Light Beating Spectrosocpy*, ed. by H.Z. Cummins, E.R. Pike (Plenum Press, New York 1974) pp. 41-74
4.10 E. Jakeman: "Photon Correlation", in *Photon Correlation and Light Beating Spectrosocpy*, ed. by H.Z. Cummins, E.R. Pike (Plenum Press, New York 1974) pp. 75-149
4.11 C.L. Mehta: "Theory of Photoelectron Counting", in *Progress in Optics*, Vol. VIII, ed. by E. Wolf (North-Holland, Amsterdam 1970) pp. 373-440
4.12 C.W. Helstrom: "Quantum Detection Theory", in *Progress in Optics*, Vol. X, ed. by E. Wolf (North-Holland, Amsterdam 1972) pp. 289-369
4.13 H. Melchior: "Demodulation and Photodetection Techniques", in *Laser Handbook*, Vol. 1, ed. by F.T. Arecchi, E.O. Schulz-Dubois (North-Holland, Amsterdam 1972) pp. 725-835
4.14 R.J. Keyes (ed.).: *Optical and Infrared Detectors*, 2nd. ed., Topics in Applied Physics, Vol.19 (Springer, Berlin, Heidelberg, New York 1980)
4.15 R.H. Kingston: *Detection of Optical and Infrared Radiation*, Springer Series in Optical Sciences, Vol. 10 (Springer, Berlin, Heidelberg, New York 1979)
4.16 R.J. Glauber: "Optical Coherence and Photon Statistics", in *Quantum Optics and Electronics*, ed. by C. de Witt, A. Blandin, C. Cohen-Tannoudji (Gordon and Breach, New York 1965) pp. 63-185
4.17 P.L. Kelley, W.H. Kleiner: Phys. Rev. *136*, A316-334 (1964)
4.18 B.R. Mollow: Phys. Rev. *168*, 1896-1919 (1968)
4.19 R.J. Glauber: "Coherence and Quantum Detection", in *Quantum Optics*, ed. by R.J. Glauber (Academic Press, New York 1969) pp. 15-56

4.20 M.O. Scully, W.E. Lamb: Phys. Rev. *179*, 368-374 (1969)
4.21 A. Selloni, P. Schwendimann, A. Quattropani, H.P. Baltes: Opt. Commun. *22*, 131-134 (1977)
4.22 A. Selloni, P. Schwendimann, A. Quattropani, H.P. Baltes: J. Phys. A *11*, 1427-1438 (1978)
4.23 A. Selloni, P. Schwendimann, A. Quattropani, H.P. Baltes: Phys. Rev. A *18*, 2234-2240 (1978)
4.24 I.R. Senitzky: "Semiclassical Radiation Theory within a Quantum Mechanical Framework", in *Progress in Optics*, Vol XVI, ed. by E. Wolf (North-Holland, Amsterdam 1978) pp. 413-448
4.25 P.W. Kruse: "The Photon Detection Process", in *Optical and Infrared Detectors*, 2nd ed., ed. by R.J. Keyes, Topics in Applied Physics, Vol. 19 (Springer, Berlin, Heidelberg, New York 1980) pp. 5-69
4.26 V. Degiorgio: "Photon Correlation Techniques", in *Photon Correlation Spectroscopy and Velocimetry*, ed. by H.Z. Cummins, E.R. Pike (Plenum Press, New York 1977) pp. 142-163
4.27 C.J. Oliver: "Correlation Techniques", in *Photon Correlation and Light Beating Spectroscopy*, ed. by H.Z. Cummins, E.R. Pike (Plenum Press, New York 1974) pp. 151-223
4.28 R. Foord, R. Jones, C.J. Oliver, E.R. Pike: Appl. Opt. *9*, 1975-1989 (1969)
4.29 E. Jakeman, E.R. Pike: J. Phys. A *1*, 690-693 (1968)
4.30 V. Degiorgio, J.B. Lastovka: Phys. Rev. A *4*, 2033-2050 (1971)
4.31 A.J. Hughes, E. Jakeman, C.J. Oliver, E.R. Pike: J. Phys. A *6*, 1327-1336 (1973)
4.32 G. Bédard: Proc. Phys. Soc. *90*, 131-141 (1967)
4.33 B.I. Cantor, M.C. Teich: J. Opt. Soc. Am. *65*, 786-791 (1975)
4.34 A. Consortini, L. Ronchi: Appl. Opt. *17*, 1286-1289 (1978)
4.35 A. Consortini, L. Ronchi: Opt. Acta *26*, 35-42 (1979)
4.36 L. Mandel: "Fluctuations of Light Beams", in *Progress in Optics*, Vol. II, ed. by E. Wolf (North-Holland, Amsterdam 1963) pp. 181-247
4.37 F.T. Arecchi, V. Degiorgio: Phys. Rev. A *3*, 1108-1123 (1971)
4.38 D. Meltzer, L. Mandel: Phys. Rev. Lett. *25*, 1151-1154 (1970)
4.39 L. Mandel, E.C.G. Sudarshan, E. Wolf: Proc. Phys. Soc. *84*, 435-444 (1964)
4.40 M. Lax: "Fluctuation and Coherence Phenomena", in *Statistical Physics, Phase Transitions and Superfluidity*, ed. by M. Chrêtien, E.P. Gross, S. Deser (Gordon and Breach, New York 1968) pp. 271-478
4.41 M. Lax, M. Zwanziger: Phys. Rev. A *7*, 750-771 (1973)
4.42 L. Mandel, E. Wolf: J. Phys. A *1*, 625-627 (1968)
4.43 E. Jakeman, E.R. Pike: J. Phys. A *1*, 627-628 (1968)
4.44 R.H. Lehmberg: Phys. Rev. *167*, 1152-1167 (1968)
4.45 F. Rocca: Phys. Rev. D *8*, 4403-4410 (1973)
4.46 A. Arneodo, F. Rocca: Z. Phys. *269*, 205-213 (1974)
4.47 L. Mandel, D. Meltzer: Phys. Rev. *188*, 198-212 (1969)
4.48 A.K. Jaiswal, G.S Agarwal: J. Opt. Soc. Am. *59*, 1446-1452 (1969)
4.49 M.C. Teich, P. Diament: J. Appl. Phys. *40*, 625-633 (1969)
4.50 M. Rousseau: J. Phys. A *8*, 1265-1276 (1975)
4.51 M. Rousseau: J. Phys. A *10*, 1043-1047 (1977)
4.52 W. Feller: *An Introduction to Probability Theory and Its Applications*, Vol. I, 3rd ed. (Wiley, New York 1970)
4.53 E. Wolf, C.L. Mehta: Phys. Rev. Lett. *13*, 705-707 (1964)
4.54 G. Bêdard: J. Opt. Soc. Am. 1201-1206 (1967)
4.55 H.A. Ferwerda: "The Phase Reconstruction Problem for Wave Amplitudes and Coherence Functions", in *Inverse Source Problems*, ed. by H.P. Baltes, Topics in Current Physics, Vol.9 (Springer, Berlin, Heidelberg, New York 1978) pp. 13-39
4.56 F.T. Arecchi, A. Berné, A. Sona: Phys. Rev. Lett. *17*, 260-263 (1966)
4.57 R.J. Glauber, O. Titulaer: Phys. Rev. *140*, B676-682 (1965)
4.58 D. Middleton: *An Introduction to Statistical Communication Theory* (McGraw-Hill, New York 1960)
4.59 S. Putterman: Phys. Rev. Lett. *39*, 585-587 (1977)
4.60 R.F. Voss: Phys. Rev. Lett. *40*, 913-916 (1978)
4.61 J.A. Abate, H.J. Kimble, L. Mandel: Phys. Rev. A *14*, 788-795 (1976)

4.62 T. Aoki: Phys. Rev. A *16*, 2432-2436 (1977)
4.63 J.R. Prescott: Nucl. Instrum. Meth. *39*, 1731-1739 (1966)
4.64 G. Lachs: IEEE Trans. QE-*10*, 590-596 (1974)
4.65 M. Corti, A. Vendramini: Rev. Sci. Instrum. *42*, 1300-1306 (1971)
4.66 I. De Lotto, P.F. Manfredi, P. Principi: Energia Nucleare *11*, 557-564 (1964)
4.67 F.A. Johnson, R. Jones, T.P. McLean, E.R. Pike: Phys. Rev. Lett. *16*, 589-592 (1966)
4.68 R.F. Chang, V. Koreman, C.O. Alley, R.W. Detenbeck: Phys. Rev. *178*, 612-621 (1969)
4.69 G. Vannucci, M.C. Teich: Opt. Commun. *25*, 267-272 (1978)
4.70 S.K. Srinivasan: J. Phys. A *11*, 2333-2340 (1978)
4.71 E. Jakeman, E.R. Pike: J. Phys. A *1*, 128-138 (1968)
4.72 E. Jakeman, C.J. Oliver, E.R. Pike: J. Phys. A *1*, 406-408 (1968)
4.73 F.T. Arecchi, A. Berné, A. Sona, P. Burlamacchi: IEEE Trans. QE-*2*, 341-350 (1966)
4.74 H.C. Kelly: J. Phys. A *5*, 104-111 (1972)
4.75 H. Sonoda, A. Kikkawa, N. Suzuki: Appl. Opt. *17*, 1006-1011 (1978)
4.76 R.L. Stratonovich: *Topics in the Theory of Random Noise*, Vol. I and II (Gordon & Breach, New York, 1963, 1967)
4.77 M.G. Kendall: *The Advanced Theory of Statistics*, Vol. I (Hafner Publishing Co., New York 1963)
4.78 H. Haken: "Cooperative Phenomena in Systems Far from Thermal Equilibrium and in Nonphysical Systems", Rev. Mod. Phys. *47*, 67-121 (1975);
H. Haken: *Synergetics, An Introduction*, 2nd ed. (Springer, Berlin, Heidelberg, New York 1978)
4.79 H. Haken: "Laser Theory", in *Encyclopedia of Physics*, Vol. XXV/2c, ed. by L. Genzel (Springer, Berlin, Heidelberg, New York 1970)
4.80 L. Mandel, E. Wolf: Rev. Mod. Phys. *37*, 231-287 (1965)
4.81 R.J. Glauber: Phys. Rev. *130*, 2529-2539 (1963)
4.82 R.J. Glauber: Phys. Rev. *131*, 2766-2788 (1963)
4.83 K.E. Cahill, R.J. Glauber: Phys. Rev. *177*, 1857-1881, 1882-1902 (1969)
4.84 W.H. Louisell: *Radiation and Noise in Quantum Electronics* (McGraw-Hill, New York 1964)
4.85 G.J. Troup, R.G. Turner: Rep. Progr. Phys. *37*, 772-816 (1974)
4.86 H.P. Baltes: Appl. Phys. *12*, 221-244 (1977)
4.87 C. Favre, Ph. A. Martin: Helv. Phys. Acta *41*, 333-361 (1968)
4.88 F. Haake: "Statistical Treatment of Open Systems by Generalized Master Equation", in *Quantum Statistics*, Springer Tracts in Modern Physics, Vol. 66 (Springer, Berlin, Heidelberg, New York 1973) pp. 98-168
4.89 R.W. Zwanzig: "Statistical Mechanics of Irreversibility", in *Lectures in Theoretical Physics*, Vol. III, ed. by W.E. Brittin, B.W. Downs, J. Downs (Interscience Publishers, New York 1961) pp. 106-141
4.90 P.N. Argyres, P.L. Kelley: Phys. Rev. *134*, A98-111 (1964)
4.91 L. Van Hove: Physica *21*, 517-540 (1955); *23*, 441-480 (1957)
4.92 E.B. Davies: Commun. Math. Phys. *39*, 91-110 (1974)
4.93 R.K. Wangsness, F. Bloch: Phys. Rev. *89*, 728-739 (1956)
4.94 W.H. Louisell: "Quantum Theory of Noise", in *Quantum Optics*, ed. by R.J. Glauber (Academic Press, New York 1969) pp. 680-742
4.95 W.H. Louisell: *Quantum Statistical Properties of Radiation* (Wiley, New York 1973)
4.96 M. Sargent III, M.O. Scully, W.E. Lamb, Jr.: *Laser Physics* (Addison-Wesley, Reading, Ma. 1974)
4.97 G. Scharf: Helv. Phys. Acta *43*, 806-828 (1970)
4.98 P. Schwendimann: Z. Phys. *251*, 244-253 (1972)
4.99 H. Mori: Progr. Theor. Phys. *33*, 423-455 (1965)
4.100 R. Kubo: Rep. Progr. Phys. *29*, 255-284 (1966)
4.101 R.H. Dicke: Phys. Rev. *93*, 99-110 (1954)
4.102 D. Wolf (ed.): *Noise in Physical Systems*, Springer Series in Electrophysics, Vol. 2 (Springer, Berlin, Heidelberg, New York 1978)

5. The Stability of Inverse Problems

M. Bertero, C. De Mol, and G. A. Viano

With 7 Figures

Many inverse problems arising in optics and other fields like geophysics, medical diagnostics and remote sensing, present numerical instability: the noise affecting the data may produce arbitrarily large errors in the solutions. In other words, these problems are *ill-posed* in the sense of Hadamard.

The basic point, in the study of ill-posed problems, is that the development of adequate computational methods, leading to stable results, requires *prior knowledge* of properties of the admissible solutions: global bounds, smoothness conditions, positivity constraints, statistical properties, etc. The problem is first to incorporate the supplementary constraints in the computational algorithm, and secondly to estimate the accuracy of the solutions for a given prior knowledge and data accuracy. General methods are available only for linear inverse problems.

This chapter begins with an outline of the main features of ill-posed problems, of their connection with inverse problems and of the basic ideas enabling one to solve them. Next we discuss *regularization theory* where the supplementary constraints are prescribed bounds on the class of admissible solutions. Then we analyze the application to ill-posed problems of the method of linear mean square estimation (*optimum filtering*), when prior knowledge of statistical properties of the solutions is available. Finally, we review the applications of the previous methods to some linear inverse problems in optics and scattering theory.

5.1 Ill-Posedness in Inverse Problems

The concept of *ill-posedness* was introduced by HADAMARD [5.1] in the field of partial differential equations. For years, ill-posed problems have been considered as mere mathematical anomalies. Indeed, it was believed that physical situations were leading only to well-posed problems like, for instance, the Dirichlet problem for elliptic equations of potential theory, or the Cauchy problem for hyperbolic equations describing wave motion. However, it appeared later that this attitude was erroneous and that many ill-posed problems, generally inverse problems, were arising from practical situations. Nowadays there is no doubt that a systematic study of these problems is of great relevance in many fields of applied physics.

5.1.1 Well-Posed and Ill-Posed Problems

It is rather difficult to give a precise and exhaustive definition of an ill-posed problem. Indeed this term covers a lot of various problems presenting many common features but also differences so important that a global and unified theory is not yet available. The best characterization is perhaps a negative one: ill-posed problems do not fulfill all the required conditions for well-posedness [5.1], i.e., *existence, uniqueness* and *continuity* of the solution on the data (requirement of stability). As clearly stated by COURANT and HILBERT [Ref.5.2, p.227], *"the third requirement, particularly incisive, is necessary if the mathematical formulation is to describe observable natural phenomena. Data in nature cannot possibly be conceived as rigidly fixed; the mere process of measuring them involves small errors. Therefore a mathematical problem cannot be considered as realistically corresponding to physical phenomena unless a variation of the given data in a sufficiently small range leads to an arbitrary small change in the solution. This requirement of "stability" is not only essential for meaningful problems in mathematical physics, but also for approximation methods".*

An example of a well-posed problem is to find a solution u of the Laplace equation

$$\frac{\partial^2 u}{\partial x^2} + \frac{\partial^2 u}{\partial y^2} = 0 \tag{5.1}$$

in some domain D of the plane, with the condition $u = g$ on the boundary of D (Dirichlet problem). It is well known that there exists a unique solution which depends continuously on the data. Indeed, the maximum principle [Ref.5.2, p.255] guarantees that when g is slightly perturbed into g', the corresponding solution u' is in a neighborhood of u. More precisely, $|g - g'| \leqq \varepsilon$ implies $|u - u'| \leqq \varepsilon$.

Any problem failing to satisfy one or more of the three requirements quoted above might be called an ill-posed (or improperly posed) problem. Nevertheless, this term is usually reserved to those problems for which the second requirement (uniqueness) is fulfilled, but not the first and the third ones. Indeed, as we shall see below, existence and continuity are in general closely related.

The first who pointed out the concepts of well- and ill-posedness was J. Hadamard. Let us recall his famous example showing the lack of continuity on the data in the Cauchy problem for elliptic partial differential equations. Consider (5.1) with the boundary conditions

$$u(x, 0) = 0 \quad , \quad \frac{\partial u}{\partial y}(x, 0) = \frac{1}{n}\sin(nx) \quad . \tag{5.2}$$

It is straightforward to verify that this problem has the following solution:

$$u(x, y) = \frac{1}{n^2}\sin(nx)\sinh(ny) \quad . \tag{5.3}$$

The term $n^{-1}\sin(nx)$ departs from zero on the x axis in an imperceptible way for n sufficiently large. However, because of the hyperbolic sine, the solution (5.3) becomes enormous at any given distance from the x axis, provided that n is sufficiently large.

Related to the Cauchy problem for the Laplace equation is the analytic continuation of functions of a complex variable. In fact, let the values of the harmonic function u, i.e., the solution of (5.1), and its normal derivative $\partial u/\partial n$ be known on some curve Γ. We denote by $f(z)$, $z = x + iy$, the analytic function $f = u + iv$, where v is the function conjugate to u. Then, on the curve Γ, v is related to u as follows

$$v(z) = \int_{z_0}^{z} \frac{\partial u}{\partial n}(z')ds + \text{constant} , \qquad (5.4)$$

where z_0 is one of the endpoints of Γ. Hence, if u and $\partial u/\partial n$ are known on Γ, one may consider that the values of the analytic function $f(z)$ on Γ are known. This shows that the solution of the Cauchy problem for the Laplace equation gives the analytic continuation of f outside Γ, which is therefore also an ill-posed problem.

Moreover, it is worth noting that the determination of an analytic function from its values on a curve Γ, inside the domain of regularity, is a problem which can be reduced to the solution of a Fredholm integral equation of the first kind, by means of the well-known Cauchy formula. Therefore, it is quite natural to guess that also integral equations of the first kind give rise to ill-posed problems. This is indeed true, as we shall show in Sect.5.1.2.

To be convinced of the practical relevance of ill-posed problems, it is sufficient to have a glance at the enormous amount of literature devoted to this field. Many references may be found for instance in the books by LAVRENTIEV [5.3], TIKHONOV and ARSENINE [5.4] and PAYNE [5.5].

5.1.2 Ill-Posedness and Numerical Instability

Let us consider the following Fredholm integral equation of the first kind:

$$\int_a^b K(x, y)\bar{f}(y)dy = \bar{g}(x) , \quad c \leq x \leq d , \qquad (5.5)$$

where the kernel $K(x, y)$ is supposed to be continuous. Assuming that there exists a unique solution $\bar{f}$ corresponding to $\bar{g}$, we might add to that solution a function $f^{(n)}(x) = C\sin(nx)$ where C is an arbitrary constant. From the Riemann-Lebesgue theorem we know that

$$\lim_{n\to+\infty} \int_a^b K(x, y)\sin(ny)dy = 0 . \qquad (5.6)$$

Hence, taking the constant C and the integer n sufficiently large, we see that widely different functions $\bar{f}$ give approximately the same $\bar{g}$. As in the case of the Cauchy problem for the Laplace equation, small modifications of $\bar{g}$ can alter radically the solution of (5.5).

Without being conscious of the ill-posedness of this problem, one could try to solve numerically (5.5) by discretizing it. By means of some N-point quadrature formula, the integral in (5.5) may be approximated by a finite sum. Then, supposing that $\bar{g}$ is given in M points, the integral equation becomes a linear algebraic system

$$[K]\underline{f} = \underline{g} \quad , \tag{5.7}$$

where [K] is a $M \times N$ matrix of components $K_{mn} = K(x_m, y_n)w_n$ (the w_n are the weight factors depending upon the quadrature formula used) while $\underline{f} = \{\bar{f}(y_n)\}$ and $\underline{g} = \{\bar{g}(x_m)\}$ are vectors in euclidean spaces of dimension N and M, respectively. At this point, let us introduce the usual euclidean scalar product between two M-dimensional vectors

$$(\underline{g}, \underline{h})_M = \sum_{m=1}^{M} g_m h_m^* \tag{5.8}$$

and the corresponding euclidean norm $\|\underline{g}\|_M^2 = (\underline{g}, \underline{g})_M$. Now, when $\underline{g}$ is affected by errors, one could always add to a given solution $\underline{f}$ a spurious vector $\underline{u}$ such that

$$\|[K]\underline{u}\|_M^2 = ([K]\underline{u}, [K]\underline{u})_M \leqq \varepsilon^2 \quad , \tag{5.9}$$

where ε is an estimate of data accuracy. Let us now investigate the shape of the set of those $\underline{u}$ satisfying (5.9). To this purpose let us put the quadratic form (5.9) in a somewhat different form

$$([K]^*[K]\underline{u}, \underline{u})_N \leqq \varepsilon^2 \quad , \tag{5.10}$$

where $[K]^*$ is a $N \times M$ matrix denoting the adjoint (or hermitian conjugate) matrix of [K]. Even if [K] is not a square matrix, $[K]^* [K]$ is a $N \times N$ symmetric, nonnegative matrix, so that it can be diagonalized. Let us denote by λ_n^2 the eigenvalues of $[K]^* [K]$ (λ_n is also called a singular value of [K]) and assume that they are all strictly positive. Of course this can happen only if $N \leqq M$. Then inequality (5.10) defines the interior of a N-dimensional nondegenerate ellipsoid with center at the origin and axes directed along the eigenvectors of $[K]^* [K]$. The length of each axis is given by $a_n = \varepsilon/\lambda_n$, $n = 1, \ldots, N$, and when the eigenvalues λ_n^2 are ordered in decreasing magnitude, the length of the greatest axis is ε/λ_N, while the length of the shortest one is ε/λ_1. The ratio between the two lengths, $\alpha = \lambda_1/\lambda_N$ is the so-called *condition number* of the matrix [K]. When α is much greater than one, the ellipsoid (5.10) contains, along certain principal directions, vectors whose euclidean norm is very large. A small change in the data vector $\underline{g}$ may produce a large error in the solution (or pseudo-solution) of (5.7). The algebraic system (5.7) is then said to be *ill-conditioned*.

In general this actually arises when discretizing Fredholm equations of the first kind. Indeed, let us consider for simplicity an integral operator whose kernel $K(x, y)$ is symmetric, and let us assume that it does not have the eigenvalue zero [of course, we also assume $a = c$ and $b = d$ in (5.5)]. Then, as it is well known, such an operator admits an infinite sequence of real eigenvalues (with finite multiplicity) accumulating to zero [Ref.5.6, Chap.2]. Hence it is easy to understand that the finer the discretization of (5.5) is (i.e., the larger N and M), the worse conditioned the resulting system (5.7) is.

5.1.3 General Formulation of Linear Inverse Problems

In order to make precise the concepts illustrated in the previous sections concerning instability, we must specify the sets to which the data and the solutions belong. Moreover, we must define what is meant by "closeness" in each set. This can be done by introducing a norm and defining a distance between two functions of the set as the norm of their difference. Particularly important in many applications is a norm induced by a scalar product (or inner product) like the norm of a vector in Euclidean space. In that way one may speak about angles and perpendiculars and perform the familiar geometrical constructions even for infinite dimensional spaces. A typical and very important example is the space of square integrable functions on some interval (a, b). This space, called $L^2(a, b)$, is equipped with the following scalar product

$$(f, g) = \int_a^b f(x)g^*(x)dx \tag{5.11}$$

and the induced norm is

$$\|f\| = (f, f)^{\frac{1}{2}} = \left(\int_a^b |f(x)|^2 dx\right)^{\frac{1}{2}} \quad . \tag{5.12}$$

The space $L^2(a, b)$ is not only a normed space, but also a *Hilbert space* [Ref.5.7, Chap.1]. This means that it is *complete* with respect to the norm, i.e., that every Cauchy sequence converges to an element of the space. Moreover, it is a separable space: there exists a countably infinite orthonormal sequence $\{u_n\}$ such that every element of the space can be indefinitely approximated in norm by linear combinations of the vectors u_n. Such a sequence is called a basis and every function f can thus be written as

$$f = \sum_{n=0}^{+\infty} f_n u_n \quad , \tag{5.13}$$

where $f_n = (f, u_n)$ are the Fourier components of f with respect to the basis $\{u_n\}$. In the following we shall often use the so-called *Parseval equality* which expresses the scalar product of two functions in terms of their Fourier components

$$(f, g) = \sum_{n=0}^{+\infty} f_n g_n^* \quad . \tag{5.14}$$

Another norm, which is often used in the case of continuous functions on the closed interval [a, b], is the so-called *uniform norm* defined as follows:

$$\| f \| = \max_{a \leq x \leq b} |f(x)| \quad , \tag{5.15}$$

i.e., the maximal value of the modulus of f on the interval [a, b]. Convergence with respect to the norm (5.15) is uniform convergence and the space of continuous functions is complete with respect to this norm.

After these few preliminaries, we can give a more precise meaning to the concept of *ill-posed linear inverse problems*.

First let us define the *direct problem*: it is a mapping of a space F of functions, called by CHADAN and SABATIER [5.8] "parameters" and by BALTES [Ref.5.9, p.1] "source functions", into a space $\bar{G}$ of functions, called "results" or "data". In the analysis of imaging systems a function of F is called an "object" and a function of $\bar{G}$ a "noiseless image". We assume that F, $\bar{G}$ are normed spaces and that the mapping is given by a linear operator A. We write A: $F \rightarrow \bar{G}$ and, in mathematical language, the space $\bar{G}$ is called the range of the operator A.

Usually the operator A is continuous. This means that to any sequence of elements of F, say $\{f^{(n)}\}$, converging to the null element, there corresponds a sequence $\{Af^{(n)}\}$ which converges to the null element of $\bar{G}$. This property ensures the stability of the direct problem: any perturbation of $\bar{g}$ vanishes when the inducing perturbation of f tends to zero. Besides it is always possible to introduce a norm in $\bar{G}$ such that $\bar{G}$ becomes a complete normed space. Let us assume now that the inverse mapping A^{-1} exists, which is equivalent to require that the equation Af = 0 has only the trivial solution f = 0. Then a theorem of Banach [Ref.5.10, p.83] implies that A^{-1} is also continuous. At this point one could try to define the inverse problem as the problem of solving the functional equation

$$A\bar{f} = \bar{g} \quad , \tag{5.16}$$

where $\bar{g}$ is a given function of $\bar{G}$. The continuity of A^{-1} would ensure the stability of the solution.

However, this approach is inadequate for the following reason. The operator A has usually a smoothing effect. Consider, for instance, the integral operator of (5.5): if the kernel K(x, y) has continuous derivatives with respect to x up to a certain order, then the same property holds for $\bar{g}(x)$. In any case the operator attenuates the higher frequencies - see (5.6). Now, in general, measurement errors or noise destroy the smoothness properties of $\bar{g}$: the "measured result" g is no longer a function of $\bar{G}$ (in the case of imaging systems g is the "noisy image"). In other words, as remarked by SABATIER [Ref.5.11, p.5], one has to extend the space $\bar{G}$ into a larger space G containing all possible results of measurements. The space G must be equipped with a norm suitable for describing experimental errors: a L^2- space with the L^2-norm for instance, when one considers mean-squared errors, or a space of continuous functions with the uniform norm (5.15), when one considers maximal absolute errors. It happens that $\bar{G}$ is no longer a complete space with respect to the norm of G and the operator A^{-1} is no longer continuous. *The inverse problem turns out to be an ill-posed problem.* Besides the equation Af = g might have no solution, because g does not necessarily belong to $\bar{G}$. We see that the questions of existence and continuity are closely connected.

We shall call F the *solution space* and G the *data space* . If we assume a simple additive model for noise and measurements errors, then we have

$$Af + h = g \quad . \tag{5.17}$$

Since both f (the solution) and h (the noise) are unknown and since the equation Af = g might have no solution, it follows that:

I) the best we can do is to search for some f reproducing the given g within a tolerable uncertainty. The problem is then reformulated as follows: find f such that

$$\|Af - g\|_G \leqq \varepsilon \quad , \tag{5.18}$$

where ε is the "size" of the noise, measured with the norm of G.
II) the previous formulation is adequate if the set H of all the functions f satisfying (5.18) is bounded and sufficiently "small" so that any element of H might be taken as an approximation of the "true" solution. However, when A^{-1} is not continuous, H is not bounded. In other words, given an arbitrary number Λ, one can find two functions $f^{(1)}$, $f^{(2)}$ satisfying (5.18) and such that $\|f^{(1)} - f^{(2)}\|_F > \Lambda$. This is precisely the meaning of Hadamard's example discussed in Sect. 5.1.1. In such a case, as we shall see below, some supplementary constraints on the solution are necessary.

The situation illustrated above is quite similar to that of ill-conditioned systems as described in Sect.5.1.2. Evidently, in the finite dimensional case the set H is always bounded, but it is very large along some directions.

Finally we want to remark that, when the inverse operator does not exist, the previous analysis can be repeated considering for (5.16) only solutions of minimal norm. These solutions can be expressed in terms of the *generalized inverse* (or pseudo-inverse) of the operator A [5.12]. The generalized inverse is an extension, for operators in functional spaces, of the Moore-Penrose inverse for matrices. When the operator A has a smoothing effect, it happens that its generalized inverse is not continuous with respect to the norm of the data space G and therefore we get again an ill-posed problem.

5.1.4 Prior Knowledge as a Remedy to Ill-Posedness

In Sect.5.1.3 we saw that, for an ill-posed problem, the set H is unbounded so that (5.18) is not sufficient for determining meaningful approximate solutions. The main idea, common to most available methods for curing ill-posedness, is to restrict the class of admissible solutions by means of suitable *prior knowledge*. In the following we shall always assume that the inverse operator A^{-1} exists.

In *regularization methods*, a subset M of the solution space F is defined in such a way that the intersection of M with the set H, should be a set K of reasonably small size (see Fig.5.1).

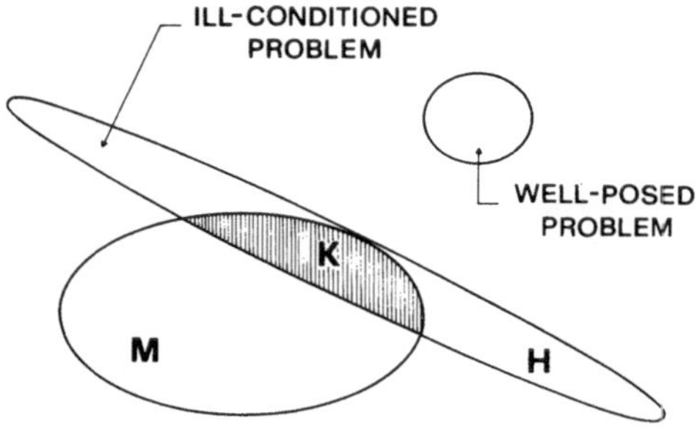

Fig. 5.1. Illustrating the difference between ill-posed and well-posed problems, and the basic idea of regularization

The problem is then said to be regularized if K collapses around the element $A^{-1}\bar{g}$, when g tends to an element $\bar{g}$ of G, i.e., when the noise tends to zero. In such a case one also says that *continuous dependence of the solution on the data has been restored*.

The set M can be defined by imposing global constraints on the class of admissible solutions. For example, one might ask for nonnegative solutions, or for solutions satisfying prescribed bounds, lying in compact sets, etc. The relevance of compactness as a way for restoring continuity was emphasized by TIKHONOV [5.13] who also introduced the concept of regularization. Besides he showed, in the case of Fredholm integral equations of the first kind, how to incorporate the constraints into the computational algorithm.

The role played by prescribed bounds in ill-posed problems for partial differential equations, has been particularly emphasized by JOHN [5.14] and PUCCI [5.15] (in Sect.5.1.5 we shall give a simple example of this approach). Later, along these lines, MILLER [5.16] formulated a regularization algorithm, having in mind the problem of analytic continuation (for an application of this method to the analytic continuation of scattering amplitudes, see also MILLER and VIANO [5.17]). Since it is formulated in the framework of Hilbert spaces, the method of MILLER [5.16] is easily adapted to the general formulation of inverse problems given in Sect.5.1.3. The prescribed bound on the solution is then expressed by means of a linear operator B as follows

$$\|Bf\|_F \leqq E \ , \tag{5.19}$$

where E is a given positive constant. The operator B is called by Miller the *constraint operator*. The method of TIKHONOV [5.13] and the method of MILLER [5.16] are not exactly equivalent even if, in many cases, they lead practically to the same results. We shall discuss these points in Sect.5.2.3.

The global bound (5.19) should express some expected properties of the solution and has to be prescribed according to the physical character of the problem one considers. If the solution represents for instance a signal, then one may know some realistic upper bound for its energy. Very often also, some smoothness condition on the solution can be prescribed by bounding its derivatives. The role of this prior knowledge is to discriminate between interesting solutions and spurious solutions generated by uncontrolled propagation of data errors. The principle of regularization methods is to include the additional conditions explicitly, at the start, instead of resorting consciously or not, during the computations, to some tricks eliminating the instability. The essential drawbacks of such apposite tricks is indeed that their implications, on the class of admissible solutions, often remain in the dark.

Another route to regularization is provided by the theory of stochastic processes. The idea is to associate random processes both to the class of admissible solutions and to the data set. Again we must have some prior knowledge about the solutions. For linear mean-square estimation (which is in general the best one can do) it is enough to know expectation values (mean values), autocorrelation and cross-correlation functions of data and solutions.

Along these lines most work has been done on ill-conditioned algebraic systems, arising from the discretization of ill-posed problems. A good review on this subject is the paper of TURCHIN et al. [5.18]. Stochastic regularization for ill-posed problems has been formulated by LAVRENTIEV [5.3] , MOROZOV [5.19] and FRANKLIN [5.20].

When data and solutions belong to Hilbert spaces, the fundamental mathematical tool is given by the theory of *weak random variables* [Ref.5.7, Chap.6]. Then linear mean-square estimation (optimum filtering) is performed, allowing a comparison with regularization theory based on the constraint (5.19), as presented in [5.21,22]. In the stochastic approach, one says that a continuous dependence of the solution on the data (i.e., stability) has been restored if the mean-square error on the solution tends to zero when the noise tends to zero. This requirement imposes conditions on the autocorrelation functions (covariance operators) of the solutions and of the noise. Stochastic regularization methods will be analyzed in Sect.5.3.

5.1.5 Hölder and Logarithmic Continuity

In Sects.5.2.4 and 5.3.3, we will review some convergence results ensuring that continuity on the data has truly been restored, by means of prior knowledge. This means that the error in the solution converges to zero when the data error ε vanishes. However, convergence theorems are not necessarily enough for practical purposes: one has to know whether the convergence is fast enough or not in order to allow efficient numerical computations. In this connection, we have to distinguish between two different types of continuity. For some problems, the error on the solution is proportional to ε^{α}, $0 < \alpha < 1$. Such a continuity, called *Hölder continuity*, may in general be considered as fairly satisfactory [5.14]. Indeed, the number of significant digits in the solution is then a fixed percentage of the number of significant digits in the data. However, there are some problems where the optimal bounds for the solution error are proportional only to $|\ln \varepsilon|^{-\beta}$, $\beta > 0$. Then a lowering of the data noise by many orders of magnitude, does not improve significantly the solution accuracy. This is *logarithmic continuity* which appears very poor for numerical computations. In other words, a real improvement of the solution is only possible if further constraints may be introduced.

As an illustration of these different types of continuity, let us consider the problem of analytic continuation of functions holomorphic in the unit disk. Suppose that the data are given on a curve Γ contained in the interior of the disk. The domain whose boundary is the unit circle and the curve Γ, may be conformally mapped in an annulus with inner radius a and outer radius R. The data are then given on the inner circle. Let f(z) be a function holomorphic in the annulus; if we denote by $M(\rho)$ the maximum of the modulus of $f[\rho \exp(i\varphi)]$ on the circle of radius ρ, $a \leq \rho \leq R$, then by Hadamard's three circle theorem [Ref.5.23, Chap.5], we have

$$M(\rho) \leq [M(a)]^{\alpha}[M(R)]^{1-\alpha} \quad , \quad \alpha = \frac{\ln(\rho/R)}{\ln(a/R)} \quad . \tag{5.20}$$

From this inequality it follows that analytic continuation to points within the annulus is stable if prior knowledge assures that the admissible functions are bounded by E on the outer circle. Indeed, let us take two such functions $f^{(1)}$, $f^{(2)}$ and consider their difference $f = f^{(1)} - f^{(2)}$. The modulus of f is bounded by 2ε on the inner circle and by 2E on the outer circle, so that, by (5.20), it is bounded by $2E(\varepsilon/E)^{\alpha}$ on any circle of radius ρ. This result implies Hölder continuity for analytic continuation to points within the domain of analyticity. However, α is equal to zero for $\rho = R$, and therefore (5.20) does not ensure the stability up to the outer circle. If one pretends to continue a function precisely up to the boundary of its analyticity domain, then a more restrictive bound is necessary. For instance, one requires that also the first derivative is bounded. In such a case it is possible to show that, at the boundary, one gets logarithmic continuity [5.14]. This fact is not surprising since analytic functions are smooth and well behaved deep inside their holomorphy domain, but may grow rough and oscillatory when approaching the boundary.

For general inverse problems, one expects that the type of restored continuity will depend upon the smoothing or filtering effect of the operator A. Consider for instance a Fredholm integral operator. Then the regularity properties of its kernel are related to the decreasing rate of its eigenvalues. In particular, for analytic kernels, the eigenvalues tend exponentially to zero [5.24] and therefore, if the constraint is not too restrictive, we get only logarithmic continuity. In other words, some information contained in the "true" solution is lost in the data. Accordingly one expects that there are severe limitations on the restoration of fine details in the solution. To check this point it is convenient to consider the reconstruction of a "blurred solution", i.e., the restoration of local weighted averages (Sect.5.2.4). Then it is possible to define a "resolution limit", practically noise independent, giving a measure of the size of the finest details which can be restored. This will be illustrated by many examples in Sect.5.4.

5.2 Regularization Theory

The concept of *regularization* was introduced by TIKHONOV [5.13] in the study of Fredholm integral equations of the first kind. The basic ideas have already been discussed in Sect.5.1.4. A similar method was developed by MILLER [5.16] for improperly posed problems in a Hilbert space setting. We chose the latter method for the following reasons. Firstly because, using the geometrical properties of Hilbert spaces, it is possible to justify, by means of elementary arguments, the main points of the theory. Secondly because Miller's theory allows precise estimations of the solution accuracy (Sect.5.2.4).

5.2.1 An Outline of Miller's Theory

As seen in Sects.5.1.3,4, the regularization of a linear inverse problem can be formulated as follows: to search for functions f satisfying both constraints

$$\|Af - g\|_G \leqq \varepsilon \tag{5.21}$$

and

$$\|Bf\|_F \leqq E \quad . \tag{5.22}$$

The spaces F and G are Hilbert spaces, $A{:}F \to G$ is a known continuous operator, ε is an estimate of the data accuracy, E is a prescribed constant and, finally, $B{:}F \to F$ is the constraint operator.

Many different choices are possible for B, according to the available prior knowledge. The simplest one is B = I (the identity operator in F); then (5.22) is a constraint on the norm of f. Another usual choice is to let B be a differential operator (see Sect.5.2.3) and then the bound (5.22) is a smoothness requirement on the solution. However, for the general formulation of the theory, it is not necessary to specify B. It is only required that B is densely defined in F and that it has a *continuous inverse* B^{-1}. In other words, there must exist a constant β such that $\| Bf \|_F \geq \beta \| f \|_F$. Therefore, the set M of the functions satisfying (5.22) is bounded, and hence also the set K of the functions satisfying (5.21,22). Of course the set K is not allowed to be empty, and this property depends on the values of the numbers ε and E. A couple $\{\varepsilon, E\}$ is said to be *permissible* [5.16] if there exists at least one function f which satisfies (5.21,22).

Let us denote by Π the set of permissible couples. It can be proved [5.16] that this set is convex, i.e., it contains the segment joining any two of its points. Its boundary can be drawn as follows. Let us introduce the functional

$$\Phi(\alpha; f) = \| Af - g \|_G^2 + \alpha \| Bf \|_F^2 , \tag{5.23}$$

where α is a positive parameter. Then, since B^{-1} is bounded, there exists a unique function $\tilde{f}_\alpha$ minimizing the functional $\Phi(\alpha; f)$ [see also the subsequent discussion from (5.25) to (5.27)]. If we write

$$\varepsilon_\alpha = \| A\tilde{f}_\alpha - g \|_G , \quad E_\alpha = \| B\tilde{f}_\alpha \|_F \tag{5.24}$$

it can be proved that ε_α is a continuously increasing and E_α a continuously decreasing function of α, when α runs from 0 to $+\infty$. Moreover, since $\tilde{f}_\alpha$ minimizes $\| Af - g \|_G$ under to the constraint $\| Bf \|_F = E_\alpha$ and likewise minimizes $\| Bf \|_F$ under the constraint $\| Af - g \|_G = \varepsilon_\alpha$, then Π is exactly the set of points which are above and to the right of the curve $\{\varepsilon_\alpha, E_\alpha\}$, $0 < \alpha < +\infty$. Since Π is a convex set, the computation of only a few points on its boundary curve, coupled with linear interpolation in between, would give a good idea of its shape.

Let us assume now that the couple $\{\varepsilon, E\}$ is permissible; then if K is not too "large", any function of K may be taken as a satisfactory estimate of the unknown solution. We are faced with the following two problems:

a) how to exhibit at least one function of K;

b) how to estimate the accuracy of the solution.

For solving these problems it is convenient to find out a simpler and more symmetric geometry than that of the set K. To this purpose, one can introduce two sets K_0 and K_1 sandwiching K [5.25,26]. Indeed, if we consider the functional defined in (5.23) with $\alpha = (\varepsilon/E)^2$

$$\Phi(f) = \| Af - g \|_G^2 + \left(\frac{\varepsilon}{E}\right)^2 \| Bf \|_F^2 , \tag{5.25}$$

then the set K_0 of the functions f satisfying the condition $\Phi(f) \leq \varepsilon^2$ is contained in K, while the set K_1 of the functions f such that $\Phi(f) \leq 2\varepsilon^2$ contains K.

In order to show that the sets K_0 and K_1 have a simpler geometrical structure than the set K, we must consider the following operator:

$$C = A^*A + (\varepsilon/E)^2 B^*B , \tag{5.26}$$

where A^* and B^* are the adjoints (hermitian conjugates) of A and B, respectively. Observe that $A^* : G \to F$, so that $A^*A : F \to F$.

The operator $C : F \to F$ is defined on the domain of B^*B and has the following properties: it is a positive definite operator, i.e., for any f in its domain

$(Cf, f)_F > 0$; it is self-adjoint, i.e., $C^* = C$; it has a continuous inverse since, from the analogous property assumed for B, it follows that there exists a positive constant γ^2 such that $\|Cf\|_F \geq \gamma^2 \|f\|_F$. The last property implies that, for *any* given g, we can introduce the function

$$\tilde{f} = C^{-1}A^*g \quad . \tag{5.27}$$

Then the functional (5.25) may be rewritten as follows:

$$\Phi(f) = (C[f - \tilde{f}] \; , \; [f - \tilde{f}])_F + \|g\|_G^2 - (g, A\tilde{f})_G \quad . \tag{5.28}$$

It is clear that the function $\tilde{f}$ minimizes the functional and that

$$\Phi(\tilde{f}) = \|g\|_G^2 - (g, A\tilde{f})_G \geq 0 \quad . \tag{5.29}$$

Now, in order to investigate the geometrical structure of the sets K_0, K_1, assume for simplicity that the operator C has a complete orthonormal set of eigenfunctions $\{u_n\}$; then the condition $\Phi(f) \leq \varepsilon^2$ can be written in the following form:

$$\sum_{n=0}^{+\infty} \gamma_n^2 |f_n - \tilde{f}_n|^2 \leq \varepsilon^2 - [\|g\|_G^2 - (g, A\tilde{f})_G] \; , \tag{5.30}$$

where $\{\gamma_n^2\}$ is the set of the eigenvalues of C and f_n, $\tilde{f}_n$ are the Fourier components of f, $\tilde{f}$ in the basis $\{u_n\}$, i.e., $f_n = (f, u_n)_F$, $\tilde{f}_n = (\tilde{f}, u_n)_F$. At this point it is clear that the sets K_0, K_1 are infinite-dimensional "ellipsoids" having the same center $\tilde{f}$ and the same principal axes, the latter being given by the eigenvectors of C.

Now, does $\tilde{f}$ belong to the set K? A sufficient condition for this is $\Phi(\tilde{f}) \leq \varepsilon^2$, which can be easily verified by numerical computation.

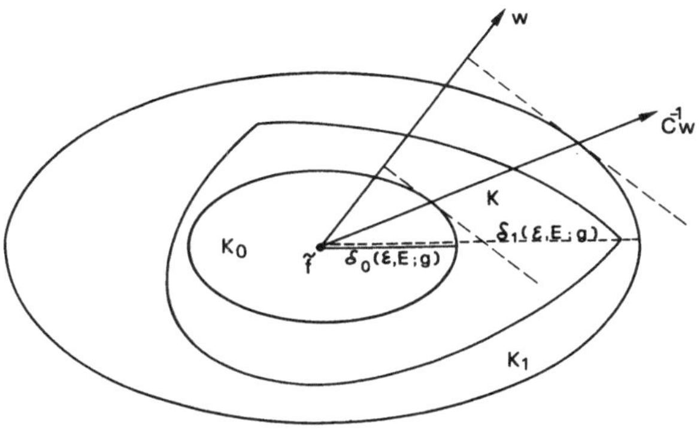

Fig. 5.2. Schematic representation of the relation between the sets K_0, K, K_1. The sets K_0, K_1 are represented as homothetic ellipses with center $\tilde{f}$

Besides the set K_0 is nonvoid if and only if this condition is satisfied. In such a case, the situation is schematically represented in Fig.5.2. It is then clear that we may take $\tilde{f}$ as an estimate of the unknown solution.

5.2.2 Eigenfunction Expansions and Numerical Filtering

Let us suppose now that A is a compact operator. A typical example is an integral operator over a finite interval and with continuous kernel. The corresponding inverse problem is the solution of a Fredholm equation like (5.5). Existence theory for these equations was developed by PICARD [5.27], using expansions in terms of the eigenfunctions of the operators A*A and AA*. Picard's theory generalizes immediately to the case of compact operators in Hilbert space. The operator A*A is a compact, self-adjoint, nonnegative operator and its inverse exists since A^{-1} exists. From the spectral theory for compact operators [Ref.5.7, Chap.3], it follows that A*A admits a countably infinite set of positive eigenvalues $\{\alpha_n^2\}$, and that the set $\{u_n\}$ of the corresponding eigenfunctions is a basis in F. Each eigenvalue has finite multiplicity and $\{\alpha_n^2\}$ can be ordered as follows: $\alpha_0^2 \geqq \alpha_1^2 \geqq \alpha_2^2 \geqq \dots$. Moreover $\alpha_n^2 \to 0$ when $n \to +\infty$. The $\alpha_n (\alpha_n > 0)$ and the u_n are called, respectively, singular values and singular functions of A. If we introduce the vectors

$$v_n = \alpha_n^{-1} A u_n \tag{5.31}$$

it is easy to check that

$$Au_n = \alpha_n u_n \ , \quad A^* v_n = \alpha_n u_n \tag{5.32}$$

and that

$$A^*Au_n = \alpha_n^2 u_n \ , \quad AA^* v_n = \alpha_n^2 v_n \quad . \tag{5.33}$$

The set $\{v_n\}$ is a complete orthonormal set in the closure of the range of the operator A [Ref.5.28, Chap.5.2], i.e., in the closure of $\bar{G}$. Therefore $\{v_n\}$ is a basis for representing noise-free data.

Solution (5.27) takes on a simple form when the u_n diagonalize B*B. In such a case, we have

$$B^*Bf = \sum_{n=0}^{+\infty} \beta_n^2 f_n u_n \tag{5.34}$$

and B has a continuous inverse if and only if $\beta_n \geq \beta > 0$ for any n. The u_n diagonalize also the operator C, and the corresponding eigenvalues are given by $\gamma_n^2 = \alpha_n^2 + (\varepsilon/E)^2 \beta_n^2$. Now, from (5.32) it follows that

$$A^*g = \sum_{n=0}^{+\infty} \alpha_n g_n u_n \ , \tag{5.35}$$

where $g_n = (g, v_n)_G$, and from (5.27) we get

$$\tilde{f} = \sum_{n=0}^{+\infty} \frac{\alpha_n}{\alpha_n^2 + (\varepsilon/E)^2\beta_n^2} g_n u_n \quad . \tag{5.36}$$

Another solution of the problem is obtained as follows. Let us assume for simplicity the the β_n^2 form a nondecreasing sequence. Then let us denote by N the greatest integer such that $\alpha_n \geq (\varepsilon/E)\beta_n$ (recall that the α_n form a nonincreasing sequence). Truncating the series (5.36) at n = N and neglecting $\varepsilon\beta_n/E$ in comparison with α_n for $n \leq N$, we obtain

$$\approx{f} = \sum_{n=0}^{N} \alpha_n^{-1} g_n u_n \quad . \tag{5.37}$$

It can be proved that $\approx{f}$ belongs to K_1 [5.16] and therefore is also an approximate solution. We recover here, with a prescribed cutoff, the well-known truncation method which is in use for eliminating the noise amplification due to eigenvalues very close to zero and which is usually called *numerical filtering* [5.29].

The preceding methods have the disadvantage that both the error bound ε and the constraint E have to be known. However, often in practice, only ε is known. Then it is possible to elaborate procedures which require the knowledge of only one element of the couple $\{\varepsilon, E\}$. For instance, let us suppose that we know a good upper bound $\bar{\varepsilon}$ for the data accuracy but none for E.

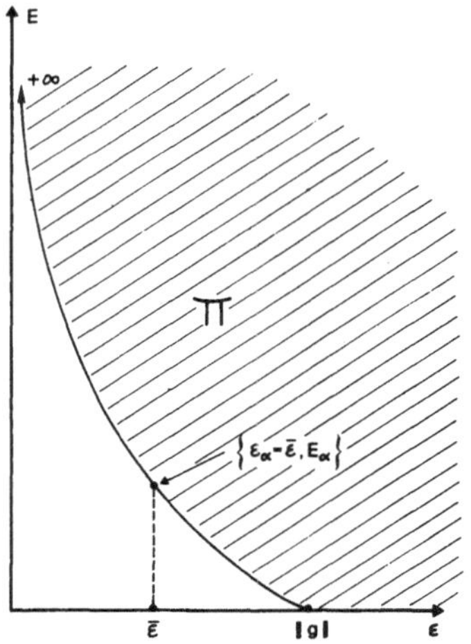

Fig. 5.3. The region Π of permissible couples $\{\varepsilon, E\}$

Nevertheless, we can obtain an estimate of the solution taking as our approximation that element of the space F which minimizes $\|Bf\|_F$ with respect to the constraint $\|Af - g\|_G = \bar{\varepsilon}$. In other words we will take as our approximation the function $\tilde{f}_\alpha$ with α determined by the condition $\varepsilon_\alpha = \bar{\varepsilon}$, as shown in Fig.5.3. This method coincides with that proposed by MOROZOV [5.30] and called the "residue method". A complete discussion of this point can be found in [5.16].

Up to now we have considered an operator A such that A*A has only a discrete spectrum. By means of a simple example, we illustrate how the regularization procedure works when the spectrum is continuous. To this purpose we consider the case of a convolution operator

$$(Af)(x) = \int_{-\infty}^{+\infty} K(x - y)f(y)dy \quad . \tag{5.38}$$

If the function K(x) is integrable over $(-\infty, +\infty)$, then its Fourier transform

$$\hat{K}(\nu) = \int_{-\infty}^{+\infty} K(x)\, e^{-2\pi i\nu x}dx \tag{5.39}$$

is a bounded continuous function on $(-\infty, +\infty)$, such that $|\hat{K}(\nu)| \to 0$ when $|\nu| \to +\infty$ (Riemann-Lebesgue theorem). If we take as solution and data space the space of square integrable functions, i.e., $F = G = L^2(-\infty, +\infty)$, then $A: F \to G$ is a continuous operator. Besides A^{-1} (which exists if $\hat{K}(\nu)$ does not vanish over some interval) is not continuous since $\hat{K}(\nu)$ tends to zero at infinity.

Now we can take as a constraint operator B

$$(Bf)(x) = \int_{-\infty}^{+\infty} \hat{\beta}(\nu)\hat{f}(\nu)\, e^{2\pi ix\nu}d\nu \quad . \tag{5.40}$$

In such a form we can write, for instance, a differential operator of order n with constant coefficients. Then $\hat{\beta}(\nu)$ is a polynomial of order n, $\hat{\beta}(\nu) = a_0 + a_1\nu + \ldots + a_n\nu^n$, and the domain of B is the set of the functions f such that $\hat{\beta}(\nu)\hat{f}(\nu)$ is square integrable over $(-\infty, +\infty)$. Besides, let us remark that B has a bounded inverse if and only if $\hat{\beta}(\nu)$ has no zeros on $(-\infty, +\infty)$.

Since the operators A*A and B*B commute, we are in a situation completely analogous to that illustrated in the preceding example. Therefore, from (5.27) we get

$$\tilde{f}(x) = \int_{-\infty}^{+\infty} \frac{\hat{K}^*(\nu)}{|\hat{K}(\nu)|^2 + (\varepsilon/E)^2|\hat{\beta}(\nu)|^2}\, \hat{g}(\nu)\, e^{2\pi ix\nu}d\nu \quad . \tag{5.41}$$

This formula corresponds to the expansion (5.36). Of course, also in this case, we can obtain a second approximation corresponding to the truncated solution (5.37). Indeed, let us denote by Λ the set of the values of ν such that $|\hat{K}(\nu)| \geq (\varepsilon/E)|\hat{\beta}(\nu)|$; since $\hat{K}(\nu)$ tends to zero at infinity while $\hat{\beta}(\nu)$ is never zero, Λ is a bounded set of $(-\infty, +\infty)$. Then the second solution is given by

$$\tilde{\tilde{f}}(x) = \int_{\Lambda} \frac{\hat{g}(\nu)}{\hat{K}(\nu)}\, e^{2\pi i\nu x}d\nu \quad . \tag{5.42}$$

5.2.3 Tikhonov Regularization Method

A general method for solving Fredholm equations of the first kind was proposed by TIKHONOV [5.13]. The method was successively developed and generalized by Tikhonov himself and by many Russian mathematicians. A copious list of references of the Russian school can be found in [5.4].

Let us consider the integral operator

$$(Af)(x) = \int_a^b K(x, y)f(y)dy \ , \quad c \leq x \leq d \tag{5.43}$$

whose inverse is assumed to exist. The solution space F is the space of the continuous functions over [a, b], normed with the uniform norm (5.15). The data space G is the space $L^2(a, b)$. The purpose is to construct a uniform approximation to the solution of (5.5).

The basic idea is to restrict the class of admissible solutions to a compact subset of F. Then a general theorem of functional analysis, (due to Tikhonov himself) assures the continuity of the inverse mapping [Ref.5.4, Chap.1, Sect.1]. The restriction to a compact subset is achieved by means of a "regularizing functional" $\Omega(f)$: the compact subsets are defined by the condition $\Omega(f) \leq E^2$, where E^2 is a given arbitrary constant.

The functional proposed by Tikhonov for Fredholm equations of the first kind is

$$\Omega(f) = \int_a^b [p(x)|f'(x)|^2 + q(x)|f(x)|^2]dx \ , \tag{5.44}$$

where the weight functions p(x) and q(x) are strictly positive. It is possible to prove, by means of the Ascoli-Arzelà theorem [Ref.5.10, Chap.1], that the set $\Omega(f) \leq E^2$ is a compact subset of F. Next a "*regularized family of approximate solutions*" $\{\tilde{f}_\alpha\}$, $\alpha > 0$, is defined as the set of functions minimizing the functionals

$$\Phi(\alpha; f) = \int_c^d |(Af)(x) - g(x)|^2dx + \alpha\Omega(f) \ , \tag{5.45}$$

where α is a free parameter.

The functions $\tilde{f}_\alpha$ are the solutions of the Euler equation for $\Phi(\alpha; f)$

$$-\alpha\{[p(x)\tilde{f}'_\alpha(x)]' + q(x)\tilde{f}_\alpha(x)\} + \int_a^b \bar{K}(x, y)\tilde{f}_\alpha(y)dy = \bar{b}(x) \ ,$$
$$\tilde{f}'_\alpha(a) = \tilde{f}'_\alpha(b) = 0 \ , \tag{5.46}$$

where

$$\bar{K}(x, y) = \int_c^d K^*(s, x)K(s, y)ds \ , \quad \bar{b}(x) = \int_c^d K^*(s, x)g(s)ds \ . \tag{5.47}$$

Denote by $\{g_\varepsilon\}$, $\varepsilon > 0$, a family of data converging to the error-free datum $\bar{g}$ when $\varepsilon \to 0$ and let $\varepsilon = \|g_\varepsilon - \bar{g}\|_G$. Besides, let $\bar{\alpha}(\varepsilon)$ be a value of the parameter α such that $c_1\varepsilon^2 \leq \bar{\alpha}(\varepsilon) \leq c_2\varepsilon^2$ (where c_1, c_2 are given arbitrary constant); i.e., $\bar{\alpha}(\varepsilon) \sim \varepsilon^2$ when $\varepsilon \to 0$. Finally, denote by $\bar{f}_\varepsilon$ the solution of (5.46) with $g = g_\varepsilon$ and $\alpha = \bar{\alpha}(\varepsilon)$. Then $\bar{f}_\varepsilon \to \bar{f}$ in the uniform norm, when $\varepsilon \to 0$ ($\bar{f}$ is the unique solution corresponding to the error-free datum $\bar{g}$) [5.13]. A simple proof of this result can be found in [5.31].

In order to compare Tikhonov's method with the method outlined in Sect.5.2.1, we can take as solution space $L^2(a, b)$. The space of continuous functions is a subspace of $L^2(a, b)$ and the set $\Omega(f) \leq E^2$ is a compact subset of $L^2(a, b)$. Then we remark that the condition $\Omega(f) \leq E^2$ can be written in the form (5.22) if the operator B is such that

$$(B^*Bf)(x) = -[p(x)f'(x)]' + q(x)f(x) \; ; \quad f'(a) = f'(b) = 0 \quad . \tag{5.48}$$

Indeed, by means of a partial integration, it is easy to see that, for any f in the domain of B^*B we have: $(B^*Bf, f)_F = \Omega(f)$. We want also to remark that the operator B^*B has a discrete spectrum with eigenvalues accumulating to infinity. Indeed, the eigenvalue equation for B^*B is nothing else but a Sturm-Liouville problem over [a, b]. It should also be remarked that the functional (5.45) coincides now with (5.23). Besides the solution $\bar{f}_\alpha$ of (5.46) can be formally written as in (5.27), with $(\varepsilon/E)^2$ replaced by α in (5.26). Clearly the method outlined in Sect.5.2.1 is essentially a generalization of Tikhonov's method to the case where F and G are Hilbert spaces and $A: F \to G$ an arbitrary linear continuous operator. Besides the parameter α is explicitly taken proportional to ε^2, the constant being given by the prescribed bound (5.22).

Numerical computations on Fredholm equations of the first kind have been done by many authors using the method described above [5.32-34]. When uniqueness does not hold, generalized inverses must be used. A simple discussion of existence and uniqueness theorems, using singular function expansions can be found in [5.35]. The clear result is that generalized inverses are not continuous. Their regularization has been analyzed using TIKHONOV's method [5.12].

5.2.4 Stability Estimates

Here we come back to the method of Sect.5.2.1. Indeed, if we want to estimate the error on the approximate solution (5.27) it is necessary to know both constants, ε and E. Of course, it is meaningless to speak about errors in the solution without specifiying how the accuracy of the solution is defined. When F is a Hilbert space, without further specifications, then there are two natural choices. The first is to measure errors in the solutions by means of the distance $\|f^{(1)} - f^{(2)}\|_F$. The second choice is to measure errors by means of the quantity (seminorm, in mathematical language) $|(f^{(1)} - f^{(2)}, w)_F|$, where w is a suitable function of F. As we shall see, this choice is convenient for the analysis of "blurred solutions". We do not consider here the case where errors are defined in terms of the uniform norm (5.15) (for a discussion of this problem see [5.22,36]).

In the first case the error may be defined as the maximum value of $\|f - \tilde{f}\|_F$ where f is any function of the set K. Recall that $\tilde{f}$ is given by (5.27) and that K is the set of the functions satisfying (5.21,22). We shall write

$$\delta(\varepsilon, E; g) = \max_{f \in K} \| f - \tilde{f} \|_F \quad . \tag{5.49}$$

In Fig.5.2 $\delta(\varepsilon, E; g)$ is the maximum distance between any point of K and $\tilde{f}$; it is clear that this maximum value is attained at the boundary of K.

Let us now denote by $\delta_0(\varepsilon, E; g)$ and $\delta_1(\varepsilon, E; g)$ the maximum length of the semi-axes of K_0 and K_1, respectively. Then, looking at Fig.5.2, we see

$$\delta_0(\varepsilon, E; g) \leqq \delta(\varepsilon, E; g) \leqq \delta_1(\varepsilon, E; g) \quad . \tag{5.50}$$

The quantities $\delta_i(\varepsilon, E; g)$, $i = 0,1$, may be easily related to the data g and to the spectrum of the operator C. Assume, for simplicity, that C has a discrete spectrum and let $\gamma^2(\varepsilon/E)$ be the smallest eigenvalue of C. Then, from (5.30), it follows that

$$\delta_i(\varepsilon, E; g) = \frac{1}{\gamma(\varepsilon/E)} \left[\varepsilon_i^2 - (\| g \|_G^2 - (g, A\tilde{f})_G)\right]^{\frac{1}{2}} , \quad i = 0,1 \ , \tag{5.51}$$

where $\varepsilon_0^2 = \varepsilon^2$ and $\varepsilon_1^2 = 2\varepsilon^2$. When C has a continuous spectrum, (5.51) is still true, $\gamma^2(\varepsilon/E)$ being the infimum of the spectrum of C. Therefore $\delta_0(\varepsilon, E; g)$ and $\delta_1(\varepsilon, E; g)$ can be computed in practical cases, and they give, respectively, a lower and an upper bound for $\delta(\varepsilon, E; g)$.

Inequality (5.29) shows that there exists an upper bound for $\delta(\varepsilon, E, g)$ independent of the data function g. More precisely: $\delta(\varepsilon, E, g) \leqq \delta(\varepsilon, E)$, where

$$\delta(\varepsilon, E) = \sqrt{2} \frac{\varepsilon}{\gamma(\varepsilon/E)} \quad . \tag{5.52}$$

The quantity $\delta(\varepsilon, E)$ is called by MILLER [5.16] *stability estimate*. Indeed, if $\delta(\varepsilon, E) \to 0$ when $\varepsilon \to 0$, for fixed E, then the error in the solution of the inverse problem also tends to zero. In other words, if the noise tends to zero and therefore $g \to \bar{g}$, where $\bar{g} = A\bar{f}$ represents noiseless data, then the estimated solution $\tilde{f}$ tends to the exact solution $\bar{f}$. This corresponds to the collapse of the ellipses K_0 and K_1 of Fig.5.2 into a point. A general condition for ensuring this is that *B has a compact inverse*. This is just a reformulation of the general result of Tikhonov (see Sect.5.2.3).

We can now state more precisely what is meant by *Hölder continuity* and *logarithmic continuity* (Sect.5.1.5). The first case corresponds to $\delta(\varepsilon, E) \sim \varepsilon^\alpha$, $0 < \alpha < 1$ (for fixed E) and the second to $\delta(\varepsilon, E) \sim |\ln\varepsilon|^{-\beta}$, $\beta > 0$. An elementary discussion of the relationship between these properties of the stability estimate and the properties of the couple of operators A, B can be done when assuming that the operators A*A and B*B commute.

Like in Sect.5.2.2, consider firstly the case where A is a compact operator. Then, from (5.52), recalling that the eigenvalues of C are given by $\gamma_n^2 = \alpha_n^2 + (\varepsilon/E)^2\beta_n^2$, it follows that

$$\delta(\varepsilon, E) = \sqrt{2}\varepsilon \sup_n \left[\alpha_n^2 + (\varepsilon/E)^2 \beta_n^2 \right]^{-1/2} . \tag{5.53}$$

First of all, let us observe that we cannot restore the continuity by choosing bounded β_n. Indeed if we take, for instance, $\beta_n = 1$ in (5.53) then we have, for any ε, $\delta(\varepsilon, E) = \sqrt{2}\, E$. In fact $\delta(\varepsilon, E) \to 0$ for $\varepsilon \to 0$ and fixed E, if B satisfies the following conditions: I) each eigenvalue of B*B has finite multiplicity; II) the β_n grow to infinity for $n \to +\infty$ [5.26]. These assumptions are equivalent to require that B^{-1} is a compact operator.

More precise results can be obtained if stronger assumptions are imposed on the β_n. For instance, if we assume that for $n \to +\infty$, $\beta_n \sim \alpha_n^{-\mu}$, $\mu > 0$, then, by computing the minimum of the function $\varphi(s) = s + (\varepsilon/E)^2 s^{-\mu}$, $s > 0$, it is easy to show that $\delta(\varepsilon, E) \sim E(\varepsilon/E)^{\alpha}$ where $\alpha = \mu(\mu+1)^{-1}$. However, the condition $\beta_n \sim \alpha_n^{-\mu}$ is too restrictive when the α_n tend to zero very rapidly. This occurs, for instance, for integral operators having an analytic kernel, since their singular values have an exponential fall-off [5.24]. In such a case, a more reasonable condition [5.22,26] is to take β_n growing as a power of n, i.e., $\beta_n \sim n^{\mu}$, $\mu > 0$. But, observing that the function $\psi(s) = \exp[-as] + (\varepsilon/E)^2 s^{2\mu}$, $s > 0$, has a minimum for $s_0 \sim |\ln(\varepsilon/E)|$ when $\varepsilon \to 0$, it follows that $\delta(\varepsilon, E) \sim E|\ln(\varepsilon/E)|^{-\mu}$. In most cases the condition $\beta_n \sim n^{\mu}$ implies a constraint upon a finite number of derivatives of the admissible solutions (see Sect.5.4).

As a second example, we consider the case of the convolution operator (5.38), the constraint operator B having the form (5.40). Noting that the infimum of the spectrum of the operator C is given by

$$\gamma^2(\varepsilon/E) = \inf_{\nu}[|\hat{K}(\nu)|^2 + (\varepsilon/E)^2 |\hat{\beta}(\nu)|^2] ,$$

from (5.52) we get

$$\delta(\varepsilon, E) = \sqrt{2}\, \varepsilon \sup_{\nu}[|\hat{K}(\nu)|^2 + (\varepsilon/E)^2 |\hat{\beta}(\nu)|^2]^{-1/2} . \tag{5.54}$$

In this case, we obtain results very similar to the previous ones. Let us suppose, for instance, that $\hat{K}(\nu)$ is a rational function (this may happen in the case of an electrical network), without zeros on the real axis, whose asymptotic behavior for $|\nu| \to +\infty$ is given by $\hat{K}(\nu) \sim \nu^{-m}$ (m is a positive integer). Then we can take as constraint operator B, a differential operator of order n, with constant coefficients; in such a case $\hat{\beta}(\nu)$ is a polynomial of order n, and therefore $\hat{\beta}(\nu) \sim \nu^n$ for $|\nu| \to +\infty$. Then, asymptotically, we have $\hat{\beta}(\nu) \sim [\hat{K}(\nu)]^{-\mu}$, $\mu = n/m$, and hence we have Hölder continuity. More precisely, $\delta(\varepsilon, E) \sim E(\varepsilon/E)^{\alpha}$ with $\alpha = \mu(\mu+1)^{-1} = n(m+n)^{-1} < 1$. Observe that $\alpha \sim 1$ when $n \gg m$, i.e., when the admissible solutions are very smooth. On the other hand, if $\hat{K}(\nu)$ decreases exponentially for $|\nu| \to +\infty$ while $\hat{\beta}(\nu)$ increases as a power, then the restored continuity is only logarithmic. As is well known, the asymptotic behaviour of $\hat{K}(\nu)$ for $|\nu| \to +\infty$ is strictly related to smoothness properties of K(x). In particular, $\hat{K}(\nu)$ decreases exponentially when K(x) is an analytic function. A list of various problems which can be reduced to the solution of integral equations of convolution type can be found, for instance, in [Ref.5.4, Chap.4, Sect.3].

As pointed out in Sect.5.1.5, when the restored continuity is of logarithmic type, it is convenient to consider the reconstruction of "blurred solutions". For the sake of simplicity, we analyze essentially the case of the convolution operator (5.38). Then we can proceed as follows. Let $w_D(x)$ be a positive, even function such that

$$\int_{-\infty}^{+\infty} w_D(x)dx = 1 , \quad \int_{-\infty}^{+\infty} x^2 w_D(x)dx = D^2 \tag{5.55}$$

and let

$$f_D(x_0) = \int_{-\infty}^{+\infty} w_D(x_0 - x)f(x)dx \quad . \tag{5.56}$$

The "blurred solution" $f_D(x_0)$ is a local weighted average of f over the distance D, and an estimate $\tilde{f}_D(x_0)$ for it is obtained by replacing f with $\tilde{f}$ in (5.56). The error in $\tilde{f}_D(x_0)$ is the maximum value of $|f_D(x_0) - \tilde{f}_D(x_0)|$ where $f_D(x_0)$ corresponds to an arbitrary function of the set K. Since we consider a convolution operator commuting with the translation operators, it is quite clear that the error in $\tilde{f}_D(x_0)$ is independent of x_0, and hence it is enough to evaluate the error in $\tilde{f}_D(0) = (\tilde{f}, w_D)_F$.

For a fixed D, we define the error by

$$\delta(\varepsilon, E; g, w_D) = \sup_{f \in K} |(f - \tilde{f}, w_D)_F| \quad . \tag{5.57}$$

In other words, it is the maximum value of the component of $f - \tilde{f}$ along the direction of the vector w_D. If we look at Fig.5.2, we clearly understand that

$$\delta_0(\varepsilon, E; g, w_D) \leqq \delta(\varepsilon, E; g, w_D) \leqq \delta_1(\varepsilon, E; g, w_D) \quad , \tag{5.58}$$

where $\delta_0(\varepsilon, E; g, w_D)$ and $\delta_1(\varepsilon, E; g, w_D)$ are quantities analogous to (5.57), the supremum being taken over the sets K_0 and K_1, respectively. Again $\delta_0(\varepsilon, E; g, w_D)$ and $\delta_1(\varepsilon, E; g, w_D)$ can be easily computed. Indeed, if we use Fig.5.2 as a schematic representation of the infinite dimensional problem, we see that the component along w_D of a vector $u = f - \tilde{f}$ of K_0 is maximal when u coincides with that point u_0 of the boundary of K_0 such that the tangent to the ellipse at u_0 is orthogonal to w_D. Now, if we write the equation of the ellipse as $(Cu, u)_F = b^2$, then the equation of the tangent in u_0 is given by $(Cu_0, u)_F = b^2$ and therefore the tangent is orthogonal to the vector Cu_0. This vector is parallel to w_D if $u_0 = aC^{-1}w_D$, where a is a constant which can be determined by requiring that u_0 belongs to the boundary of K_0. It follows that $a = b(C^{-1}w_D, w_D)_F^{-\frac{1}{2}}$, so that $\delta_0(\varepsilon, E; g, w_D) = |(u_0, w_D)_F| = b(C^{-1}w_D, w_D)_F^{\frac{1}{2}}$. Using a similar argument for K_1 and recalling (5.30), we have

$$\delta_i(\varepsilon, E; g, w_D) = [\varepsilon_i^2 - (\|g\|_G^2 - (g, A\tilde{f})_G)]^{\frac{1}{2}}(C^{-1}w_D, w_D)_F^{\frac{1}{2}} \quad , \tag{5.59}$$

where $\varepsilon_0^2 = \varepsilon^2$ and $\varepsilon_1^2 = 2\varepsilon^2$. For infinite dimensional "ellipsoids" the previous argument can be made completely rigorous using the Schwarz inequality.

Again we can find an upper bound on the error, which is independent of g, i.e., $\delta(\varepsilon, E; g, w_D) \leqq \delta(\varepsilon, E; w_D)$ where [5.16]

$$\delta(\varepsilon, E; w_D) = \sqrt{2}\,\varepsilon(C^{-1}w_D, w_D)_F^{\frac{1}{2}} \quad . \tag{5.60}$$

It is possible to prove that $\delta(\varepsilon, E; w_D)$ tends to zero when $\varepsilon \to 0$, provided that the constraint operator B would have a bounded inverse. Hence in this case, we are allowed to take B = I. This type of continuity can be called *weak continuity*.

More generally, the problem of restoring "blurred solutions" can just be formulated as the problem of restoring a family of linear functionals like $(f, w_\lambda)_F$ where λ is some suitable parameter (like the center of the averaging interval). Such a point of view is usually adopted for inverse problems in geophysics [5.37,38].

5.3 Optimum Filtering

When statistical properties of data and solutions are available, filtering methods provide an alternative way for regularizing ill-posed problems.

5.3.1 Random Variables in a Hilbert Space

Since we will use a perhaps unfamiliar description of stochastic processes, we begin with a brief sketch of the working frame. For further details the reader may consult standard books on random processes like [5.39-41] or, for the more specific questions concerning Hilbert space valued random variables, [Ref.5.7, Chap.6] and the review article [5.42].

Let us first recall that a *random variable* (in short: r.v.) X is an application of some set Ω (the set of the outcomes ω of an experiment) on the set of the real numbers, i.e., $X(\omega)$ is a real number called a value of the r.v. X. A probability measure P is defined on the subsets of Ω, called events, and X is described by the distribution function

$$F_X(x) = P[X \leqq x] = P\{\omega | X(\omega) \leqq x\} \quad , \tag{5.61}$$

where one must read the r.h.s. as the "probability of the event containing all outcomes ω such that $X(\omega) \leqq x$". The *mean value* of X will be denoted by $m_X = E[X]$, where E means *mathematical expectation*, and its *variance* by $\sigma_X^2 = E[(X - m_X)^2]$. Two (or more) r.v. are said to be jointly distributed if they are defined on the same space Ω; then they may be described by the joint distribution function

$$F_{XY}(x, y) = P[X \leqq x, Y \leqq y] = P\{\omega | X(\omega) \leqq x, Y(\omega) \leqq y\} \quad . \tag{5.62}$$

Their *covariance coefficient* is $\mu_{XY} = E[(X - m_X)(Y - m_Y)]$. The two r.v. are *uncorrelated* if $\mu_{XY} = 0$. A complex r.v. $Z = X + iY$ is viewed as a pair of jointly distributed real r.v. X, Y.

A *Hilbert space valued random variable* ξ is an application of Ω on a Hilbert space F, i.e., $\xi(\omega)$ is an element of F. When F is a space of functions, then ξ is a stochastic process. We prefer, however, to use the previous appellation, since stochastic processes are not necessarily defined in a Hilbert space. It is quite obvious that, if F is a real (complex) Hilbert space, then, for any w in F, $\xi_w = (\xi, w)_F$ is a real (complex) r.v. Accordingly, the Fourier components of ξ in a basis $\{u_n\}$ of F are an infinite set of jointly distributed r.v. $\xi_n = (\xi, u_n)_F$. Simplifying somehow, we can define ξ by requiring that, for any sequence $\{a_n\}$ of real numbers, the following probability makes sense:

$$P[\xi_1 \leqq a_1, \xi_2 \leqq a_2, \ldots, \xi_n \leqq a_n, \ldots] \quad . \tag{5.63}$$

However, in order to include processes like white noise, one has to consider *weak random variables* (in short: w.r.v.). In this case it is required that (5.63) be defined only for any sequence having a finite number of elements different from $+\infty$. Such a probability measure on F is also called a *cylinder set measure*, because it is only defined on the "cylinders" of F, i.e., the sets which are bounded only along a finite number of directions [5.7,42]. Indeed, as it is the case for white noise, the probability that ξ takes values in a bounded set is not necessarily defined. Let us remark that, in writing (5.63), we have implicitly assumed that F was a real Hilbert space; the extension to complex Hilbert spaces and to corresponding w.r.v. is quite obvious. Next, let us only mention that an alternative way to the introduction of w.r.v. is to define like GEL'FAND and VILENKIN [5.43] generalized processes, i.e., applications of Ω into a space of distributions.

In the following we will assume, for simplicity, that all w.r.v. ξ have zero mean; in other words, for any w in F, $E[(\xi, w)_F] = 0$. This is not a restrictive hypothesis since, when the mean is not zero, one can always consider, instead of ξ, the reduced w.r.v. $\xi' = \xi - E[\xi]$. At this point we still have to introduce the concept of covariance operator $R_{\xi\xi}$ of the w.r.v. ξ. Such an operator is strictly related to the so-called autocovariance function of a stochastic process (for zero mean processes autocorrelation and autocovariance functions coincide). Indeed, let us assume for a moment that F is $L^2(a, b)$ and that it is possible to define in some way the complex r.v. $\xi(x)$, where x is a point of [a, b]. Then the *autocovariance function* of ξ is given by

$$R_{\xi\xi}(x, y) = E[\xi(x)\xi^*(y)] \tag{5.64}$$

and we call *covariance operator* $R_{\xi\xi}$ the integral operator whose kernel is (5.64)

$$(R_{\xi\xi}f)(x) = \int_a^b R_{\xi\xi}(x, y)f(y)dy \quad . \tag{5.65}$$

"White noise" is by definition a Gaussian process ζ for which, formally, $R_{\zeta\zeta}(x, y) = \varepsilon^2\delta(x - y)$, and hence $R_{\zeta\zeta} = \varepsilon^2 I$, where I is the identity operator in F. From (5.64,65) and the definition of the scalar product in $L^2(a, b)$ it is easy to check that, for any f, w in F

$$(R_{\xi\xi}f, w)_F = E[(f, \xi)_F(\xi, w)_F] \quad . \tag{5.66}$$

In the theory of w.r.v. (5.66) is adopted as a definition of $R_{\xi\xi}$, a definition which remains valid in any Hilbert space F. Indeed, we will always assume that the w.r.v. has a finite second moment, i.e., we require that $E[|(\xi, f)_F|^2]$ is finite for any f in F, and is a continuous function of f. Then the r.h.s. of (5.66) is a continuous bilinear form over F and hence there exists a bounded, linear, self-adjoint, non-negative operator $R_{\xi\xi}$ fulfilling (5.66) (see, e.g. [5.7]).

For two stochastic processes ξ, η the *cross-covariance function* is defined by

$$R_{\xi\eta}(x, y) = E[\xi(x)\eta^*(y)] \tag{5.67}$$

and the cross-covariance operator $R_{\xi\eta}$ is the integral operator whose kernel is (5.67). If ξ takes values in the Hilbert space F and η in the Hilbert space G, then $R_{\xi\eta} : G \to F$, and it is easy to check that

$$(R_{\xi\eta}g, f)_F = E[(g, \eta)_G(\xi, f)_F] \quad . \tag{5.68}$$

Equation (5.68) can be taken as a definition of the cross-covariance operator for processes having a finite second moment. Besides the following relation holds: $R_{\xi\eta} = R^*_{\eta\xi}$.

5.3.2 Best Linear Estimates

With the previous background, let us turn back to our linear inverse problem. The basic equation is (5.17) and the functions f, g, h will be considered as values of jointly distributed w.r.v., respectively ξ, η, ζ. The w.r.v. ξ takes values in the Hilbert space F, while η and ζ take values in the Hilbert space G. The w.r.v. ξ, η, ζ are assumed to satisfy the following equation

$$A\xi + \zeta = \eta \ , \tag{5.69}$$

where the linear operator $A : F \to G$ is continuous, and its inverse A^{-1} is supposed to exist. The inverse problem consists in estimating a value of ξ, given an observed value g of η. Prior knowledge would be knowledge of the joint distribution of the w.r.v. ξ and ζ (solution and noise). This is usually too much for linear estimations. It is enough to assume a knowledge of the mean values of the w.r.v. ξ, ζ and of the appropriate covariance and cross-covariance operators. The following assumptions are usually introduced:

I) *ξ and ζ have zero mean;*
II) *ξ and ζ are uncorrelated, i.e.,* $R_{\xi\zeta} = 0$;
III) $R^{-1}_{\zeta\zeta}$ *exists.*

The third assumption is the mathematical formulation of the fact that all components of the data function are affected by noise, or in other words that no component of the noise is equal to zero with probability one. Thanks to the assumptions I), II) the covariance operator of η is given by (see, e.g. [5.20])

$$R_{\eta\eta} = AR_{\xi\xi}A^* + R_{\zeta\zeta} \tag{5.70}$$

and the cross-covariance operator $R_{\xi\eta}$ is

$$R_{\xi\eta} = R_{\xi\xi}A^* \quad . \tag{5.71}$$

We will also assume that $R_{\zeta\zeta}$ contains a parameter ε, which tends to zero when the noise vanishes, i.e.,

$$R_{\zeta\zeta} = \varepsilon^2 N \ , \tag{5.72}$$

where N is a given operator (for white noise N = I).

The classical procedure of linear mean-square estimation can now be formulated as follows. A *linear estimator* of ξ will be any w.r.v. $\tilde{\xi}_L = L\eta$ where $L : G \to F$ is an arbitrary linear continuous operator. From a value g of η one obtains then a *linear estimate* of the possible values of ξ, $\tilde{f}_L = Lg$. Now we have to find some way of evaluating the validity of such an estimator. For instance, we can measure its validity in estimating the scalar r.v. $(\xi, w)_F$ (for any given element w in F) by the mean-square error

$$\delta^2(\varepsilon; w; L) = E\{|(\xi - L\eta, w)_F|^2\} \quad . \tag{5.73}$$

It is then natural to ask whether there exists an operator L_0 minimizing the error (5.73). If the covariance operator $R_{\zeta\zeta}$ has a bounded inverse, L_0 exists and is unique for any w in F. It is given by

$$L_0 = R_{\xi\eta}R_{\eta\eta}^{-1} = R_{\xi\xi}A^*[AR_{\xi\xi}A^* + R_{\zeta\zeta}]^{-1} \quad . \tag{5.74}$$

The w.r.v. $\tilde{\xi} = L_0\eta$ is called the *best linear estimator* of ξ and, given a value g of η, the *best linear estimate* $\tilde{f}$ for the value of ξ is

$$\tilde{f} = R_{\xi\xi}A^*[AR_{\xi\xi}A^* + R_{\zeta\zeta}]^{-1}g \quad . \tag{5.75}$$

Let us just sketch the proof of this result (see, e.g. [Ref.5.7, Chap.6] or [5.20]). Since $R_{\zeta\zeta}$ has a bounded inverse, $R_{\eta\eta}$ has also a bounded inverse and $L_0 = R_{\xi\eta}R_{\eta\eta}^{-1}$ is a linear continuous operator from G into F. On the other hand, using (5.66,68), one can write

$$\begin{aligned} E\{|(\xi - L\eta, w)_F|^2\} &= ([R_{\xi\xi} - R_{\xi\eta}L^* - LR^*_{\xi\eta} + LR_{\eta\eta}L^*]w, w)_F \\ &= ([L - L_0]R_{\eta\eta}[L^* - L_0^*]w, w)_F + ([R_{\xi\xi} - L_0R_{\eta\eta}L_0^*]w, w)_F \end{aligned} \tag{5.76}$$

and, since the operator $(L - L_0)R_{\eta\eta}(L^* - L_0^*)$ is positive definite when $L \neq L_0$, the minimum is attained if and only if $L = L_0$. Let us still remark that the previous result can be extended to the case where $R_{\zeta\zeta}^{-1}$ is not bounded: there exists a unique continuous operator L_0 minimizing (5.73) if and only if the operator $R_{\xi\eta}R_{\eta\eta}^{-1}$ is bounded on its domain [5.21].

We consider also the case of a w.r.v. ξ with a finite *variance* defined by

$$E\{\|\xi\|_F^2\} = E\,\{\sum_{n=0}^{+\infty}(u_n, \xi)_F(\xi, u_n)_F\} < +\infty \quad , \tag{5.77}$$

where $\{u_n\}$ is a basis in F [note that (5.77) does not depend upon the choice of a particular basis]. Let us remark that (5.77) can also be written with the help of the covariance operator $R_{\xi\xi}$, using (5.66)

$$E\{\|\xi\|_F^2\} = \sum_{n=0}^{+\infty}(R_{\xi\xi}u_n, u_n) = \mathrm{Trace}\,(R_{\xi\xi}) \quad . \tag{5.78}$$

Hence we see that ξ has finite variance if and only if $R_{\xi\xi}$ has a finite trace (one says then that $R_{\xi\xi}$ is a nuclear or trace class operator). When ξ has finite variance, we may define the following "global" mean-square error (for the estimator $\tilde{\xi}_L = L\eta$)

$$\delta^2(\varepsilon; L) = E\{\|\xi - L\eta\|_F^2\} \ , \tag{5.79}$$

which will be finite if and only if $L\eta$ has also a finite variance. When it exists, the operator L_0, which minimizes (5.73), minimizes also (5.79) if and only if $L_0\eta$ has a finite variance. When $R_{\zeta\zeta}$ has a bounded inverse, the previous condition is satisfied if the operator $L_0 = R_{\xi\eta}R_{\eta\eta}^{-1}$ is of the Schmidt class, i.e., it satisfies the condition trace $(L_0^*L_0) < +\infty$.

Now, as in Sect.5.2.3, let us briefly discuss the situations where (5.75) can be conveniently represented by means of eigenfunction (singular function) expansions. We consider first the case of a compact operator A, using the same notations as in Sect.5.2.3. We expand ξ and ζ in terms of the eigenfunctions of the operators A*A and AA*, respectively; their Fourier components are the random variables $\xi_n = (\xi, u_n)_F$, $\zeta_m = (\zeta, v_m)_G$. Then we assume [in addition to I) - III)] that

IV) *the Fourier components of ξ are mutually uncorrelated as well as the Fourier components of ζ.*

Equivalently, the following representations for $R_{\xi\xi}$ and $R_{\zeta\zeta}$ are valid:

$$R_{\xi\xi}f = \sum_{n=0}^{+\infty}\rho_n^2 f_n u_n \ , \quad R_{\zeta\zeta}g = \varepsilon^2\sum_{m=0}^{+\infty}\nu_m^2 g_m v_m \ , \tag{5.80}$$

where $f_n = (f, u_n)_F$, $g_m = (g, v_m)_G$, ρ_n^2 is the variance of ξ_n and $\varepsilon^2\nu_m^2$ the variance of ζ_m [recall (5.72)].

Then the best linear estimate (5.75) becomes

$$\tilde{f} = \sum_{n=0}^{+\infty}\frac{\alpha_n\rho_n^2}{\alpha_n^2\rho_n^2+\varepsilon^2\nu_n^2}\,g_n u_n \tag{5.81}$$

and it results that the operator L_0 is bounded if and only if $\sup(\alpha_n\rho_n^2\nu_n^{-2}) < +\infty$ [5.21,22].

Equation (5.81) can also be written as follows:

$$\tilde{f} = \sum_{n=0}^{+\infty} [1 - \exp(-2J_n)] \frac{g_n}{\alpha_n} u_n , \tag{5.82}$$

where

$$J_n = 1/2 \ln\left(1 + \frac{\alpha_n^2 \rho_n^2}{\varepsilon^2 \nu_n^2}\right) . \tag{5.83}$$

This form is interesting because, in the case of Gaussian processes, we recognize in J_n the average mutual information contained in the scalar random variables $\xi_n = (\xi, u_n)_F$ and $\eta_n = (\eta, \nu_n)_G$ (see, e.g. [5.44]). Indeed, we have

$$J_n = -1/2 \ln(1 - r_n^2) , \tag{5.84}$$

where r_n, given by

$$r_n^2 = \frac{\alpha_n^2 \rho_n^2}{\alpha_n^2 \rho_n^2 + \varepsilon^2 \nu_n^2} = \frac{|E[\xi_n \eta_n^*]|^2}{E[|\xi_n|^2] E[|\eta_n|^2]} \tag{5.85}$$

is precisely the correlation coefficient of ξ_n and η_n. The best linear estimate $\tilde{f}$ hence appears as a penalized version of the unstable formal solution $A^{-1}g = \sum_n \alpha_n^{-1} g_n u_n$, where the penalized components are those components g_n containing too little information about the components f_n.

A truncated solution, similar to (5.37), can be obtained by introducing the set $I(\varepsilon)$ of the values of the index n such that $\alpha_n \rho_n \geq \varepsilon \nu_n$. This condition is equivalent to require $r_n^2 \geq 1/2$ or $J_n \geq (\ln 2)/2$. If $r_n \to 0$ when $n \to +\infty$ [this condition is assured by the condition $\sup(\alpha_n \rho_n^2/\nu_n^2) < +\infty$ which implies the convergence of (5.81) for any g in G], then the set $I(\varepsilon)$ is finite and we can consider the finite sum

$$\sum_{n \in I(\varepsilon)} \frac{g_n}{\alpha_n} u_n . \tag{5.86}$$

It can be proved [5.22] that $\tilde{L}_0$ minimizes (5.73) when we consider linear estimators with only a finite number of components different from zero.

As a second example, let us consider the case where A is a convolution operator — see (5.38). Besides we assume that both ξ and ζ are *stationary processes* with autocovariance functions $R_{\xi\xi}(x - y)$ and $R_{\zeta\zeta}(x - y)$, respectively. Let $S_{\xi\xi}(\nu)$ and $S_{\zeta\zeta}(\nu)$ be the *power spectra* (spectral density) of ξ and ζ, respectively; then (5.75) takes the usual form of a *Wiener filter* [5.40,41]

$$\tilde{f}(x) = \int_{-\infty}^{+\infty} \frac{\hat{K}^*(\nu) S_{\xi\xi}(\nu)}{|\hat{K}(\nu)|^2 S_{\xi\xi}(\nu) + S_{\zeta\zeta}(\nu)} \hat{g}(\nu)\, e^{2\pi i x \nu} d\nu . \tag{5.87}$$

5.3.3 Mean-Square Errors

The r.v. $(\xi - L_0\eta, w)_F$ is the error we commit when taking $(L_0\eta, w)_F$ as an estimator of $(\xi, w)_F$. Its variance is

$$\delta^2(\varepsilon; w) = E[|(\xi - L_0\eta, w)_F|^2] \tag{5.88}$$

and therefore $\delta(\varepsilon; w)$ is the *mean-square error* in the estimation of $(\xi, w)_F$. The parameter ε is defined by (5.72). For simplicity let us consider only the case where $R_{\zeta\zeta}$ has a bounded inverse. Then the optimum filter L_0 certainly exists and, from (5.74,76) it follows that

$$\delta(\varepsilon; w) = ([R_{\xi\xi} - L_0R_{\eta\eta}L_0^*]w, w)_F^{\frac{1}{2}} \quad . \tag{5.89}$$

Furthermore, it is possible to prove that, when both inverse operators $R_{\xi\xi}^{-1}$ and A^{-1} exist, then $\delta(\varepsilon; w) \to 0$ when $\varepsilon \to 0$, for any w in F [5.21].

It is now natural to define a *relative mean-square error* as being the ratio between the variance of the error and the variance of the estimated r.v. $(\xi, w)_F$ [5.20]. Since $E[|(\xi, w)_F|^2] = (R_{\xi\xi}w, w)_F$, we get from (5.89)

$$\delta_{rel}(\varepsilon; w) = \frac{([R_{\xi\xi}-L_0R_{\eta\eta}L_0^*]w,w)_F^{\frac{1}{2}}}{(R_{\xi\xi}w,w)_F^{\frac{1}{2}}} \quad . \tag{5.90}$$

This quantity gives a precise measure of the *reliability* of the estimate. It is interesting to remark [5.20] that, when A^{-1} is not continuous, one can find a sequence $\{w^{(n)}\}$ such that $\delta_{rel}(\varepsilon, w^{(n)}) \to 1$ when $n \to +\infty$, for fixed ε. In other words, in the case of an ill-posed problem, for any value of $\varepsilon > 0$, there will be vectors w in F for which the r.v. $(\xi, w)_F$ cannot be reliably estimated.

When $R_{\xi\xi}$ is of the trace class, i.e., ξ has a finite variance - see (5.78) - and $R_{\zeta\zeta}$ has a bounded inverse, then the optimum filter L_0 is of the Schmidt class and one can define a "global" mean-square error as $\delta(\varepsilon) = \delta(\varepsilon; L_0)$ - see (5.79). An expression for $\delta(\varepsilon)$, similar to (5.89), can be derived remarking that $\delta^2(\varepsilon)$ is the trace of the covariance operator of $\xi - L_0\eta$

$$\delta(\varepsilon) = [\mathrm{Trace}\{R_{\xi\xi} - L_0R_{\eta\eta}L_0^*\}]^{\frac{1}{2}} \quad . \tag{5.91}$$

When the inverse operators $R_{\xi\xi}^{-1}$ and A^{-1} both exist, then one can prove that $\delta(\varepsilon) \to 0$, when $\varepsilon \to 0$ [5.21].

In the case of a compact operator A, when assumption IV) of Sect.5.3.2 is satisfied, (5.89) becomes

$$\delta(\varepsilon; w) = \varepsilon\left(\sum_{n=0}^{+\infty} \frac{\rho_n^2\nu_n^2}{\alpha_n^2\rho_n^2+\varepsilon^2\nu_n^2} |w_n|^2\right)^{\frac{1}{2}} , \tag{5.92}$$

where $w_n = (w, u_n)_F$. It is quite easy to show that $\delta(\varepsilon; w) \to 0$ when $\varepsilon \to 0$ [5.21], under the sole condition that the operators $R_{\xi\xi}^{-1}$ and $R_{\zeta\zeta}^{-1}$ exist [5.21,22].

As regards the "global" mean-square error (5.91), it becomes

$$\delta(\varepsilon) = \varepsilon \left(\sum_{n=0}^{+\infty} \frac{\rho_n^2 \nu_n^2}{\alpha_n^2 \rho_n^2 + \varepsilon^2 \nu_n^2} \right)^{\frac{1}{2}} . \tag{5.93}$$

It is also easy to show that $\delta(\varepsilon) \to 0$, when $\varepsilon \to 0$, provided that ξ has a finite variance, i.e., $\Sigma_n \rho_n^2 < +\infty$ [5.21].

In the case of a convolution operator A and of stationary processes ξ, ζ, one can only define the mean-square error (5.89). Indeed, the covariance operator of a stationary process is never of the trace class. An expression for $\delta(\varepsilon; w)$ can be easily derived using (5.87,89).

5.3.4 Comparison with Miller's Regularization Method

In Miller's method, the estimates of the solution of the problem (5.17) have to belong to the set K defined by (5.21,22) and this is a "rigid" condition, in the sense that all the functions outside K are rejected as meaningless. In probabilistic methods, the restrictions are less categorical since one considers the probability distributions of the solutions and of the errors. In fact, the knowledge of $R_{\zeta\zeta}$ corresponds to the bound (5.21) for the error, while the knowledge of $R_{\xi\xi}$ corresponds to the bound (5.22) for the solution. Moreover, both Miller's method and optimum filtering are least square methods. Hence, it is not surprising to find similarities between the solutions provided by the two methods. Indeed, thanks to the following identity, valid when $R_{\xi\xi}^{-1}$ and $R_{\zeta\zeta}^{-1}$ exist

$$(A^* R_{\zeta\zeta}^{-1} A + R_{\xi\xi}^{-1}) R_{\xi\xi} A^* = A^* R_{\zeta\zeta}^{-1} (A R_{\xi\xi} A^* + R_{\zeta\zeta}) \tag{5.94}$$

it is easy to show that (5.27) and (5.75) coincide formally when putting

$$R_{\zeta\zeta} = \varepsilon^2 I \ , \quad R_{\xi\xi} = E^2 (B^* B)^{-1} \quad . \tag{5.95}$$

In fact, the condition introduced by Miller in order to restore continuous dependence on the data (i.e., that the constraint operator B should have a bounded inverse) corresponds to the condition that the w.r.v. ξ has finite second moment, i.e., there exists a bounded operator $R_{\xi\xi}$ defined by (5.66). Moreover, the mean-square error (5.89) can be considered as the analogue of the stability estimate (5.60) for the restoration of "blurred solutions". Thanks to the identification (5.95), they coincide up to a factor $\sqrt{2}$. Looking at the relative error (5.90), we are tempted to define its fellow in regularization theory by

$$\delta_{rel}(\varepsilon, E; w) = \frac{\varepsilon}{E} \frac{(C^{-1}w,w)_F^{\frac{1}{2}}}{([B^*B]^{-1}w,w)_F^{\frac{1}{2}}} \quad . \tag{5.96}$$

In fact, it is also possible to derive this formula in an intrinsic way, without reference to its probabilistic analogue [5.26]. In spite of different starting points, the similarities of the solutions and of the error estimates are very interesting and therefore the conjunction of both points of view can provide complementary insights on the regularization of linear inverse problems.

For the reader's convenience we summarize the main analogies between Miller's regularization method and optimum filtering in the following scheme.

	Regularization method
Data	the function g; $g = Af + h$
Prior knowledge	$\|h\|_G = \|Af - g\|_G \leq \varepsilon$, $\|Bf\|_F \leq E$; knowledge of ε, E and of the operator B
Requirement	an estimate of the function f
Solution	$\tilde{f} = [A^*A + (\frac{\varepsilon}{E})^2 B^*B]^{-1}A^*g$

	Optimum filtering
Data	a value g of the r.v. $\eta = A\xi + \zeta$
Prior knowledge	ξ, ζ are zero mean, uncorrelated r.v.; knowledge of the covariance operators $R_{\xi\xi}$, $R_{\zeta\zeta}$
Requirement	an estimate of a value of ξ
Solution	$\tilde{f} = R_{\xi\xi}A^*[AR_{\xi\xi}A^* + R_{\zeta\zeta}]^{-1}g$

5.4 Linear Inverse Problems in Optics

Surveys of inverse problems in optics and electromagnetics can be found in [5.9,45]. Due to the rapid growth of research in this field, we do not attempt a complete review. Our aim is to focus on stability problems and therefore we select only a few examples, using for simplicity the scalar theory of light. The harmonic time dependence $\exp(-i\omega t)$ is assumed, and by wave functions we mean scalar complex amplitudes.

5.4.1 Inverse Problems in Fourier Optics

The problem of restoring data that have been degraded by a linear bandlimited system has for long received much attention both in optics [5.46] and in radio astronomy [5.47,48]. For simplicity we will consider only a one-dimensional system. Then, in the absence of noise, such a system is described by a linear equation like

$$\int_{-X/2}^{X/2} S(x - y)\bar{f}(y)dy = \bar{g}(x) \ , \tag{5.97}$$

where S(x), the *point spread function*, has a Fourier transform which vanishes outside a finite interval $[-\Omega/2, \Omega/2]$, $\bar{f}$ is the wave function in the object plane and $\bar{g}$ the wave function in the image plane. We call $\bar{f}$ the object and $\bar{g}$ the noiseless image.

Since $\bar{f}$ is zero outside the interval $[-X/2, X/2]$, its Fourier transform is an entire analytic function. So it was observed [5.46,48] that analytic continuation in the frequency domain will in principle allow for restoration of unlimited details of $\bar{f}$. As remarked by many authors [5.49,50], this result seems to be in contradiction with the concept of *number of degrees of freedom of an image* [5.51-53], which essentially means that the image never contains enough information to reconstruct the object unambiguously. The contradiction disappears if one takes into account the noise and *logarithmic continuity*, which arise for object restoration.

Prolate Spheroidal Wave Functions (PSWF)

We summarize here the main properties of the prolate spheroidal wave functions $\psi_n(c, x)$ [5.54-56], which are a fundamental tool for the analysis of bandlimited systems. The $\psi_n(c, x)$ can be defined as the continuous solutions, on the closed interval [-1, 1], of the differential equation

$$-[(1 - x^2)\psi'(x)]' + c^2x^2\psi(x) = \chi\psi(x) \quad . \tag{5.98}$$

Continuous solutions exist only for certain discrete positive values χ_n of the parameter $\chi : 0 < \chi_0 < \chi_1 < \dots$. Then $\psi_n(c, x)$ is just the solution of (5.98) corresponding to the eigenvalue χ_n. The behavior of χ_n when $n \to +\infty$ is [5.57]

$$\chi_n = n(n + 1) + \frac{1}{2} c^2 + O\left(\frac{1}{n^2}\right) \quad . \tag{5.99}$$

The $\psi_n(c, x)$ can be uniquely extended to entire analytic functions, and they will be normalized as follows:

$$\int_{-\infty}^{+\infty} |\psi_n(c, x)|^2 dx = 1 \ ; \quad n = 0,1,2, \dots \quad . \tag{5.100}$$

The PSWF are also solutions of the eigenvalue equation

$$\int_{-1}^{1} \frac{\sin[c(x-y)]}{\pi(x-y)} \psi_n(c, y)dy = \lambda_n \psi_n(c, x) \quad . \tag{5.101}$$

The eigenvalues λ_n form a decreasing sequence: $1 > \lambda_0 > \lambda_1 > \ldots > 0$ and have a step behavior: they are approximately equal to one for values of the index less than $N_0 = 2c/\pi$ and then fall off to zero exponentially. More precisely, their behavior for $n \to +\infty$ is [5.58]

$$\lambda_n = O\{\frac{1}{n} \exp[-2n \ln(\frac{n}{ec})]\} \quad . \tag{5.102}$$

The eigenvalues λ_n are also the normalization constants of the PSWF on the interval [-1, 1]

$$\int_{-1}^{1} |\psi_n (c, x)|^2 dx = \lambda_n \quad . \tag{5.103}$$

The fundamental properties of the PSWF are:

a) The $\psi_n(c, x)$ are bandlimited functions; their Fourier transforms vanish outside the interval $[-c/2\pi, c/2\pi]$

$$\int_{-\infty}^{+\infty} e^{-2\pi i \nu x} \psi_n(c, x)dx = (-i)^n \sqrt{\frac{2\pi}{c\lambda_n}} \psi_n(c, \frac{2\pi\nu}{c}) \theta(\frac{2\pi\nu}{c}) \quad , \tag{5.104}$$

where $\theta(s) = 1$ for $|s| < 1$ and $\theta(s) = 0$ for $|s| > 1$ (see, e.g. [5.59]).

b) The $\psi_n(c, x)$ are a basis in the space of the square-integrable bandlimited functions.

c) The functions $u_n(x) = \lambda_n^{-\frac{1}{2}} \psi_n(c, x)$ are a basis in $L^2(-1, 1)$.

Statements b) and c) exhibit a remarkable property of the PSWF: they are orthogonal over two different intervals. This property is fundamental for the extrapolation of bandlimited functions.

Perfect Lowpass Filter

We consider first (5.97) with the point spread function $S(x) = (\pi x)^{-1} \sin(\pi\Omega x)$ (perfect lowpass filter through the band $[-\Omega/2, \Omega/2]$). The connection with the general formulation of a linear inverse problem, given in Sect.5.1.3, is as follows. Since the object radiates a finite power, we can take $L^2(-X/2, X/2)$ as solution space F. Assuming that the noisy image g is known only on the interval $[-X/2, X/2]$, we can also take $L^2(-X/2, X/2)$ as data space G. Then, object restoration consists in inverting the integral operator

$$(Af)(x) = \int_{-X/2}^{X/2} \frac{\sin[\pi\Omega(x - y)]}{\pi(x - y)} f(y)dy \quad . \tag{5.105}$$

The operator A is self-adjoint, nonnegative and compact. The quantity $R = \Omega^{-1}$ is the *Rayleigh resolution distance* and $N_0 = \Omega X$ is the *number of degrees of freedom* of the image. Observe that N_0 is the number of eigenvalues of A which are approximately equal to one and also that $N_0 = \mathrm{Trace}(A)$ [5.60]. In fact the eigenvalues of A are the λ_n associated to the PSWF with $c = \pi\Omega X/2$ and the corresponding eigenfunctions are

$$u_n(x) = \left(\frac{2}{X\lambda_n}\right)^{\frac{1}{2}}\psi_n\left(c, \frac{2x}{X}\right) \; , \quad c = \pi\Omega X/2 \quad . \tag{5.106}$$

In order to apply to this problem the general results of Sect.5.2.2, we need a constraint operator B commuting with A. This requirement is satisfied by the differential operator

$$(B^*Bf)(x) = -[(\tfrac{1}{4}X^2 - x^2)f'(x)]' + c^2x^2f(x) \tag{5.107}$$

since, from the definition of the PSWF, it follows that the u_n, defined by (5.106), are the eigenfunctions of B^*B and the χ_n are the corresponding eigenvalues. Furthermore, by means of a partial integration one gets

$$(B^*Bf, f)_F = \int_{-X/2}^{X/2} (\tfrac{1}{4}X^2 - x^2)|f'(x)|^2dx + c^2 \int_{-X/2}^{X/2} x^2|f(x)|^2dx \tag{5.108}$$

so that condition (5.22) is a constraint on the first derivative of f. Note that (5.108) has the same form as (5.44); however, the functions p(x) and q(x) in (5.108) are not strictly positive. Hence the set defined by (5.22,108) is not compact with respect to the uniform norm (5.15) (see [Ref.5.36, p.195]) but it is compact with respect to the L^2-norm.

Now the restored object is given by (5.36) with $\alpha_n = \lambda_n$, $\beta_n = \sqrt{\chi_n}$ and

$$g_n = \int_{-X/2}^{X/2} g(x)u_n(x)dx \tag{5.109}$$

or by the truncated solution (5.37). It is interesting to remark that, since the λ_n decrease exponentially fast when $n > N_0$, while $\beta_n \sim n$, the number of terms N in (5.37) is equal to N_0 (the number of degrees of freedom) plus a number of terms which is roughly proportional to $|\ln \varepsilon|$, which number is therefore rather insensitive to the noise. This fact is strictly related to logarithmic continuity [5.61]. Indeed, if we consider the stability estimate (5.53), with $\alpha_n = \lambda_n$ and $\beta_{n_1} = \sqrt{\chi_n}$, from the behavior (5.102) and (5.99), we conclude that $\delta(\varepsilon, E) \sim E|\ln(\varepsilon/E)|^{-1}$. This result can be extended to the case where a finite number of derivatives of f are bounded [5.22, 36]. The following statements are justified.

I) The error on the restored object tends to zero when $\varepsilon \to 0$ and therefore, *in principle*, unlimited resolution of details is possible (at least in the framework of classical optics, since the previous analysis does not take into account the quantum-mechanical limitations on measurement of the light field [5.62]).

II) When the object is not extremely smooth (what is equivalent to say that its Fourier transform is not negligible outside the band $[-\Omega/2, \Omega/2]$ – see [5.22,61]), the error on the restored object tends to zero so slowly that, *in practice*, resolution beyond the limit corresponding to the number of degrees of freedom becomes impossible. This conclusion agrees with the results of earlier analysis of object restoration [5.63,64].

More precise results about resolution can be obtained by considering the restoration of "blurred solutions" as sketched in Sect.5.2.4. The "blurred object" is given by (5.56) (the integration ranging now over $[-X/2, X/2]$) at least when $D \ll X$ and x_0 is sufficiently far from the borders. Then the error in the restoration of $f_D(0) = (f, w_D)_F$ is an estimate of the error we commit in the restoration of details whose size is D. One should expect a trade-off between resolution and error [5.63]: the restoration error has to be greater for smaller values of D.

In order to analyze the effect on resolution of different types of noise [5.63], we focus on optimum filtering methods. The relative error in the restoration of $f_D(0)$ is given by (5.90) with $w = w_D$. Let us assume, for simplicity, that the stochastic processes describing object and noise satisfy the assumptions I) - IV) of Sect.5.3.2. Besides we assume that the variances of the Fourier components of the object [with respect to the basis (5.106)] are constant, i.e., $\rho_n^2 = E^2$. This assumption is reasonable if the correlation distance δ for the stochastic process representing the object (δ gives the size of the finest details that should be resolved) is much smaller than the Rayleigh distance R. Then, from (5.90,92) we get

$$\delta_{rel}\left(\frac{\varepsilon}{E}; w_D\right) = \left(1 - \frac{1}{\|w_D\|^2} \sum_{n=0}^{+\infty} \frac{\lambda_n^2}{\lambda_n^2 + (\varepsilon/E)^2 \nu_n^2} |w_{D,n}|^2\right)^{\frac{1}{2}} , \qquad (5.110)$$

where $\|w_D\|$ is the norm of w_D in $L^2(-X/2, X/2)$ and the $w_{D,n}$ are the Fourier coefficients of w_D in the basis (5.106). We consider two types of noise in the image plane [5.63]: *white measurement noise*, i.e., $\nu_n^2 = 1$, and *band-limited measurement noise*, i.e., $\nu_n^2 = \lambda_n$. In both cases it is easy to show [5.26] that $\delta_{rel}(\varepsilon/E; w_D) \to 1$ (100% error) when $D \to 0$, i.e., when w_D tends to the Dirac delta measure.

In Fig.5.4 we give the results of numerical computations (the numerical method is described in [5.26]) for $c = 10$, $X = 2$ and $w_D(x) = N\theta(x/d)\,\mathrm{sinc}^2(x/d)$ [N is a normalization constant, $\theta(s)$ is the characteristic function of the interval $[-1, 1]$ and d is a parameter related to D through (5.55)]. In the case of white measurement noise, the curves are rapidly decreasing up to a value of D/R of about 0.5 and then become rather flat. Besides a lowering of ε from 10^{-2} to 10^{-6} does not modify the situation in a significant way. If we accept only an error of a few percent, then it is difficul to get a resolution better than R. One expects that superresolution should become even more difficult for greater values of c [Ref.5.65, p.470]. Figure 5.4 also shows that a smoothing of the noise (band-limited measurement noise) [5.63] is equivalent to a lowering of the white noise from 10^{-2} to 10^{-6}.

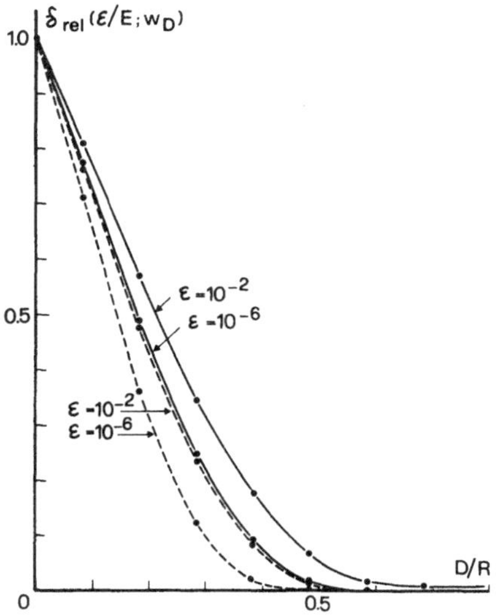

Fig. 5.4. Relative errors vs the resolution parameter D/R, where $R = \pi X/2c$ ($X = 2$, $c = 10$). Undotted curves correspond to white measurement noise and dotted curves to bandlimited measurement noise. In both cases $E = 1$

The case of *incoherent illumination* in the object plane [point spread function $S(x) = \Omega(\pi\Omega x)^{-2} \sin^2(\pi\Omega x)$, frequency band $[-\Omega, \Omega]$] has been analyzed by many authors [5.66-68]. In this case, as well as for any bandlimited system, the restored continuity is also logarithmic. Indeed, thanks to the Paley-Wiener theorem [5.69], the point spread function of a bandlimited system, with bandwidth Ω, is an entire analytic function of order ρ, $0 < \rho \leq 1$, and type $\tau \leq \pi\Omega$. Then, it follows from general results of HILLE and TAMARKIN [5.24] that the singular values of the integral operator defined by (5.97) have the behavior $\alpha_n \sim C \exp(-Dn \ln n)$, $n \to +\infty$, where C, D are suitable constants. As we remarked in Sect. 5.2.4, this behavior implies logarithmic continuity. Therefore it should be possible to define, for any bandlimited system, a resolution limit and a number of degrees of freedom, practically noise independent. For instance, in the case of incoherent illumination, this number has been estimated to be of the order of $2\Omega X$ (the bandwidth is 2Ω), corresponding to a resolution limit of about R/2 [5.68]. This result agrees with an analysis of the restoration of "blurred objects" [5.26]. However it must be remarked that, in the case of incoherent illumination the solutions have to satisfy a positivity constraint which could improve the resolution limit (for a brief discussion of this point see Sect.5.4.6).

Bandwidth Extrapolation

A method which has been proposed for the analysis of an arbitrary bandlimited system is the following [5.46,48,70-72]: by Fourier transforming the image one can obtain the Fourier transform of the object over the band $[-\Omega/2, \Omega/2]$ of the system; since this Fourier transform is an entire analytic function, by analytic continuation one could restore its values for all frequencies and therefore also restore the object. We remark that this problem is mathematically equivalent to the extrapolation of optical images beyond borders [5.59,73].

Let f be the object and $\hat{f}$ its Fourier transform

$$\hat{f}(\nu) = \int_{-X/2}^{X/2} e^{-2\pi i\nu x} f(x)dx \quad . \qquad (5.111)$$

The data $\hat{g}$ are the values (affected by errors of $\hat{f}$ on the band $[-\Omega/2, \Omega/2]$ and we can take $L^2(-\Omega/2, \Omega/2)$ as data space G. The solution space F is the set of the functions $\hat{f}$ like (5.111), normed with the norm of $L^2(-\infty, +\infty)$. Then F is a Hilbert space of analytic functions. The direct problem is merely the restriction of $\hat{f}(\nu)$ to $[-\Omega/2, \Omega/2]$, i.e., $(A\hat{f})(\nu) = \hat{f}(\nu)$ when $|\nu| < \Omega/2$, $(A\hat{f})(\nu) = 0$ elsewhere. Then $A^* : G \to F$ is the integral operator [5.22]

$$(A^*\hat{g})(\nu) = \int_{-\Omega/2}^{\Omega/2} \frac{\sin[\pi X(\nu-\nu')]}{\pi(\nu-\nu')} \hat{g}(\nu')d\nu' \quad . \qquad (5.112)$$

Observing that AA^* coincides with the integral operator (5.105), except for an exchange of Ω and X, one can write for these operators A, A^* equations like (5.32) where $\alpha_n = \sqrt{\lambda_n}$ (the eigenvalues of the PSWF) and u_n, v_n are replaced by ($c = \pi\Omega X/2$)

$$\hat{u}_n(\nu) = \sqrt{\frac{2}{\Omega}}\, \psi_n(c, \frac{2\nu}{\Omega}) \;, \quad \hat{v}_n(\nu) = \sqrt{\frac{2}{\Omega\lambda_n}}\, \psi_n(c, \frac{2\nu}{\Omega})\theta(\frac{2\nu}{\Omega}) \quad . \qquad (5.113)$$

From properties II), III) of the PSWF it follows that $\{\hat{u}_n\}$ is a basis in F while $\{\hat{v}_n\}$ is a basis in G. Remark that the expansion of $\hat{f}(\nu)$ as a series of the $\hat{u}_n$ is equivalent to the expansion of f(x), in (5.111), as a series of the u_n given in (5.106). Indeed $i^n u_n$ and $\hat{u}_n$ are related by (5.104).

BUCK and GUSTINCIC [5.70] assumed that the stochastic processes, representing the object and the noise, satisfy conditions I) - IV) of Sect. 5.3.2 and that $\rho_n^2 = E^2$ (where the ρ_n^2 are the variances of the components of the object in the basis $\{u_n\}$). In the case of white noise, their solution to the problem of analytic continuation is given by (5.85) with $\alpha_n = \sqrt{\lambda_n}$, $\rho_n^2 = E^2$, $\nu_n^2 = 1$, $g_n = (\hat{g}, \hat{v}_n)_G$ and u_n replaced by $\hat{u}_n$ (we denote this estimate by $\hat{\hat{f}}$). It has been remarked by these authors that an increase by a factor of 10 in the signal-to-noise ratio E/ε adds only one more significant term in the series (5.85) for $\hat{\hat{f}}$ (and therefore, for large apertures, the improvement in resolution is negligible). The same estimate f was obtained by VIANO [5.73] in the framework of regularization theory, considering the constraint operator B = I [compare with (5.36) where $\alpha_n = \sqrt{\lambda_n}$, $\beta_n = 1$, $g_n = (\hat{g},\hat{v}_n)_G$ and u_n replaced by $\hat{u}_n$]. This assumption is equivalent to requiring that the object radiates a finite power. VIANO [5.73] proved that such an estimate converges to the "true solution", when $\varepsilon \to 0$, uniformly over any finite interval containing the band $[-\Omega/2, \Omega/2]$. In order to have stability with respect to the norm of F, stronger conditions on the objects are required. If we introduce, for instance, the constraint operator (5.107), then it is easy to prove (as in the case of the perfect lowpass filter) that the analytic continuation of the Fourier transform of the object outside the band $[-\Omega/2, \Omega/2]$ is stable with respect to the norm of F, but that we get only logarithmic continuity. This result agrees with the conclusions of [5.70].

5.4.2 Inverse Diffraction

According to SHERMAN [5.76] and SHEWELL and WOLF [5.74], inverse diffraction can be defined as the problem of determining the field distribution on a boundary surface from the knowledge of the distribution on a surface situated within the domain where the wave propagates. An extensive treatment of uniqueness in inverse diffraction is given by HOENDERS [5.75] both in the scalar and in the vector case.

The fundamental reason of the instability of inverse diffraction is that space acts like a filter for the higher modes. For instance, in the scattering of a plane wave, with wave number k, by a body whose largest dimension is R, only kR modes are propagated up to the far zone, while the others are attenuated.

Inverse Diffraction from Plane to Plane

The direct problem is to determine a wave function u, solution of Helmholtz equation in the half-space $z \geqq z_0$

$$\nabla^2 u + k^2 u = 0 \tag{5.114}$$

satisfying Sommerfeld's condition at infinity and the condition $u = u_0$ (u_0 being a given function) on the plane $z = z_0$. This solution can be most conveniently expressed in terms of Fourier transforms [5.74,76]. If we write

$$\hat{u}(p, q; z) = \iint_{-\infty}^{+\infty} e^{-ik(px + qy)} u(x, y, z) dx\, dy \ , \tag{5.115}$$

then

$$\hat{u}(p, q; z) = \exp[ikm(z - z_0)]\hat{u}(p, q; z_0) \ , \tag{5.116}$$

where

$$m = (1 - p^2 - q^2)^{\frac{1}{2}} \ , \quad \mathrm{Im}(m) \geqq 0 \ . \tag{5.117}$$

The inverse problem is the following: given the values g (affected by errors) of the wave function u on the plane $z = z_1 > z_0$, estimate the values of u on any z plane between z_0 and z_1. If the radiating power is finite, then u is square integrable over any z plane and therefore we can take $L^2(R^2)$ both as solution and as data space. Writing $\bar{f}(x, y) = u(x, y, z)$, $\bar{g}(x, y) = u(x, y, z_1)$ ($\bar{g}$ is the noiseless wave function), from (5.116) we derive that $\bar{g} = A\bar{f}$, where

$$(Af)\hat{}(p, q) = \exp[+ im(z_1 - z)]\hat{f}(p, q) \ . \tag{5.118}$$

Instability is due to the effect of inhomogeneous (evanescent) waves $(p^2 + q^2 > 1)$. Assuming that f is generated by a field distribution on the plane $z = z_0$ (with an L^2-norm bounded by E^2), we get a class of admissible solutions defined by the constraint operator

$$(Bf)\hat{}(p, q) = \exp[-im(z - z_0)]\hat{f}(p, q) \ . \tag{5.119}$$

The operators A, B have the form (5.38) and (5.40), respectively, with $\hat{K}(p, q) = \exp[im(z_1 - z)]$ and $\hat{\beta}(p, q) = \exp[-im(z - z_0)]$ so that the regularized solution is given by (5.41) or (5.42). Besides, observing that $\hat{\beta}(p, q) = [\hat{K}(p, q)]^{-\mu}$, $\mu = (z - z_0)/(z_1 - z)$, from (5.54) $\delta(\varepsilon, E) \sim E(\varepsilon/E)^{\alpha}$ follows, where $\alpha = (z - z_0)/(z_1 - z_0)$, $0 < \alpha < 1$.

This result is very similar to "three line theorems" derived by MILLER [5.77] in the case of the backward heat equation and the Cauchy problem for the Laplace equation. As we see, we get Hölder continuity if $z > z_0$, while we do not have stability for $z = z_0$ (in this case B = I). To restore the stability even there, we have to take stronger constraints. If we assume, for instance, that the wave function on the plane $z = z_0$ has also square integrable first derivatives, then we have stability up to the plane $z = z_0$. In this case, however, the restored continuity is only logarithmic (see Sect.5.2.4). Finally, if the wave function on the plane $z = z_0$ is assumed to contain only spatial frequencies below the wave number k, i.e., $\hat{u}(p, q; z_0) = 0$ if $p^2 + q^2 > 1$, then a well-behaved inversion formula can be derived [5.74]. In other words, inverse diffraction can be formulated as well-posed problem when the effect of evanescent waves can be disregarded. It has recently been shown [5.117-122] that the total field due to the inhomogeneous waves does not decay exponentially with distance z, but much slower ($z^{-3/2}$ or z^{-2}). In view of these results, the inverse diffraction problem seems to deserve reconsideration.

Inverse Diffraction for Cylindrical Waves

We consider a wave function $u = u(\rho, \varphi)$ (ρ, φ are circular cylinder coordinates), solution of (5.114), satisfying Sommerfeld's radiation condition at infinity and the condition $u = u_0$ on the circular cylinder of radius ρ_0. The solution of this problem (direct problem) is represented by the Fourier series

$$u(\rho, \varphi) = \sum_{n=-\infty}^{+\infty} \frac{H_n^{(1)}(k\rho)}{H_n^{(1)}(k\rho_0)} c_n e^{in\varphi} , \tag{5.120}$$

where the $H_n^{(1)}$ are the Hankel functions of the first kind and the c_n are the Fourier coefficients of $u_0(\varphi) = u(\rho_0, \varphi)$

$$c_n = \frac{1}{2\pi} \int_{-\pi}^{\pi} u(\rho_0, \varphi)\, e^{-in\varphi} d\varphi \quad . \tag{5.121}$$

The inverse problem is as follows: given the values g (affected by errors) of u on the cylinder of radius $\rho_1 > \rho_0$, estimate the wave function over any cylinder of radius ρ, $\rho_0 \leq \rho \leq \rho_1$. If we denote by $\bar{g}(\varphi) = u(\rho_1, \varphi)$ the noiseless data and by $\bar{f}(\varphi) = u(\rho, \varphi)$ ($\rho < \rho_1$) the unknown solution, then, from (5.120), $\bar{g} = A\bar{f}$ where

$$(A\bar{f})(\varphi) = \sum_{n=-\infty}^{+\infty} \frac{H_n^{(1)}(k\rho_1)}{H_n^{(1)}(k\rho)} \bar{f}_n\, e^{in\varphi} , \tag{5.122}$$

and the $\bar{f}_n = c_n H_n^{(1)}(k\rho)/H_n^{(1)}(k\rho_0)$ are the Fourier coefficients of $\bar{f}$. We take $L^2(-\pi, \pi)$ both as solution and data space. Then, if $\bar{f}(\varphi) = u(\rho, \varphi)$ is generated by a wave function on the cylinder of radius ρ_0, the class of admissible solutions on the cylinder of radius ρ is characterized by the constraint operator

$$(B\bar{f})(\varphi) = \sum_{n=-\infty}^{+\infty} \frac{H_n^{(1)}(k\rho_0)}{H_n^{(1)}(k\rho)} \bar{f}_n e^{in\varphi} , \tag{5.123}$$

which is derived from (5.120) and the equality $(B\bar{f})(\varphi) = u(\rho_0, \varphi)$. If we write $\alpha_n = |H_n^{(1)}(k\rho_1)/H_n^{(1)}(k\rho)|$, $\beta_n = |H_n^{(1)}(k\rho_0)/H_n^{(1)}(k\rho)|$, then the estimated wave function is given by (5.36) or (5.37) (remark that now the index n takes values from $-\infty$ to $+\infty$). From the behavior of Hankel functions when $|n| \to +\infty$, it follows that $\alpha_n \sim \exp[-|n| \ln(\rho_1/\rho)]$ while $\beta_n \sim \exp[|n| \ln(\rho/\rho_0)]$. These behaviors imply $\beta_n \sim \alpha_n^{-\mu}$ with $\mu = \ln(\rho/\rho_0)/\ln(\rho_1/\rho)$ and from (5.53) one has $\delta(\varepsilon, E) \sim E(\varepsilon/E)^{\alpha}$ with $\alpha = \ln(\rho/\rho_0)/ \ln(\rho_1/\rho_0)$ (for a more precise estimate see [Ref.5.78, Chap.3]). It is interesting to compare this result with the stability estimate for analytic continuation implied by Hadamard's "three circle theorem" (Sect.5.1.5). A similar result holds for harmonic continuation in a disc [5.77].

The constraint operator (5.123) does not imply stability in the L^2-norm when $\rho = \rho_0$ (in this case B = I). Then we have stability if, for instance,

$$\| B\bar{f}\|^2 = \|\frac{\partial \bar{f}}{\partial\varphi}\|^2 + \|\bar{f}\|^2 = \sum_{n=-\infty}^{+\infty} (n^2 + 1)|\bar{f}_n|^2 \leq E^2 . \tag{5.124}$$

Let us remark that, if the wave function u represents a z-polarized electric field, i.e., $E_\rho = E_\varphi = 0$, $E_z = u$, then the components H_ρ, H_φ of the magnetic field are proportional to $\rho^{-1}\partial u/\partial\varphi$ and $\partial u/\partial\rho$, respectively. Therefore, in this case, (5.124) implies a bound on both E_z and H_ρ. However this constraint gives only logarithmic continuity, since $\beta_n \sim n$ while α_n tends to zero exponentially (Sect.5.2.4).

The previous results can be easily extended to the case of spherical surfaces, using the expansion of the wave function as a series of spherical harmonics $Y_\ell^m(\theta, \varphi)$.

Inverse Diffraction from Far-Field Data

For simplicity we consider again cylindrical waves. Then the wave function (5.120) has the behavior

$$u(\rho, \varphi) \sim \sqrt{\frac{2}{i\pi k\rho}}\, e^{ik\rho}\bar{g}(\varphi), \quad \rho \to +\infty , \tag{5.125}$$

where $\bar{g}$ is the *scattering amplitude* (or *pattern function*)

$$\bar{g}(\varphi) = \sum_{n=-\infty}^{+\infty} \frac{(-i)^n}{H_n^{(1)}(k\rho_0)} c_n e^{in\varphi} . \tag{5.126}$$

The direct problem is to compute $\bar{g}$, given the wave function u_0 on the cylinder of radius ρ_0. The inverse problem is to estimate u on any cylinder of radius ρ, $\rho_0 \leq \rho < \infty$, given a noisy scattering amplitude g. If we write $\bar{f}(\varphi) = u(\rho, \varphi)$, then from (5.120,126) we get $\bar{g} = A\bar{f}$ where

$$(A\bar{f})(\varphi) = \sum_{n=-\infty}^{+\infty} \frac{(-i)^n}{H_n^{(1)}(k\rho)} \bar{f}_n e^{in\varphi} \quad . \tag{5.127}$$

We can use again (5.123) as a constraint operator. Then we have $\alpha_n = |H_n^{(1)}(k\rho)|^{-1}$, $\beta_n = |H_n^{(1)}(k\rho_0)/H_n^{(1)}(k\rho)|$ and it is easy to see (using the results of Sect.5.2.4) that the regularized solution (5.36) or (5.37) is stable with respect to the L^2-norm when $\rho > \rho$. For $\rho = \rho_0$ the constraint (5.124) implies at most logarithmic continuity. It is interesting to understand in which cases one can get Hölder continuity. Let us consider, for instance, a perfectly conducting circular cylinder of radius ρ_0, illuminated by a plane wave. Then, at the surface of the cylinder the scattered wave takes the values $u_0(\varphi) = -\exp(ik\rho_0 \cos\varphi) = u(\rho_0, \varphi)$. If we write $\bar{f}(\varphi) = u(\rho_0, \varphi)$, we have the Fourier expansion

$$\bar{f}(\varphi) = - \sum_{n=-\infty}^{+\infty} i^n J_n(k\rho_0) e^{in\varphi} \quad . \tag{5.128}$$

Since $|H_n^{(1)}(k\rho_0)J_n(k\rho_0)| \sim (\pi|n|)^{-1}$, $\bar{f}$ satisfies the condition

$$\| B\bar{f}\|^2 = \sum_{n=-\infty}^{+\infty} |H_n^{(1)}(k\rho_0)|^2 |\bar{f}_n|^2 \leq E^2 \quad , \tag{5.129}$$

where E is a suitable constant. Equation (5.129) implies $\beta_n = \alpha_n^{-1}$ and from (5.53) we get $\delta(\varepsilon, E) \sim E(\varepsilon/E)^{1/2}$, i.e., a rather good Hölder continuity. This arises also for scatterers with very smooth shape.

Finally, we discuss the angular resolution which can be obtained in the restoration of the wave function on the cylinder of radius ρ_0. We consider the restoration of a "blurred wave function" (Sect.5.2.4) and we use the constraint (5.123) with $\rho = \rho_0$ (i.e., B = I). The "blurring function" is $w_D(\varphi) = N\theta(\varphi/d)\,\mathrm{sinc}^2(\varphi/d)$. The constants N, D are given by (5.55) (where the integration ranges over [-d, d]). Then the relative error, for the restoration of the "blurred wave function" at $\varphi = 0$, can be computed by means of (5.96) (with B = I). The numerical method is described in [Ref.5.78, Chap.3]. In Fig.5.5 we give the values of the relative error as a function of the parameter D/D_0, where $D_0 = \lambda/(2\pi\rho_0) = (k\rho_0)^{-1}$ [Ref.5.78, Chap.3], for $\lambda = 2\rho_0$ and $\lambda = \rho_0/2$. As we see superresolution, i.e., restoration of details of the order of a wavelength and below, is easier when the wavelength is greater than the radius ρ_0 of the cylinder.

5.4.3 An Inverse Scattering Problem for Perfectly Conducting Bodies

An interesting combination of analytical and numerical techniques, involving the solution of linear problems, has been proposed by IMBRIALE and MITTRA [5.79] in the case of the inverse scattering problem for perfectly conducting bodies, and applied to the restoration of circular and elliptic cylinders.

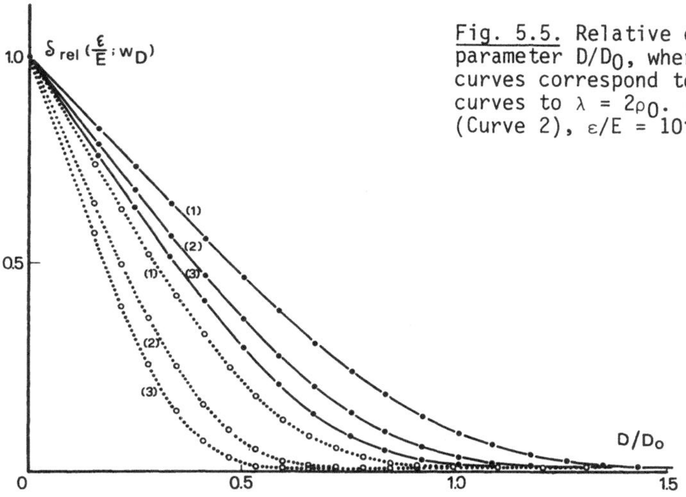

Fig. 5.5. Relative errors vs the resolution parameter D/D_0, where $D_0 = \lambda/2\pi\rho_0$. Undotted curves correspond to $\lambda = \rho_0/2$ and dotted curves to $\lambda = 2\rho_0$. (Curve 1), $\varepsilon/E = 10^{-2}$; (Curve 2), $\varepsilon/E = 10^{-4}$; (Curve 3), $\varepsilon/E = 10^{-6}$

We consider only plane wave incidence and we assume that the plane wave is z polarized. The perfectly conducting surfaces are assumed to be parallel to the z axis, so that the electromagnetic field may be derived from the single quantity $E_z = u$. The incident wave function is given by $u_0(\rho, \varphi) = \exp(ik\rho_0 \cos\varphi)$ and the associated scattered wave function u_s has the asymptotic behavior (5.125). The total field $u = u_0 + u_s$ satisfies the Helmholtz equation (5.114) in the free region and is subjected to the boundary condition $u = 0$ on the surfaces of the bodies.

The datum of the problem is the noisy scattering amplitude g and the main idea of the method is the following: reconstruct the wave function near the obstacle (from the knowledge of g) and locate points where the total wave function is zero, in order to identify points of the surface of the scatterer. This program can be accomplished in two steps. The *first step* is essentially the problem of inverse diffraction from far-field data discussed in Sect.5.4.2. Indeed, if ρ_0 is the radius of the circle tangent to the surface of the body (see Fig.5.6), at the exterior of this circle the field can be represented by the series (5.120). However, the radius ρ_0 is not known and must be determined: one must solve the inverse diffraction problem for various values of ρ_0 and choose the value for which the restored field has a zero. This zero gives a point of the surface of the scatterer. Of course, the accuracy in the determination of the zero depends on the accuracy in the restoration of the near field. As we have remarked in Sect.5.4.2, if the scatterer is a circular cylinder the accuracy can be very good (Hölder continuity). One can conjecture that, generally, the accuracy in the restoration of the near field is good when the surface of the scatterer is very smooth and poor when the surface of the scatterer is rough. The *second step* is the analytic continuation of the wave function into the region of nonconvergence of the series (5.120), i.e., $\rho < \rho_0$. If we know that the

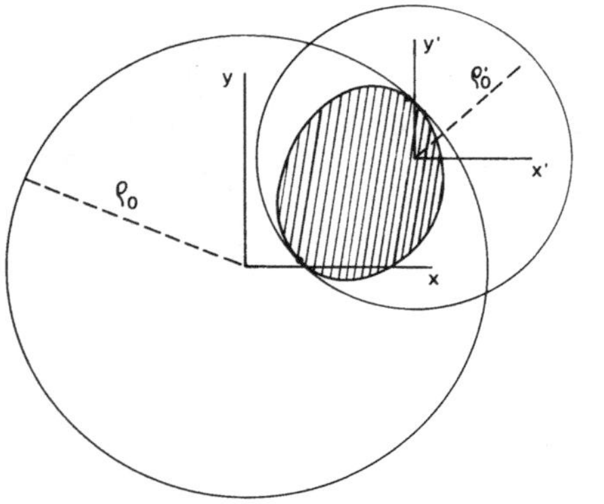

Fig. 5.6. Geometry for analytic continuation into region $\rho < \rho_0$ in the case of a convex body

scatterer is a *convex* body, then the analytic continuation can be accomplished by a simple technique of shifting the origin of the coordinate system [5.79] (see Fig. 5.6). Since the new coordinate system is obtained by translating the original one, then the scattering amplitude in the new system can be easily obtained from g [5.79]. The solution of the problem of inverse diffraction, with the new scattering amplitude as data, gives another point of the surface of the body. The procedure can be repeated and, since the body is convex, a few points can be sufficient in order to characterize the shape of the scatterer. An accurate analysis of this method in the case of a perfectly conducting *circular cylinder* has been done by CABAYAN et al. [5.80], using a stability criterion due to TWOMEY [5.29], which is equivalent to use the truncated solution (5.37) in the special case B = I. They show that even a "coarse" near-field map can give some information on the size and center position of the scatterer. Of course, the results of the method are very good because in this case, as shown in Sect.5.4.2, we have Hölder continuity.

When the body is *not convex*, analytic continuation of the wave function can be done as follows [5.79]. Once the total field, outside the minimum circle enclosing the scatterer, has been restored, then one can take a point in the exterior region as the origin of a new coordinate system ρ', φ' and represent the total field in the neighborhood of this point as a series of Bessel functions

$$u(\rho', \varphi') = \sum_{n=-\infty}^{+\infty} \frac{J_n(k\rho')}{J_n(k\rho_0')} c_n' e^{in\varphi'} \quad . \tag{5.130}$$

The circle $\rho' = \rho_0'$ is interior to the region $\rho > \rho_0$ (see Fig.5.7) and the c_n' are the Fourier coefficients of $u(\rho_0', \varphi')$. Of course, the summation of this series is an ill-posed problem if $\rho' > \rho_0'$, since $|J_n(k\rho')/J_n(k\rho_0')| \sim \exp[|n| \ln(\rho'/\rho_0')]$ when $|n| \to +\infty$. A discussion of the regularization of (5.130) can be found in [Ref.5.78, Chap.3]. The results are very similar to those for inverse diffraction. Anyway one can estimate the series (5.130) which, in principle, converges in the interior of the circle $\rho_1' > \rho_0'$, tangent to the body surface. Since ρ_1' is not known, one should estimate the series (5.130) for various values of $\rho' > \rho_0'$ and choose the smallest value of ρ' for which the restored field has a zero. So a new point of the scatterer has been determined. By a series of overlapping circles it is then possible, in principle, to ob-

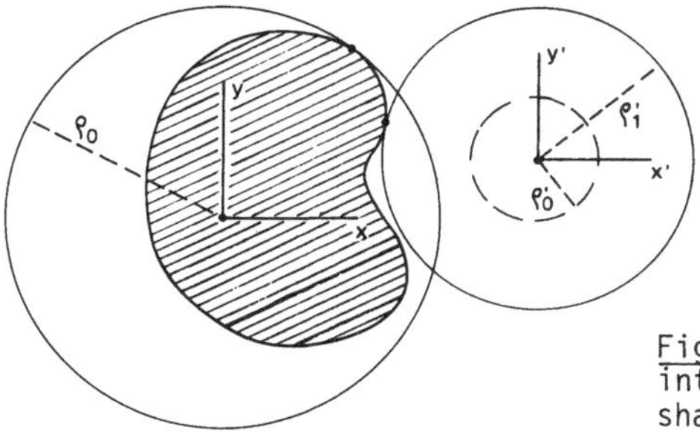

Fig. 5.7. Geometry for analytic continuation into region $\rho < \rho_0$ in the case of an arbitrarily shaped body

tain values of u in all the points of the space external to the scatterer. Of course error propagation can prevent having a satisfactory estimate of the shape of the scatterer.

The previous method of analytic continuation has also been used by AHLUWALIA and BOERNER [5.81] for recovering the electrical size, the surface locus and the averaged local surface impedance in the case of circular cylindrical monobody and twobody shapes. The same authors have also extended the method to the case of spherical surfaces [5.82].

5.4.4 Inverse Scattering Problems in the Born Approximation

We consider two problems: the determination of the shape of a perfectly conducting body and the reconstruction of the refractive index of a semi-transparent object.

When a *perfectly conducting object* is illuminated by an electromagnetic wave, the scattered field can be determined using the Born approximation (also known as Kirchhoff's or physical optics approximation) if the wavelengths λ are small compared with the characteristic dimensions of the scatterer. Assuming that the measured data are the values of the backscattered far field, then the determination of the shape of the body is a linear inverse problem [5.83,84]. Consider a plane wave, with electric field $\underline{E}_i(\underline{r}) = \underline{E}_0 \exp(ik\underline{s}_0 \cdot \underline{r})$, scattered by a smooth, convex and bounded target τ. The backscattered field $\underline{E}_b(\underline{r})$, i.e., the field observed in the direction $\underline{s}_b = -\underline{s}_0$, is given by $\underline{E}_b(\underline{r}) \sim \rho(\underline{k})(2\sqrt{\pi}r)^{-1} \exp(ikr)\underline{E}_0$, $r \to +\infty$, $\rho(\underline{k})$ being proportional to the backward scattering amplitude. Then one can prove that, in the Born approximation

$$\Gamma(\underline{k}) = \frac{2\sqrt{\pi}}{k^2} [\rho(\underline{k}) + \rho^*(-\underline{k})] = \int \gamma(\underline{r}) \exp(2i\underline{k} \cdot \underline{r}) d\underline{r} \ , \tag{5.131}$$

where $\underline{k} = k\underline{s}_0$ and $\gamma(\underline{r})$ is the characteristic function of the target, i.e., $\gamma(\underline{r}) = 1$ when $\underline{r}$ is in τ, $\gamma(\underline{r}) = 0$ otherwise [5.83,84] (see also [Ref.5.75, Sect.3.2.5]). If the backscattered field could be measured for all frequencies and for all directions of incidence, then the Fourier transform of $\gamma(\underline{r})$ would be known and $\gamma(\underline{r})$ could be determined. In practice, $\Gamma(\underline{k})$ is measurable only in a restricted domain of the $\underline{k}$ space. This does not seem to be a serious restriction, as regards the possibility of a unique reconstruction of $\gamma(\underline{r})$, since $\Gamma(\underline{k})$ is an analytic function of $\underline{k}$ [Ref. 5.75, Sect.3.2.5]. However, this reconstruction, which is equivalent from the mathematical point of view to bandwidth extrapolation (Sect.5.4.1), is not stable.

In radar applications, $\Gamma(\underline{k})$ is measured only in a set of points interior to the annular region $m \leq |\underline{k}| \leq M$, where $k_1 = m/2$ and $k_2 = M/2$ correspond to minimum and maximum values of the usable frequency band. It must also be remarked that the lower bound has to be large enough in order to assure the validity of (5.131): the largest wavelength in the incident field, $\lambda_1 = 4\pi/m$, must be short compared to the target shape. The inaccessibility of low frequency information is essentially related to the intrinsic limitations of the Born approximation. The instability due to this fact has been investigated by many authors [5.85-88]. PERRY [5.86], for instance, applied Tikhonov's regularization method to the one-dimensional case, assuming that $\Gamma(\underline{k})$ is known for $|\underline{k}| \geq m$ (perfect high-pass filter). In other words, he did not care about the limitations due to the lack of information at high frequencies. If the object is located within the interval [-X/2, X/2], then $\gamma(x)$ can be determined as the solution of the Fredholm integral equation of the second kind

$$\gamma(x) - \int_{-X/2}^{X/2} \frac{\sin[m(x-y)]}{\pi(x-y)}\, \gamma(y)dy = g(x) \ , \tag{5.132}$$

where $g(x)$ is the inverse Fourier transform of the available values of $\Gamma(\underline{k})$. Equation (5.132) has the form $(1-A)\gamma = g$ where A is the integral operator (5.105) with $\pi\Omega = m$. Therefore solving (5.132) is not an ill-posed problem in the strict sense but, when m is large, it is an ill-conditioned problem. Indeed, the condition number of $1-A$ is $\alpha = (1-\lambda_0)^{-1}$, where λ_0 is the largest eigenvalue of A. From the behavior of λ_0 for large m [5.89] it follows: $\alpha \sim (2\pi/mX)^{\frac{1}{2}} \exp(mX)/4$. In radar applications mX is relatively large and therefore α can be rather large. After discretization, an ill-conditioned problem shows the same features as an ill-posed problem and therefore regularization methods can be useful.

In order to circumvent the lack of information at low frequencies, another technique has been first suggested by BOJARSKI [5.83], and further developed by MAGER and BLEISTEIN [5.88]. The essence of the method is to examine not the characteristic function of the target, but rather the directional derivative of this function. Indeed, if $\underline{s}\cdot\underline{\nabla}\gamma$ is the derivative of γ in the direction of the unit vector $\underline{s}$, then its Fourier transform is the product of $\Gamma(\underline{k})$ by the factor $\underline{s}\cdot\underline{k}$. In this way, one simultaneously attenuates low-frequency data while enhancing the effect of high-frequency data. One expects that the limitations of the method are essentially due to the lack of information at high frequencies. The function $\underline{s}\cdot\underline{\nabla}\gamma$ is highly singular; more precisely $\underline{s}\cdot\underline{\nabla}\gamma = \underline{s}\cdot\underline{n}\delta$ where $\underline{n}$ is the unit outward normal to the surface of the body and δ is a Dirac delta measure concentrated on the surface of the body. As proved by MAGER and BLEISTEIN [5.88], similar features are shown by the function

$$h(\underline{r}, \underline{s}) = \frac{1}{(2\pi)^2} \int i\underline{s}\cdot\underline{k}a(\underline{k})\Gamma(\underline{k})\, e^{-i\underline{k}\cdot\underline{r}}d\underline{k} \ . \tag{5.133}$$

Here $a(\underline{k})$ is the characteristic function of the domain in $\underline{k}$-space (interior to the annular region $m \leq |\underline{k}| \leq M$) where the backscattered field is measured. In the high-frequency limit ($m \gg 1$), if $\underline{r}_0$ is a point on the surface of the scatterer, then

$$\mathrm{Re}\{h(\underline{r}, \underline{s})\} \sim (\mathrm{constant})|\underline{r}-\underline{r}_0|^{-1}[\sin(M|\underline{r}-\underline{r}_0|) - \sin(m|\underline{r}-\underline{r}_0|)] \qquad (5.134)$$

provided that the vector $\underline{r}-\underline{r}_0$ is orthogonal to the surface at $\underline{r}_0$ and its direction coincides with a direction of incidence. Therefore the function $\mathrm{Re}\{h(\underline{r}, \underline{s})\}$ has a central lobe which peaks on the target surface in regions with surface normals parallel to directions of incidence. The height of the central lobe is proportional to $M-m$ and its width is approximately equal to $2\pi/M$ if $M \gg m$. Remark that $2\pi/M = \lambda_2/2$ where λ_2 is the smallest wavelength in the incident field. It must also be observed that this result has been derived by MAGER and BLEISTEIN [5.88] for noiseless data. Now, the factor $\underline{s}\cdot\underline{k}$ does not only enhance the effect of high-frequency data but also the effect of the noise on these data. Error propagation in the determination of $\mathrm{Re}\{h(\underline{r}, \underline{s})\}$ is controlled by the quantity (condition number) $\alpha = M/m$. When this parameter is large, the resolution limit is certainly worse than $\lambda_2/2$. For the numerical examples presented by MAGER and BLEISTEIN [5.88], α is of the order of 2 and therefore the results are quite good.

Finally we want to remark that an improvement of the resolution limit intrinsic to $\mathrm{Re}\{h(\underline{r}, \underline{s})\}$ would require an analytic continuation of the backscattered field in the region $|\underline{k}| > M$. From the analysis of bandwidth extrapolation done in Sect. 5.4.1, it clearly appears that this problem is affected by logarithmic continuity and therefore an improvement of the resolution limit $\lambda_2/2$ is practically impossible.

The reconstruction of the *refractive index* of weakly scattering semi-transparent objects, using the Born approximation, has been widely discussed [5.90-94], with special attention to the problem of uniqueness of the solution [Ref.5.75, Sect. 3.4.3]. As shown by WOLF [5.91,95], modulus and phase of the scattering amplitude can be derived, using holographic data, from the homogeneous part of the angular spectrum of the scattered field. DÄNDLIKER and WEISS [5.92] stressed that appropriate variation of the direction of the incident wave is crucial for holographic 3D reconstruction.

The wave function u satisfies the equation

$$\nabla^2 u + k_0^2 n^2(\underline{r}) u = 0 \quad , \qquad (5.135)$$

where $n(\underline{r})$ is the (possibly complex) refractive index at the point $\underline{r}$. If the object is situated in free space, then $n(\underline{r}) = 1$ outside the object. Equation (5.135) can be recasted in the following form

$$\nabla^2 u + k_0^2 u = F(\underline{r}) u \quad , \qquad (5.136)$$

where

$$F(\underline{r}) = -k_0^2[n^2(\underline{r}) - 1] \quad . \tag{5.137}$$

The function $F(\underline{r})$ is called the *scattering potential* and it is evidently zero at all points outside the object. Consider an incident plane wave $u_i(\underline{r}) = \exp(ik_0\underline{s}_0 \cdot \underline{r})$; then the Born approximation can be used for the determination of the scattered wave function $u_s(\underline{r})$, if the object scatters weakly, i.e., if $|u_s| \ll |u_i|$. When this condition is satisfied, in the far zone we have

$$u_s(\underline{r}, k_0\underline{s}_0) \sim -\frac{\exp(ik_0r)}{4\pi r} A_B(k_0\underline{s}_0, k_0\underline{s}) \ , \tag{5.138}$$

where

$$A_B(k_0\underline{s}_0, k_0\underline{s}) = \int F(\underline{r}') \exp[-ik_0(\underline{s} - \underline{s}_0) \cdot \underline{r}']d\underline{r}' \quad . \tag{5.139}$$

Therefore the Born approximation to the scattering amplitude is essentially given by the Fourier transform $\hat{F}(\underline{k})$ of the scattering potential $F(\underline{r})$. Inspection of (5.139) shows that, for a fixed direction of incidence $\underline{s}_0$, A_B gives those Fourier components of $F(\underline{r})$ which correspond to points on the surface of the sphere with center $k_0\underline{s}_0$ and radius k_0. By varying the direction of incidence $\underline{s}_0$, a (theoretically infinite) number of experiments would allow one to determine $\hat{F}(\underline{k})$ for all values of $\underline{k}$ lying within the sphere of radius $2k_0$. Then a bandlimited approximation $F_{b\ell}(\underline{r})$ to the scattering potential is given by

$$F_{b\ell}(\underline{r}) = \frac{1}{(2\pi)^3} \int_{|\underline{k}| \leq 2k_0} \hat{F}(\underline{k}) \, e^{i\underline{k} \cdot \underline{r}} \, d\underline{k} \quad . \tag{5.140}$$

A rough measure of the limit of resolution of WOLF's approach [5.91], intrinsic to (5.140), is given by $\lambda_0/2 = \pi/k_0$ (when the scattered field is determined by side-band holography the limit of resolution is about $9\lambda_0$ [5.95]). An improvement beyond these limits is, in principle, possible since $F(\underline{k})$ is an analytic function when the object is localized within a finite volume τ. We encounter, once more, a problem which is equivalent, from the mathematical point of view, to bandwidth extrapolation. Therefore a significant improvement of the resolution limit seems to be, in practice, impossible. Besides, the effect of the noise can be very important (it is necessary to detect a weak scattered field in the presence of a strong unscattered field), so that even the theoretical limit of resolution cannot be reached. A new approach to the optical inverse scattering problem, based on interference with three variations of a spherical reference wave, has been proposed by LAM et al. [5.123].

Experiments have been undertaken in order to investigate the use of the technique suggested by Wolf's theory [5.91,95] and computational reconstruction of objects from holograms has been attempted [5.96-98]. Very simple objects have been considered, i.e., rectangular (homogeneous and inhomogeneous) and cylindrical bars. In these cases, because of the symmetry of the objects only one hologram is needed. The reported numerical results show spurious oscillations which probably can be smoothed by a filtering of the Fourier transform of $F(\underline{r})$. The 3D scattering potential of microscopic objects (40 μm diameter) has recently been reconstructed by FERCHER et al. [5.124].

5.4.5 Object Reconstruction from Projections and Abel Equation

Object reconstruction from projections and Abel equations are two examples of inverse problems which arise when the variations of the dynamical functions over a given wavelength are so small that diffraction can be neglected. For both problems there exists an enormous amount of literature. Our purpose is only to point out that, as a consequence of the physical approximation intrinsic to these problems, the restored continuity is quite good (Hölder continuity).

Assuming straight line ray propagation with the amplitude (or the phase) of the ray controlled by the line integral of a density function, a projection of the object onto a plane is measured. Typical examples are X-ray shadowgraphs. In two dimensions the mathematical formulation of the problem is as follows. Let $\bar{f}(\underline{r})$ be a density function which has support in the circle $|\underline{r}| \leqq R$ and let L be the straight line defined by $\underline{s} \cdot \underline{r} = p$. Here $\underline{r} \equiv \{x, y\}$ is a point of the plane, $\underline{s} \equiv \{\cos\varphi, \sin\varphi\}$, $0 \leqq \varphi < \pi$, is the unit vector orthogonal to L and p is the distance of L from the origin, $-\infty < p < +\infty$. If $\underline{t} \equiv \{-\sin\varphi, \cos\varphi\}$ is the unit vector of the direction of L, consider the line integral

$$\bar{g}(p, \varphi) = \int_{-\infty}^{+\infty} \bar{f}(p\underline{s} + q\underline{t})dq \quad . \qquad (5.141)$$

Obviously $\bar{g}(p, \varphi) = 0$ when $|p| \geqq R$. The function $\bar{g}(p, \varphi)$ is known as the *Radon transform* of $\bar{f}(\underline{r})$ and, for fixed φ, it gives the projection of $\bar{f}(\underline{r})$ onto a straight line parallel to $\underline{s}$. Thus, in two dimensions, object reconstruction from projections is exactly the inversion of the Radon transform. The solution of this problem was given by RADON [5.99] in 1917. Nowadays there are many fields of application: it is sufficient to mention computerized tomography (see, e.g. [5.100,101]), radio astronomy [5.102], electron microscopy [5.103], radar target shape estimation [5.104] and so on (see also [Ref.5.45, Sect.2.3]). Solving (5.141) is an ill-posed problem. This fact clearly appears from the Radon inversion formula [5.100] since it contains the derivative of the noiseless data $\bar{g}$. In order to investigate error propagation, let us take as solution space F the space of the square integrable functions, which have support in the circle $|\underline{r}| \leqq R$, and as data space G the space of the square integrable functions over the rectangle $0 \leqq \varphi < \pi$, $|p| \leqq R$. Then the linear operator A defined by (5.141) is a continuous operator form F into G [Ref.5.100, Sect.12].

Using Parseval's equality for Fourier transform, the norm of f can be written as follows:

$$\| f \|_F^2 = \int_{|\underline{r}| \leq R} |f(\underline{r})|^2 d\underline{r} = \frac{1}{(2\pi)^2} \int_0^{\pi} d\varphi \int_{-\infty}^{+\infty} |h(\nu, \varphi)|^2 d\nu \ , \tag{5.142}$$

where $h(\nu, \varphi) = |\nu|^{\frac{1}{2}} \hat{f}(\nu \underline{s})$ and $\underline{s} \equiv \{\cos\varphi, \sin\varphi\}$. Then, from the "projection slice theorem" (see, e.g. [5.105] or [Ref.5.45, Sect.2.3.4])

$$(Af)(p, \varphi) = \frac{1}{2\pi} \int_{-\infty}^{+\infty} e^{ip\nu} \hat{f}(\nu \underline{s}) d\nu \tag{5.143}$$

and from Parseval's equality it follows

$$\| Af \|_G^2 = \int_0^{\pi} d\varphi \int_{-R}^{R} |(Af)(p, \varphi)|^2 dp = \frac{1}{2\pi} \int_0^{\pi} d\varphi \int_{-\infty}^{+\infty} \frac{|h(\nu,\varphi)|^2}{|\nu|} d\nu \quad . \tag{5.144}$$

Finally we can require, as a constraint, a bound on the first derivatives of $f(\underline{r})$. This bound is not very realistic in many applications of the Radon transform, since one should also reconstruct discontinuous functions. However our purpose is only to discuss, in the simplest way, the stability of the inversion procedure. Then, using again Parseval's equality we have

$$\| Bf \|_F^2 = \int_{|\underline{r}| \leq R} \left(\left|\frac{\partial f}{\partial x}\right|^2 + \left|\frac{\partial f}{\partial y}\right|^2 \right) d\underline{r} = \frac{1}{(2\pi)^2} \int_0^{\pi} d\varphi \int_{-\infty}^{+\infty} |\nu|^2 |h(\nu, \varphi)|^2 d\nu \quad . \tag{5.145}$$

It is now easy to recognize that the stability estimate for the Radon inverse problems is given by (5.54) if we put $|\hat{K}(\nu)| = |\nu|^{-\frac{1}{2}}$ and $|\hat{\beta}(\nu)| = |\nu|$. Therefore $|\hat{\beta}(\nu)| = |\hat{K}(\nu)|^{-\mu}$, $\mu = 2$ and $\delta(\varepsilon, E) \sim E(\varepsilon/E)^{\alpha}$ with $\alpha = 2/3$. The restored stability is quite good and this result should be related to the fact that, in this mathematical model, diffraction has been neglected.

When the object has circular symmetry, then $\bar{g}(p, \varphi)$, given by (5.141), is the same for all φ. Put $\varphi = 0$ in (5.141), so that $p = x$, $q = y$. In terms of the variable $\rho = (x^2 + y^2)^{\frac{1}{2}}$ we have $(x, y > 0)$

$$\bar{g}(x) = 2 \int_x^{+\infty} \frac{\rho \bar{f}(\rho)}{(\rho^2 - x^2)^{\frac{1}{2}}} d\rho \quad . \tag{5.146}$$

This is a form of the Abel integral equation which is also fundamental whenever Fermat's principle can be used for the calculation of the rays. Many applications of the Abel equation are discussed in [5.106]. From the previous remark we expect that the same kind of stability holds for object reconstruction from projection and the Abel equation. However it is interesting to derive directly this result. To this purpose let us consider the inversion of the following integral operator (the various forms of the Abel equation can be treated in a similar way):

$$(Af)(x) = \int_0^x \frac{f(y)}{\sqrt{\pi x(x-y)}}\, dy \ , \tag{5.147}$$

taking $L^2(0, +\infty)$ both as solution and data space. Then $A: F \to G$ is a linear continuous operator. Introducing the Mellin transform of f,

$$\hat{f}_M(\nu) = \int_0^{+\infty} x^{-\frac{1}{2}+i\nu} f(x)dx \ , \tag{5.148}$$

it is easy to show that the Mellin transform of Af is given by $[\Gamma(1/2 - i\nu)/\Gamma(1 - i\nu)]$ $\hat{f}_M(\nu)$. Using Parseval's equality for Mellin transform [Ref.5.107, pp.94-95] we get

$$\| Af \|^2 = \int_0^{+\infty} |(Af)(x)|^2 dx = \frac{1}{2\pi} \int_{-\infty}^{+\infty} \frac{\mathrm{tgh}(\pi\nu)}{\nu} |\hat{f}_M(\nu)|^2 d\nu \quad . \tag{5.149}$$

Take the constraint operator defined by $(Bf)(x) = xf'(x)$. Observing that the Mellin transform of Bf is given by $-(1/2 + i\nu)\hat{f}_M(\nu)$ and using again Parseval's equality we get

$$\| Bf \|^2 = \frac{1}{2\pi} \int_{-\infty}^{+\infty} (\nu^2 + \tfrac{1}{4}) |\hat{f}_M(\nu)|^2 d\nu \quad . \tag{5.150}$$

From (5.149,150) it is easy to recognize that the stability estimate $\delta(\varepsilon, E)$ for the Abel equation is given by (5.54) with $|\hat{K}(\nu)|^2 = \nu^{-1}\mathrm{tgh}(\pi\nu) \sim |\nu|^{-1}$, $|\nu| \to +\infty$, and $|\hat{\beta}(\nu)|^2 = \nu^2 + 1/4 \sim \nu^2$, $|\nu| \to +\infty$. Since $|\hat{\beta}(\nu)| \sim |\hat{K}(\nu)|^{-\mu}$, $|\nu| \to +\infty$, with $\mu = 2$, it follows $\delta(\varepsilon, E) \sim E(\varepsilon/E)^{2/3}$, i.e., the same result as for the inversion of the Radon transform.

5.4.6 Concluding Remarks and Open Problems

In this review of a continuously expanding field, we restricted ourselves to some typical linear inverse problems, and so we omitted many important topics where regularization methods apply as well. Let us mention for instance, polarization utilization in electromagnetic inverse scattering [5.45, and Chapt. 7 of this volume] and laser anemometry data analysis [5.108,109]. Regularization methods can also be useful for synthesis problems [5.110]. The main difference between inverse and synthesis problems can be easily understood in the case of an antenna: the inverse problem is the identification of an actual antenna from measurement of its radiation pattern, while the synthesis problem is the design of an antenna producing a given radiation pattern. In the latter case, rather than in stability, one is interested in "sensitivity": if the computed antenna is not exactly realized, how much will its pattern function be modified?

Since the mathematical pathology of all those problems is quite similar whatever the particular field one considers, we think that regularization theory should provide a unified framework for treating linear inverse problems and for investigating thoroughly their stability. However, regularization methods are by no means a cure-all. Indeed, as we have seen, the faster the decay of the eigenvalues (singular values) of the operator A, the greater the loss of information due to the smoothing effect of A. Hence it is expected that in some cases, the available data are truly insufficient and no meaningful prior knowledge can provide a satisfactory solution. Very important in this connection is the precise valuation of error propagation in the regularized inversion procedure. This enables us to estimate in practical cases, the accuracy of the solution for a given noise level and a given prior knowledge. We also showed that this error analysis enlightens theoretical questions like the problem of superresolution. Summarizing the results of Sect.5.4.1, we can say that superresolution appears practically impossible for imaging systems with a large aperture, because it would require unrealistically high signal-to-noise ratios. This is due to a very fast increase of the relative error on blurred solutions beyond the Rayleigh limit. A similar feature arises in near-field reconstruction from the scattering amplitudes (see Sect.5.4.2): the reconstruction of source details of the order of a wavelength and below (for a review on this subject see [5.111]) appears very difficult when the wavelength λ is much smaller than the characteristic dimension ℓ of the source. Superresolution becomes however easier when λ is of the same order as ℓ.

More fundamental questions are still open in the field of ill-posed problems. The first point, very important, is to develop a sound theory when data are given only at a finite number of points (in regularization theory, as described in this chapter, one assumes that the data are functions defined everywhere). Some results in this direction have been obtained for the problems of analytic continuation [5.112,113] harmonic continuation [5.114] and numerical differentiation [5.115]. In these cases, the main ideas of regularization theory (prior knowledge, least-square methods, stability estimates) have been maintained.

A second point would be to extend regularization theory beyond its actual frame: linear problems and prior knowledge expressed in the form (5.22). For instance, a positivity constraint, which appears naturally in some problems, cannot be expressed in this way. For particular ill-posed problems (harmonic continuation, backward heat equation), it is known that positive solutions are necessarily stable [5.116]. However, the requirement of positivity alone is not sufficient for stabilizing Fredholm integral equations of the first kind. Some algorithms, reviewed in [5.72], have been developed for introducing the positivity constraint in the analysis of imaging systems but, to our knowledge, no theoretical analysis of the solution accuracy has been done. It is clear that a supplementary constraint of positivity improves the solution, but the quantitative estimation of this improvement is still an open question.

Acknowledgements

We are deeply indebted to Dr. McWhirter who has kindly supplied us with a long list of references. We also want to thank Prof. G. Talenti for many long and informative discussions concerning the mathematics of ill-posed problems. One of us (C.D.M.) is indebted to Prof. J. Reignier for critical and helpful remarks. Last but not least, it is a pleasure for the authors to thank Miss B. Basiglio for the preparation of the manuscript and careful typing.

References

5.1 J. Hadamard: *Lectures on the Cauchy Problem in Linear Partial Differential Equations* (Yale University Press, New Haven 1923)
5.2 R. Courant, D. Hilbert: *Methods of Mathematical Physics*, Vol. 2 (Interscience, New York 1962)
5.3 M.M. Lavrentiev: *Some Improperly Posed Problems of Mathematical Physics*, Springer Tracts in Natural Philosophy, Vol. 11 (Springer, Berlin, Heidelberg, New York 1967)
5.4 A. Tikhonov, V. Arsenine: *Méthodes de Résolution de Problèmes Mal Posés* (Mir, Moscow 1976)
5.5 L.E. Payne: *Improperly Posed Problems in Partial Differential Equations* (SIAM, Philadelphia 1975)
5.6 S.G. Mikhlin: *Integral Equations* (Pergamon, London 1957)
5.7 A.V. Balakrishnan: *Applied Functional Analysis*, Applications of Mathematics, Vol.3 (Springer, Berlin, Heidelberg, New York 1976)
5.8 K. Chadan, P.C. Sabatier: *Inverse Problems in Quantum Scattering Theory* (Springer Berlin, Heidelberg, New York 1977)
5.9 H.P. Baltes (ed.): *Inverse Source Problems in Optics*, Topics in Current Physics, Vol. 9 (Springer, Berlin, Heidelberg, New York 1978)
5.10 M. Reed, B. Simon: *Methods of Modern Mathematical Physics* (Academic Press, New York 1972)
5.11 P.C. Sabatier (ed.): *Applied Inverse Problems*, Lecture Notes in Physics, Vol. 85 (Springer, Berlin, Heidelberg, New York 1978)
5.12 M.Z. Nashed (ed.): *Generalized Inverses and Applications* (Academic Press, New York 1976)
5.13 A.N. Tikhonov: Sov. Math. Dokl. *4*, 1035-1038 (1963)
5.14 F. John: Commun. Pure Appl. Math. *13*, 551-585 (1960)
5.15 C. Pucci: Atti Accad. Naz. Lincei *18*, 473-477 (1955)
5.16 K. Miller: SIAM J. Math. Anal. *1*, 52-74 (1970)
5.17 K. Miller, G.A. Viano: J. Math. Phys. *14*, 1037-1048 (1973)
5.18 V.F. Turchin, V.P. Kozlov, M.S. Malkevich: Sov. Phys.-Usp. *13*, 681-703 (1971)
5.19 V.A. Morozov: U.S.S.R. Comp. Math. and Math. Phys. *10*, N.4, 10-25 (1970)
5.20 J.N. Franklin: J. Math. Anal. Appl. *31*, 682-716 (1970)
5.21 M. Bertero, G.A. Viano: Bollettino U.M.I. *15*-B, 483-508 (1978)
5.22 M. Bertero, C. De Mol, G.A. Viano: J. Math. Phys. *20*, 509-521 (1979)
5.23 E.C. Titchmarsh: *The Theory of Functions* (Oxford University Press, Oxford 1939)
5.24 E. Hille, J.D. Tamarkin: Acta Math. *57*, 1-76 (1931)
5.25 G. Talenti: Bollettino U.M.I. *15*-A, 1-29 (1978)
5.26 M. Bertero, C. De Mol, G.A. Viano: Opt. Acta *27*, 307-320 (1980)
5.27 E. Picard: R.C. Mat. Palermo *29*, 615-619 (1910)
5.28 T. Kato: *Perturbation Theory for Linear Operators* (Springer, Berlin, Heidelberg, New York 1966)
5.29 S. Twomey: J. Franklin Inst. *279*, 95-109 (1965)
5.30 V.A. Morozov: Sov. Math. Dokl. *8*, 1000-1003 (1967)
5.31 J.N. Franklin: Math. Comp. *28*, 889-907 (1974)
5.32 D.L. Phillips: J. ACM *9*, 84-96 (1962)
5.33 S. Twomey: J. ACM *10*, 97-101 (1963)
5.34 L. Eldén: Ph. D. Thesis, Linköping University (1977)

5.35 G.F. Miller: "Fredholm Equations of the First Kind", in *Numerical Solution of Integral Equations*, ed. by L.M. Delves, J. Walsh (Claredon Press, Oxford 1974) pp.175-188
5.36 M. Bertero, C. De Mol, G.A. Viano: "On the Regularization of Linear Inverse Problems in Optics", in *Applied Inverse Problems*, ed. by P.C. Sabatier, Lecture Notes in Physics, Vol. 85 (Springer, Berlin, Heidelberg, New York 1978) pp.180-199
5.37 G. Backus, F. Gilbert: Geophys. J. R. Astron. Soc. *16*, 169-205 (1968)
5.38 G. Backus, F. Gilbert: Phil. Trans. Roy. Soc. A-*266*, 123-192 (1970)
5.39 J.L. Doob: *Stochastic Processes* (Wiley, New York 1953)
5.40 L.E. Franks: *Signal Theory* (Prentice-Hall, Englewood Cliffs 1969)
5.41 A. Papoulis: *Probability, Random Variables and Stochastic Processes* (McGraw-Hill, New York 1965)
5.42 G.E. Backus: "Inference from Inadequate and Inaccurate Data", in *Mathematical Problems in the Geophysical Sciences*, ed. by W.H. Reid, Lectures in Applied Math., Vol 14 (AMS, Providence 1971)
5.43 I.M. Gel'fand, N.Y. Vilenkin: *Generalized Functions*, Vol. 4 (Academic Press, New York 1964)
5.44 I.M. Gel'fand, A.M. Yaglom: Am. Math. Soc. Trans. *12*, 199-246 (1959)
5.45 W.M. Boerner: "Polarization Utilization in Electromagnetic Inverse Scattering"; Communications Laboratory Rpt. 78-3, University of Illinois at Chicago Circle (1978)
5.46 H. Wolter: "On Basic Analogies and Principal Differences between Optical and Electronic Information", in *Progress in Optics*, Vol. I, ed. by E. Wolf (North Holland, Amsterdam 1961) pp.155-210
5.47 R.N. Bracewell, J.A. Roberts: Aust. J. Phys. *7*, 615-640 (1954)
5.48 Y.T. Lo: J. Appl. Phys. *32*, 2052-2054 (1961)
5.49 H. Wolter: Physica *24*, 457-475 (1958)
5.50 B.J. Hoenders, H.A. Ferwerda: Optik *37*, 542-556 (1973)
5.51 C. Shannon: *The Mathematical Theory of Communication* (University of Illinois Press, Urbana 1949)
5.52 D. Gabor: "Light and Information", in *Astronomical Optics and Related Subjects*, ed. by Z. Kopal (North-Holland, Amsterdam 1956) pp.17-30
5.53 G. Toraldo di Francia: J. Opt. Soc. Am. *59*, 799-804 (1969)
5.54 D. Slepian, H.O. Pollack: Bell System Tech. J. *40*, 43-63 (1961)
5.55 D. Slepian: J. Math. & Phys. *44*, 99-140 (1965)
5.56 D. Slepian, E. Sonnenblick: Bell Syst. Tech. J. *44*, 1745-1759 (1965)
5.57 C. Flammer: *Spheroidal Wave Functions* (Stanford University Press, Stanford 1957)
5.58 H.J. Landau: Trans. Am. Math. Soc. *115*, 242-256 (1965)
5.59 B.R. Frieden: "Evaluation, Design and Extrapolation Methods for Optical Signals, Based on Use of the Prolate Functions", in *Progress in Optics*, Vol. 9, ed. by E. Wolf (North-Holland, Amsterdam 1971) pp.311-407
5.60 F. Gori, G. Guattari: Opt. Commun. *7*, 163-165 (1973)
5.61 M. Bertero, C. De Mol, G.A. Viano: Opt. Lett. *3*, 51-53 (1978)
5.62 C.W. Helstrom: J. Opt. Soc. Am. *67*, 833-838 (1977)
5.63 C.K. Rushforth, R.W. Harris: J. Opt. Soc. Am. *58*, 539-545 (1968)
5.64 J.W. Goodman: "Synthetic Aperture Optics", in *Progress in Optics*, Vol. 8, ed. by E. Wolf (North-Holland, Amsterdam 1970) pp. 1-50
5.65 G. Toraldo di Francia: Riv. Nuovo Cimento *1* (Numero Speciale), 460-484 (1969)
5.66 C.W. Helstrom: J. Opt. Soc. Am. *57*, 297-303 (1967)
5.67 C.L. Rino: J. Opt. Soc. Am. *59*, 547-553 (1969)
5.68 M. Bendinelli, A. Consortini, L. Ronchi, B.R. Frieden: J. Opt. Soc. Am. *64*, 1498-1502 (1974)
5.69 H. Dym, H.P. McKean: *Fourier Series and Integrals* (Academic Press, New York 1972)
5.70 G.J. Buck, J.J. Gustincic: IEEE Trans. AP-*15*, 376-381 (1967)
5.71 B.R. Frieden: J. Opt. Soc. Am. *57*, 1013-1019 (1967)
5.72 B.R. Frieden: "Image Enhancement and Restoration", in *Picture Processing and Digital Filtering*, 2nd ed., ed. by T.S. Huang, Topics in Applied Physics, Vol. 6 (Springer, Berlin, Heidelberg, New York 1979) pp.177-248

5.73 G.A. Viano: J. Math. Phys. *17*, 1160-1165 (1976)
5.74 J.R. Shewell, E. Wolf: J. Opt. Soc. Am. *58*, 1596-1603 (1968)
5.75 B.J. Hoenders: "The Uniqueness of Inverse Problems", in *Inverse Source Problems in Optics*, ed. by H.P. Baltes, Topics in Current Physics, Vol. 9 (Springer, Berlin, Heidelberg, New York 1978) pp.41-82
5.76 G.C. Sherman: J. Opt. Soc. Am. *57*, 1490-1498 (1967)
5.77 K. Miller: Arch. Rational Mech. Anal. *16*, 126-154 (1964)
5.78 C. De Mol: "Sur la Regularisation des Problèmes Inverses Linéaires"; Ph.D. Thesis, Université Libre de Bruxelles (1979)
5.79 W.A. Imbriale, R. Mittra: IEEE Trans. AP-*18*, 633-642 (1970)
5.80 H.S. Cabayan, R.C. Murphy, T.J.F. Pavlásek: IEEE Trans. AP-*21*, 346-351 (1973)
5.81 H.P.S. Ahluwalia, W.M. Boerner: IEEE Trans. AP-*21*, 663-672 (1973)
5.82 H.P.S. Ahluwalia, W.M. Boerner: IEEE Trans. AP-*22*, 673-682 (1974)
5.83 N.N. Bojarski: "A Survey of Electromagnetic Inverse Scattering"; Syracuse Univ. Res. Corp., Special Projects Lab. Rpt., DDC #AD-813-851 (1966)
5.84 R.M. Lewis: IEEE Trans. AP-*17*, 308-314 (1969)
5.85 W. Tabbara: IEEE Trans. AP-*21*, 245-247 (1973)
5.86 W.L. Perry: IEEE Trans. AP-*22*, 826-829 (1974)
5.87 W. Tabbara: IEEE Trans. AP-*23*, 446-448 (1975)
5.88 R.D. Mager, N. Bleistein: IEEE Trans. AP-*26*, 695-699 (1978)
5.89 W.H.J. Fuchs: J. Math. Anal. Appl. *9*, 317-330 (1964)
5.90 R.W. Hart, E.P. Gray: J. Appl. Phys. *35*, 1408-1415 (1964)
5.91 E. Wolf: Opt. Commun. *1*, 153-156 (1969)
5.92 R. Dändliker, K. Weiss: Opt. Commun. *1*, 323-328 (1969)
5.93 H.A. Ferwerda, B.J. Hoenders: Optik *39*, 317-326 (1974)
5.94 A.J. Devaney: J. Math. Phys. *19*, 1526-1531 (1978)
5.95 E. Wolf: J. Opt. Soc. Am. *60*, 18-20 (1970)
5.96 W.H. Carter: J. Opt. Soc. Am. *60*, 306-314 (1970)
5.97 W.H. Carter, P.C. Ho: Appl. Opt. *13*, 162-172 (1974)
5.98 P.C. Ho, W.H. Carter: Appl. Opt.*15*, 313-314 (1976)
5.99 J. Radon: Ber. Verh. Sächs. Akad. Wiss. Leipzig, Math. Phys. Kl. *69*, 262-271 (1917)
5.100 K.T. Smith, D.C. Solomon, S.L. Wagner: Bull. AMS *83*, 1227-1270 (1977)
5.101 L.A. Shepp, J.B. Kruskal: Am. Math. Monthly *85*, 420-439 (1978)
5.102 R.N. Bracewell, A.C. Riddle: The Astrophys. J. *150*, 427-434 (1967)
5.103 R.A. Crowther, D.J. Rosier, A. Klug: Proc. Roy. Soc. London A-*317*, 319-340 (1970)
5.104 Y. Das, W.M. Boerner: IEEE Trans. AP-*26*, 274-279 (1978)
5.105 M.V. Berry, D.F. Gibbs: Proc. Roy. Soc. London A-*314*, 143-152 (1970)
5.106 L. Colin (ed): "Mathematics of Profile Inversion", NASA TM-X-62.150 (1972)
5.107 E.C. Titchmarsh: *Introduction to the Theory of Fourier Integrals*, 2nd ed. (Oxford Universtiy Press, Oxford 1948)
5.108 J.G. McWhirter, E.R. Pike: J. Phys. A-*11*, 1729-1745 (1978)
5.109 J.G. McWhirter, E.R. Pike: Phys. Scr. *19*, 417-425 (1979)
5.110 G.A. Deschamps, H.S. Cabayan: IEEE Trans. AP-*20*, 268-274 (1972)
5.111 H.G. Schmidt-Weinmar: "Spatial Resolution of Subwavelength Sources from Optical Far-Zone Data", in *Inverse Source Problems in Optics*, ed. by H.P. Baltes, Topics in Current Physics, Vol. 9 (Springer, Berlin, Heidelberg, New York 1978) pp. 83-118
5.112 J.R. Cannon, K. Miller: J. SIAM Numer. Anal. B-*2*, 87-98 (1965)
5.113 K. Miller, G.A. Viano: Nucl. Phys. B-*25*, 460-470 (1971)
5.114 G. Alessandrini: "An Extrapolation Problem for Harmonic Functions", Bollettino U.M.I. (to be published)
5.115 G. Alessandrini: "On Differentiation of Approximately Given Functions", preprint Istituto Matematico U. Dini, Firenze (1979)
5.116 F. John: Ann. Mat. Pura Appl. *40*, 129-142 (1955)
5.117 G.C. Sherman, J.J. Stamnes, E. Lalor: J. Math. Phys. *17*, 760-776 (1976)
5.118 H.P. Baltes, H.G. Schmidt-Weinmar: Phys. Lett. *60A*, 275-277 (1977)
5.119 H.G. Schmidt-Weinmar, W.B. Ramsay: Appl. Phys. *14*, 175-181 (1977)
5.120 H.G. Schmidt-Weinmar: Can. J. Phys. *55*, 1102-1114 (1977)
5.121 J.T. Foley, E. Wolf: J. Opt. Soc. Am. *69*, 761-764 (1979)

5.122 H.P. Baltes, B. Steinle: J. Opt. Soc. Am. *69*, 910 (1979)
5.123 D.K. Lam, H.G. Schmidt-Weinmar, A. Wouk: Can. J. Phys. *54*, 1925-1936 (1976)
5.124 A.F. Fercher, H. Bartelt, H. Becker, E. Wiltschko: Appl. Opt. *18*, 2427-2439 (1979)

6. Combustion Diagnostics by Multiangular Absorption

R. Goulard and P. J. Emmerman

With 10 Figures

Absorption techniques are being applied to three-dimensional combustion diagnostics. By using multiangular scanning, the traditional "onion peeling" method can be extended from axisymmetrical flows to arbitrary distributions of radicals and pollutants in the flow. Since scattering "point" techniques are limited by their very small cross section, such an extension would be a step improvement in sensitivity, down to low radical concentrations. Convolution Fourier transforms and iterative algorithms have already been proven in X-ray absorption tomography and interferometric applications. They are currently tested and compared on typical pollutant and radical concentration as they appear in flames or exhausts. The effect of the number of scans is analyzed for parallel beams. A trade-off exists between accuracy and the number of viewing angles. A five-angle procedure gives 10% accuracy with a moderately filtered convolution algorithm. An experiment feasibility study shows that near time-continuous three-dimensional maps of low concentrations (1 ppm) can be obtained at repetition rates up to 20 kHz. Specific applications to radical and pollutant mapping are discussed, as well as multiangular scanning strategies.

There exists a considerable body of literature concerning combustion diagnostics. Many methods have been developed which make use of the dependence of optical properties of the flow on its thermodynamic properties. In this fashion, maps of concentrations and temperature can be obtained by such techniques as interferometry, scattering, fluorescence, emission or absorption. All these methods have their place in the array of diagnostic devices needed by the experimentalist. What makes absorption diagnostics especially suitable to chemical analysis is the large cross section of most flame components: hence its ability to measure low concentrations or to track rapidly fluctuating signals. Its main limitation is that it is a path-integrated measurement, which cannot resolve a nonuniform property profile along the line of sight. Hence, until recently, its limited success in combustion applications where three-dimensional flows prevail.

The purpose of this chapter is to first briefly review the state of the art of spectrographic absorption techniques for homogeneous samples (Sect.6.1). Then we shall review and extend the work done outside the combustion community to apply absorption techniques to three-dimensional nonhomogeneous cases (Sect.6.2). Finally, recent theoretical results on three-dimensional absorption diagnostics in flames will be presented in Sect.6.3, followed by a proposed experiment (Sect.6.4).

6.1 Absorption in Homogeneous Media

The concentration N_i of a species i is determined by setting up an absorption experiment where the incoming intensity I_0 is attenuated by a sample of length L until it emerges with the value I. If the absorption cross section at the chosen frequency is $Q_{\nu i}$, the Bouguer-Lambert-Beer law states that

$$\frac{I}{I_0} = \exp(-N_i Q_{\nu i} L) \ . \tag{6.1}$$

Also, if one picks two frequencies ν_1, and ν_2 corresponding to two transitions whose lower state populations are N_1 and N_2, and if ΔE is the energy differential between them, Boltzmann's equation

$$\frac{N_2}{N_1} = \frac{g_2}{g_1} \exp\left(-\frac{h(\nu_2 - \nu_1)}{kT}\right) \tag{6.2}$$

will yield the temperature T. This knowledge may remove any dependence the cross section Q_ν may have on temperature. Another option, if one is not concerned about the temperature field per se, is to select a rotational line whose population is fairly constant with temperature changes.

If one is interested in low concentration measurements, the product $N_i Q_{\nu i} L$ is usually small and (6.1) can be written

$$\frac{\Delta I}{I_0} = \frac{I - I_0}{I_0} = -N_i Q_{\nu i} L \ . \tag{6.3}$$

It can be seen from this equation that a low threshold concentration N_i will be measureable if the path is long ($L \gg 1$), the cross section large ($Q_{\nu i} \gg 1$) and if very small values of $\Delta I/I_0$ can be detected.

Therefore it appears that low concentrations are more likely to be detected in atmospheric applications ($L \simeq 1$ km) than in such in situ measurements as flames diagnostics ($L \simeq 10$ cm). The cross sections $Q_{\nu i}$ of most pollutants and radicals correspond to their electronic transitions (near UV) or vibration-rotation (IR). PATEL [6.1] presented a synopsis of the spectral range of radicals, pollutants, and tunable laser sources relevant to some of the combustion chemical constituents. The corresponding values of $Q_{\nu i}$ are in the 10^{-18} to 10^{-17} cm^2 range (see also [Ref.6.2, p. 274]).

The most decisive parameter in securing a low threshold seems to be the ratio $\Delta I/I_0$. Different sources give very different results:

I) Noncoherent Sources. Most noncoherent sources [continuous spectra (Xe discharge), or bands (NO-Rf or discharge)] tend to be noisy. Experiments with NO by DAVIS et al. [6.3] indicated a value of $\Delta I/I_0$ of the order of 1%. Very stable sources (e.g., Globars) can improve this to 0.2% [6.4]. This relatively poor threshold limits also the dynamic range of the measurement. By substitution into (6.3), one estimates in situ thresholds N_i of the order of 10^2 ppm (with $N_{700\ K,\ 1\ atm} \cong 10^{19}\ cc^{-1}$).

II) Lasers. The advent of laser sources, especially tunable ones, produced sources which could concentrate much more power on the spectral lines of interest. Thus the information obtained is freer from the interference of other chemicals and background noise; it also supplies additional information related to line shape. PATEL [6.1] has reviewed the potential of lasers for pollutant detection: dye lasers, tunable diode lasers and discretely tunable lasers. HANST [6.5] had already suggested a match between a number of pollutants and available laser sources. Recently REID et al. [6.6] reported a NH_3 absorption experiment. It is felt that tunable UV lasers have not evolved far enough yet to be usable in routine diagnostics [6.7].

To summarize Patel's conclusion, it appears that the availability of very narrow laser lines and the use of differential analysers brings down the $\Delta I/I_0$ ratio to 10^{-4}. This, for a path of 10 cm, corresponds to a threshold of 1 ppm. In atmospheric applications ($L = 10^5$ cm), less than 1 ppb can be detected.

The conclusion of this very brief review is that there exists now a technology with which in situ measurements of 1 ppm are possible, provided the sample is homogeneous. In the next section, we shall examine what additional techniques are needed for the homogeneity restriction to be lifted.

6.2 Multiangular Scanning

There exist two ways to generate the information necessary to obtain the profile of a property along a line (n points) or for a planar cross section ($M \times N$ grid). One is *frequency scanning* where n frequencies are measured along a given line of sight. This method has been evolved with great success in satellite meteorology and stellar atmospheres [6.8-10]. The other is *multiangular scanning* where the absorption of M equally-spaced parallel beams is measured in a given direction θ (M measurements); the procedure is then repeated for N other angles (see Fig.6.1). From the resulting set of $M \times N$ data, it is possible to retrieve a grid of properties.

The frequency scanning method has been proposed in the case of combustion diagnostics [6.11], but the inherent difficulty of converting optical lengths (i.e., density $\times$ length) to physical lengths, makes it cumbersome, except for those cases where additional information is available (e.g., known mixing ratio, atmospheric density law, etc.). The multiangular method has been applied to interferometry [6.12] and to medical X-ray tomography, the latter with great commercial success [6.13,14]. The procedure is outlined below.

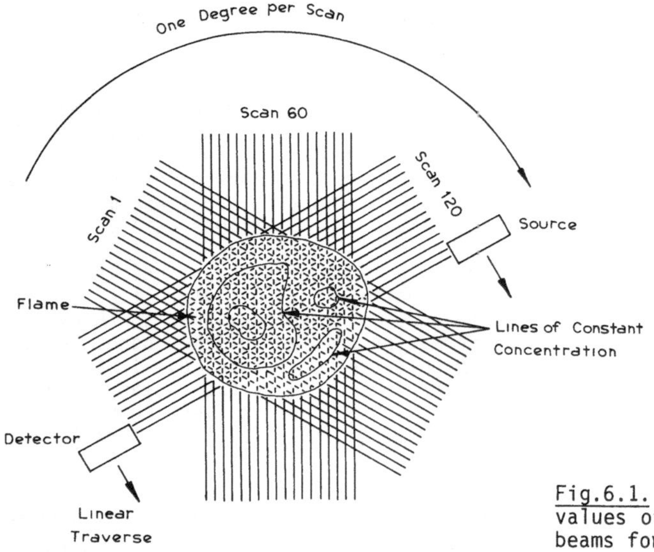

Fig.6.1. Typical scanner. Here 180 values of θ are used and 160 parallel beams for each value of θ

6.2.1 Basic Equation

The multiangular scanning technique is based (Fig.6.2) on the image reconstruction of a property field F (x, y) from its projection in the θ direction

$$P(r,\theta) = \int_{-\infty}^{+\infty} F(r,s)\, ds \quad , \tag{6.4}$$

where $r = x \cos\theta + y \sin\theta$ and $s = -x \sin\theta - y \cos\theta$.

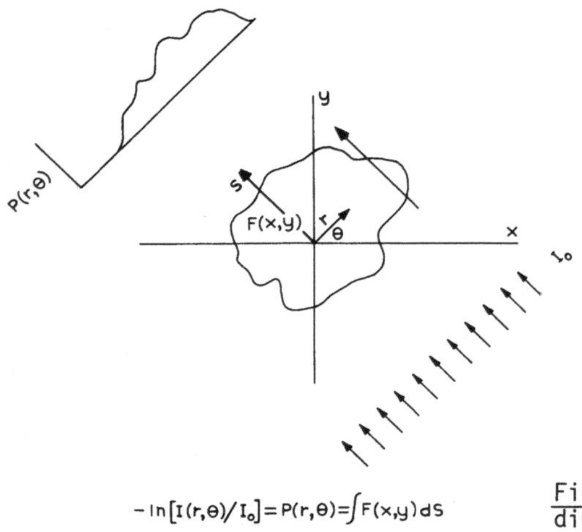

Fig.6.2. Projection P(r, θ) in polar coordinates

There exists a number of mathematical methods which use the set of $(M \times N)$ values of $P(r, \theta)$ obtained experimentally (by scanning the field of interest for N values of θ and M values of r), and then convert this data into a grid of $F(x, y)$ values. At the limit, an infinite number of measurements would give an exact map of $F(x, y)$. Once the map $F(x, y)$ has been established for a section z, measurements are performed at other sections z and eventually the three-dimensional map $F(x, y, z)$ is obtained.

In the case of flame absorption data, if a beam of initial intensity I_0 emerges from the flame with an intensity I, the absorption equation takes the form

$$\frac{I}{I_0} = \exp\left[- \int_{-\infty}^{+\infty} (N_i Q_i ds)\right] \quad , \tag{6.5}$$

where the product (cross section Q_i x concentration N_i) for the component i can be considered as a local function $F(r, s)$ of the thermodynamic properties and concentration of i at the point (r, s) or (x, y). Since the ratio I/I_0 can be measured experimentally, it is seen that by writing (6.5) in the form

$$- \ln \frac{I}{I_0} = \int_{-\infty}^{+\infty} (N_i Q_i)\, ds \quad , \tag{6.6}$$

the techniques developed in other fields of application to solve (6.4) can be readily extended to the three-dimensional measurement of $(N_i Q_i)$ in flames.

In practice, except for the "generalized onion peeling" method proposed by CHEN and GOULARD [6.15], the multiangular scanning method has not been applied to absorption. Since "onion peeling" methods amplify experimental errors, the available algorithms developed for interferometry and tomography are worth closer inspection in terms of possible use in absorption experiments. They fall in three general categories which have been evaluated and compared, especially in both contexts of X-ray tomography [6.16] and interferometry [6.12,17]. The three cases are discussed in Sects.6.2.2-4.

6.2.2 Two-Dimensional Fourier Transform

This method is based on the fact that by taking the Fourier transforms of (6.4), one obtains a "spectrum" $\hat{P}(\omega,\theta)$ of the experimentally available function $P(r,\theta)$

$$\hat{P}(\omega,\theta) \equiv \int_{-\infty}^{+\infty} P(r,\theta) e^{-i\omega r}\, dr \quad , \tag{6.7}$$

where $\omega = 2\pi\rho$, and ρ is the spatial frequency. If (6.4) is substituted into (6.7), then it is known that the one-dimensional Fourier transform $\hat{P}(\omega,\theta)$ of the projection $P(r,\theta)$ is equal to the two-dimensional Fourier transform of $F(r,s)$

$$\hat{F}(\omega,\theta) \equiv \int_{-\infty}^{+\infty}\int_{-\infty}^{+\infty} F(r,s)e^{-i(\omega r + \eta s)}drds\Big|_{\eta = 0} \tag{6.8}$$

when it is evaluated along the line $\eta = 0$ in the two-dimensional frequency plane. This is the "central-slice theorem" (see, e.g. [6.14]). As θ is changed, the line $\eta = 0$ rotates and a 180° θ-range sweeps *all* values $\hat{F}(\omega,\theta)$ in the frequency plane (see Fig.6.3). It is therefore possible to retrieve the function $F(x,y)$ as the inverse Fourier transform of $\hat{F}(\omega,\theta)$. The procedure is discussed later in this chapter. Numerical methods, using finite series rather than integrals, have been adapted to this case. Their accuracy improves with the number of measurements and they tend to be successfull mostly for situations where access is free from all angles.

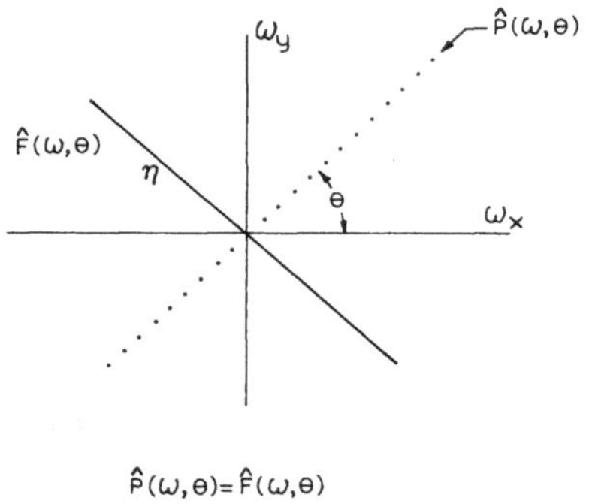

Fig.6.3. The central-slice theorem

Fast Fourier transform techniques (FFT) have accelerated this process substantially [6.18,19], but still faster procedures have evolved. In the *convolution methods* developed by RAMACHANDRAN and LAKSHMENARAYANAN [6.20], the profiles $P(r,\theta)$ collected for each θ are immediately "back projected" on the beam (θ,r), with the use of appropriate filter functions (see Sect.6.3). The picture matrix is thus constructed in a cumulative fashion as the angular scanning proceeds over θ. Typically, an ACTA scanner will deliver a 64,000 element map of (N_iQ_i) in 4½ minutes or less [6.21]. This method has been extremely successful in detecting lesions and abnormalities in brain or limb cross sections, based only on the small variations of the product N_iQ_i for body tissues and fluids in the X-ray range (~50 kev).

A method akin to Fourier transforms relies on the use of different series expansions than Fourier, which happen to match by their first terms the main features of the property profiles at hand. For instance, the Hermitian polynomials $H_{mn}(x,y)$ have been shown to be well suited to near-Gaussian density profiles [6.22].

6.2.3 Linear Superposition Techniques

These techniques are based on straightforward algorithms which repeatedly adjust the estimated $F(x,y)$ map until convergence is accomplished. This is done [6.14] by comparing for each beam (θ,r), the value of $P(r,\theta)$ calculated on the basis of the estimated values of $F(x,y)$, with the actual measurement of $P(r,\theta)$. An algorithm increases then the estimated $F(x,y)$ values along the beam (r,θ) by the *same* averaged increment, in such a way as to reduce the difference between measured and calculated $P(r,\theta)$ to zero. The procedure is repeated until the predicted values of $P(r,\theta)$ coincide with actual measurements *for all* θ*'s*. As in many other forms of image processing, the use of filters is necessary.

6.2.4 Algebraic Reconstruction Techniques (ART)

Here, one divides the (x,y) plane in a grid of $M\times N$ elements, which divide each beam (r,θ) in a set of n segments of known length a_n. Thus the projection $P(r,\theta)$ is equal to

$$P(r,\theta) = \sum_n a_n(r,\theta)\, F_n(x,y) \quad . \tag{6.9}$$

Systems of 200×200 linear equations can be solved for that many values of $F(x,y)$. Iterative methods are quite effective for this application [6.13,23].

6.2.5 Applications and Results

There is no theory which can predict with any generality the relative performance of these methods in any particular application. It seems however that most successes have been met with convolution methods when rapid detailed mapping is required, whereas algebraic reconstruction techniques (ART) seem to be most effective where only a few viewing angles are available [6.24].

In general, the accuracy claimed by X-ray manufacturers for the product N_iQ_i is 0.5% [6.14], which is ample to distinguish between such similar fluids as water and blood. This explains the fine resolution which medical tomography is able to provide in the mapping of tumors, for instance. Another X-ray application which is being considered is that of materials testing, where the density N_i of gaseous flaws is orders of magnitude less than that of the surrounding solid [6.25]. Recent efforts have also been made to capture the dynamics of some internal organs (heart, lungs); WOOD et al. [6.26] have devised a three-dimensional diagnostic device which fires 28 separate X-ray sources and records the corresponding images within 10 ms.

In most of these applications, the objective is to map volumes containing uniform samples of different NQ's. The emphasis is not so much on determining an accurate value of NQ itself, so long as one can tell one substance from another; the emphasis is rather on one's ability to map accurately the contours of these discontinuous samples (a tumor, for instance). This puts a large requirement on the number of terms involved in the Fourier transformation (i.e., on the number of viewing angles). Such a constraint is likely to be considerably relaxed in *combustion applications* where diffusion phenomena yield fairly smooth maps of concentrations N_i, for a fairly constant $Q_{\nu i}$. It should be possible then to diminish the number of viewing angles, a vital advantage since the characteristic times of combustion are often too short to permit the required number of angular scans with a single instrument. Such a device is proposed in Sect.6.4.

In Sect.6.3 we discuss the factors which condition the accuracy of the convolution method which we propose to use in combustion applications.

6.3 The Reconstruction Procedure

It is a mathematical truism that as the number of equally spaced noiseless projections $P(r,\theta)$ (see Fig.6.2) and their corresponding number of infinitesimal rays approach infinity, the difference between the reconstruction of any bandlimited function and the function itself approaches zero. It is just as true that any actual implementation of reconstruction instrumentation and computational algorithms will yield results that significantly deviate from this limiting case if the proper hardware and software trade-offs are not generated.

6.3.1 Reconstruction Errors

Reconstruction errors can generally be attributed to one or more of the following causes:

1) Inadequate number of equally spaced projection angles θ.
2) Inadequate number of equally spaced lateral scanning positions r (fixed θ).
3) Measurement times exceeding the characteristic times during which the function being measured does not vary appreciably.
4) Incomplete optical access to the function or measurement space.
5) Offsets and drift of source or detector gain.
6) Imperfect geometrical positioning of source and detector.
7) Lack of adequate signal-to-noise ratio S/N in the absorption measurement.

This section is concerned primarily with the effects of the limitation in number of projection angles θ and scanning positions r. However, a short review of these error sources is provided for background information.

Incomplete optical access to the function space occurs when the field of view of the instrumentation does not extend fully to the periphery of the system under test or when there are opaque objects within the function space. The ability of both the algebraic and Fourier based reconstruction algorithms to compensate for this limitation in measurement data has been explored empirically by OPPENHEIM [6.27]. Gain offsets and drift and imperfect geometrical positioning of source and detector are common problems in many measurement systems. Their effects on reconstruction solutions have been extensively analyzed [6.28,29]. Although these sources of error will tend to decrease the accuracy of reconstruction, it is worth noting that the measurement speeds necessitated by combustion dynamics would limit error due to drift. Deviations caused by imperfect calibration may be corrected if the calibration offsets are ascertained prior to the processing of the projection data. The effect of finite signal-to-noise ratios S/N on the absorption of projection measurements has been explored for a variety of noise conditions [6.30]. Due to the dynamics of the combustion process the standard methods of increasing the S/N ratio by signal averaging are not generally applicable. The absorption coefficient N_iQ_i distribution can be expected to be neither repetitive nor static and the time limitation for projection measurements precludes multiple sampling. However, the common noise rejection techniques of narrowing aperture times and integration windows do apply to reconstruction instrumentation. Reconstruction error is dependent on function bandlimit, on system resolution, as well as on S/N ratio. Resolution and spatial bandwidth requirements must be specified before the effect of S/N ratios can be quantified.

For simplicity, the relationships between resolution, spatial bandlimit and projection data will be defined for parallel-ray projections. Let the measurement or function space have a radius R such that $F(x,y) = 0$ for $r \geq R$, where $r^2 = x^2 + y^2$, and assume the function to be bandlimited, i. e., $\hat{F}(x,y) \equiv \hat{F}_\theta(\rho) = 0$ for $\rho > \rho_{max}$ where ^ indicates a Fourier transform operation. From the sampling theorem, a function can be uniquely recovered from its samples if it is sampled at a rate grater than twice the highest frequency component of the function (sampling rate $\geq 2\rho_{max}$). It is now possible to determine the number of measurements required to accurately reconstruct the bandlimited function F(x,y). One axis of the two-dimensional transform plane of the function is specified by $4\rho_{max}R = M$ samples, where 2R is the diameter of the physical space. The minimum number of equally spaced (in angle) projections required to reconstruct F(x,y) has been derived[1], [6.36,37] as $N \simeq 2\ R\rho_{max}$.

---

[1] A simplified derivation of $N \simeq 2\pi R\rho_{max}$ can be given as follows: $M = 4(\rho_{max})R$ samples define the bandlimited projection in both the spatial and frequency domains. The frequency domain of the projection extends from $-\rho_{max}$ to $+\rho_{max}$ and therefore the frequency spacing between samples is $2\rho_{max}/4\rho_{max}R = 1/2R$. The two-dimensional frequency domain of a function with bandlimit ρ_{max} and spatial limit 2R, can be defined by a square matrix of sample points separated by 1/2R. If 1/2R is assumed to be the maximum allowable separation of sample axes in the frequency domain diverging at angle $\Delta\theta$, then $2\pi\rho_{max}/(1/2R) \simeq 2\pi/\Delta\theta$ or $\Delta\theta \simeq (2R\rho_{max})^{-1}$ and $N = \pi/\Delta\theta \simeq 2\pi R\rho_{max}$.

Therefore, if ρ_{max} is isotropic $(2\pi R\rho_{max})(4\rho_{max}R) = 2\pi(2R\rho_{max})^2$ measurements are needed to accurately reconstruct the spatially and frequency limited function F(x,y). If the resolution of this measurement technique is defined to be the smallest spacing between measurements, it is thereby simply related to the bandlimit by the equation

$$\text{Resolution} = \frac{1}{2\rho_{max}} \quad . \tag{6.10}$$

To summarize, if a function is bandlimited at ρ_{max} it will be accurately reconstructed if $4\rho_{max}R$ equally spaced rays are measured at $2\pi\rho_{max}R$ equally spaced angles. It should be noted that since Fourier theory does not allow a function to be both bandlimited and spatially limited, this use of bandlimit is interpreted as approximate.

6.3.2 An Observation of the Oversampling Requirement of Reconstruction

It had been stated previously, that the number of uniformly distributed samples M required to adequately define a function of bandlimit ρ_{max} would be $4\rho_{max}R$ and the transform of this sampled function would define a central axis in the transform domain of the function. This central axis would then be defined over its range $-\rho_{max}$ to $+\rho_{max}$ by the same number of samples M. Therefore the spatial frequency spacing between samples would be $2\rho_{max}/4\rho_{max}R = 1/2R$. Since the two-dimensional Fourier transform of a bandlimited (at ρ_{max}) function can be specified by a square matrix of $M \times M$ samples, the number of samples within a circle of radius ρ_{max} would be approximately $\pi(M/2)^2$ or $\pi(4\rho_{max}R/2)^2$. In comparison with the sampling requirement of reconstruction derived earlier: $2\pi(2R\rho_{max})^2$, there appears to be oversampling by a factor 2. This oversampling in the frequency domain becomes apparent by noting the change in density of sampled frequencies from the center of the frequency domain to its periphery. In general, oversampling results in improved S/N ratio and its effects should be noticed when noise is added to the simulated projection data.

6.3.3 Number of Measurements $M \times N$ in Combustion Application

Meaningful distributions of absorption coefficients in any dynamic process can only be obtained if the aperture time of the measurement is small compared to the time required for significant change of the measured quantities (characteristic time). This time constraint, coupled with the difficulties and expense of rapidly scanning detector and source arrays, force the minimization of the number of required measurements. Before this minimization is attempted, it is worthwhile to relate the abstract term bandlimit to actual physical processes and real source and detector limitations. Most spatial distributions of physical properties do have an approximate if not absolute bandlimit and therefore can be represented by their truncated transform with negligible error. This is partly due to the fact that any measurement represents an integration or averaging over a finite interval of time and space which inherently bandlimits the measured physical properties. In reconstruction, the finite width of the transmitted ray and detector act as a low pass filter by averaging the signal over their spatial dimensions.

If the number of required measurements is separated into two components, the first being the number of samples collected at one angle $M = (4\rho_{max}R)$, and second being the number of angles $N = (2\pi R\rho_{max})$, it can be observed that these two numbers differ only by a factor of $\pi/2$ and have the same dependence on ρ_{max} and R. Since R defines the boundary of the function and measurement space required by the experiment, it is thereby fixed. Now ρ_{max} is the approximate spatial bandlimit of the function and can only be anticipated by some a priori knowledge of the function itself. From these statements it can be ascertained that neither the number of rays along one axis nor the number of angles can be decreased if the full accuracy of the reconstruction is to be guaranteed. However, the functional dependence of reconstruction accuracy on either M or N has not been defined except for this limiting case of fulfilling the sampling theorem on a bandlimited function.

From a hardware standpoint much more can be gained by minimizing N rather than M. The detectors required for combustion diagnostics are available in single and two-dimensional array configurations of up to 100 elements in a single dimension. The energy sources can be optically structured into a fan or plane beam. This means that no motion is required to obtain the projection values, with a high bandlimit, from a single angle. Due to the fact that the sources and detectors are of finite size and cannot be collocated in reconstruction instrumentation, to obtain a large number of angular views, physical rotation of source and detector pairs is required. If the number of angular views could be constrained to a reasonably small number, it becomes physically and economically feasible to have a stationary, almost instantaneous reconstruction measurement system.

6.3.4 The Convolution Algorithm

The effect of limiting the number of angular views will be determined on a variety of symmetric and nonsymmetric, continuous and discontinuous functions. The convolution algorithm, which is representative of the class of Fourier based reconstruction techniques, will be used in this analysis because of its advantages of speed of implementation and of being analytically tractable. A simplified derivation of the algorithm will be presented here with much more detailed development available in the literature [6.20].

Let $P(r,\theta)$ be the integral of the function to be reconstructed along the line s defined by r and θ (Fig.6.2)

$$P(r,\theta) = \int F(x,y)\, ds \quad , \tag{6.11}$$

where $r = x\cos\theta + y\sin\theta$. The Fourier transform of $P(r,\theta)$ is defined as

$$\hat{P}(\omega,\theta) = \int_{-\infty}^{\infty} e^{-i\omega r} P(r,\theta)\,dr \quad , \text{ where } \omega = 2\pi\rho \quad , \tag{6.12}$$

and

$$\hat{F}(\omega,\theta) = \int_{-\infty}^{\infty}\int_{-\infty}^{\infty} F(x,y)\exp[-i\omega(x\cos\theta + y\sin\theta)]dxdy \quad . \tag{6.13}$$

From the central slice theorem of Fourier transforms we have

$$\hat{P}(\omega,\theta) = \hat{F}(\omega,\theta) \quad , \tag{6.14}$$

hence,

$$F(x,y) = \frac{1}{4\pi^2}\int_0^{\pi} d\theta \int_{-\infty}^{\infty} \hat{P}(\omega,\theta)\exp[i\omega(x\cos\theta + y\sin\theta)]|\omega|d\omega \quad . \tag{6.15}$$

Let V be equal to the inner integral

$$V(r,\theta) = \frac{1}{2\pi}\int_{-\infty}^{\infty} \hat{P}(\omega,\theta)|\omega|e^{i\omega r}d\omega \quad . \tag{6.16}$$

Define a function ϕ such that

$$\hat{\phi}(\omega) = |\omega| \quad , \tag{6.17}$$

then from convolution

$$V(r,\theta) = \frac{1}{2\pi}\int_{-\infty}^{\infty} \hat{P}(\omega,\theta)\hat{\phi}(\omega)e^{i\omega r}d\omega = \int_{-\infty}^{\infty} P(\tau,\theta)\phi(r-\tau)d\tau \quad , \tag{6.18}$$

hence,

$$F(x,y) = \frac{1}{2\pi}\int_0^{\pi} d\theta \int_{-\infty}^{\infty} P(\tau,\theta)\phi(x\cos\theta + y\sin\theta - \tau)d\tau \quad . \tag{6.19}$$

Since $P(r,\theta)$ will only be known in the sampled domain, it is necessary to replace the integrals by their discrete summation approximations

$$F(x,y) = \frac{R}{MN}\sum_{j=1}^{N}\sum_{k=1}^{M} P(r_k,\theta_j)\phi(x\cos\theta_j + y\sin\theta_j - r_k) \quad , \tag{6.20}$$

where $R = 1$, $\theta_j = (j-1)\pi/N$, $r_k = -R + k(2/M)$.

The choice of $\phi(r)$ greatly affects the computational requirements as well as the accuracy of the convolution algorithm. Now $\phi(r)$ can be considered a filter function and thereby its selection must be dependent on the noise of the instrumentation system as well as the bandlimit of the function to be reconstructed. KWOH et al. [6.31] provided a comparison and optimization of $\phi(r)$. Two commonly used $\phi(r_k)$'s were implemented in this study. These are

$$\phi(0) = \frac{M^2}{\pi} \quad , \; \phi(r_k) = -\frac{M^2}{\pi}(4k^2 - 1)^{-1} \quad , \; k = 1,2,3,\ldots. \qquad (6.21)$$

(Shepp and Logan filter), where ϕ is linearly interpolated when evaluated between r_k's, and

$$\bar{\phi}(r_k) = 0.4\phi(r_k) + 0.3\phi(r_{k+1}) + 0.3\phi(r_{k-1}) \quad , \qquad (6.22)$$

where $\phi(r_k)$ is defined in the preceding equation (modified Shepp and Logan filter).

6.3.5 Simulated Test Functions and Results

Functions $F(x,y)$ composed of summations of Gaussian fields have been tested, because of their resemblance to smooth NQ profiles in diffusion flames. To test the occurence of sharp discontinuities and point sources, a summation of cylindrical and Gaussian functions was reconstructed. The projection values of these computer simulated functions are determined by numerically integrating along the required paths in the function space. In reconstructing this group of functions the value of M is 100. The original Shepp and Logan filter $\phi(r_k)$ is utilized in all cases except where otherwise designated. In the following figures, N is defined as the number of angular scans. The functions tested and the corresponding figures can be summarized as follows:

$$\left.\begin{array}{l} F_1 = e^{-20(x^2 + y^2)} \\ N = 2,\ 3,\ 6,\ 10,\ 20 \end{array}\right\} \text{Fig.6.4}$$

$$\left.\begin{array}{l} F_2 = 0.5\, e^{-20[(x + 0.3)^2 + y^2]} + 0.8\, e^{-20[(x - 0.3)^2 + y^2]} \\ N = 2,\ 3,\ 6,\ 10,\ 20 \end{array}\right\} \text{Fig.6.5}$$

$$\left.\begin{array}{l} F_3 = 0.5\, e^{-20[(x + 0.3)^2 + y^2]} + 0.8\, e^{-20[(x - 0.3)^2 + y^2]} \\ \quad + 0.7\, e^{-25[(x + 0.5)^2 + (y - 0.5)]^2} \\ N = 2,\ 3,\ 6,\ 10,\ 20 \end{array}\right\} \text{Fig.6.6}$$

$$\left.\begin{array}{l} F_4 = F_3 \text{ for } x^2 + y^2 \leq 0.7,\ F_4 = 0.3 \text{ for } 0.7 < x^2 + y^2 < 0.9 \\ N = 20 \text{ with and without modified Shepp and Logan filter.} \end{array}\right\} \text{Fig.6.7}$$

Several significant characteristics of the Fourier based reconstruction algorithm can be obtained from an inspection of Figs.6.4-7. Increasing N causes the reconstruct-

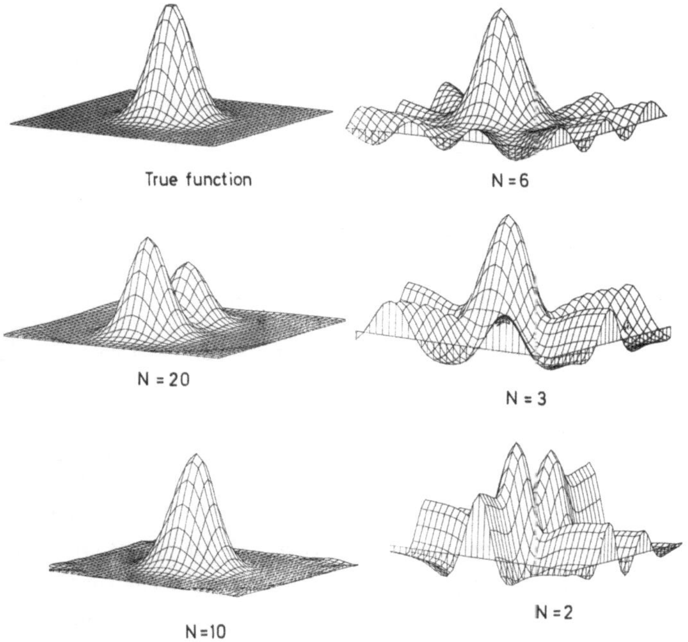

Fig.6.4. Reconstruction of one Gaussian F_1 using N viewing angles

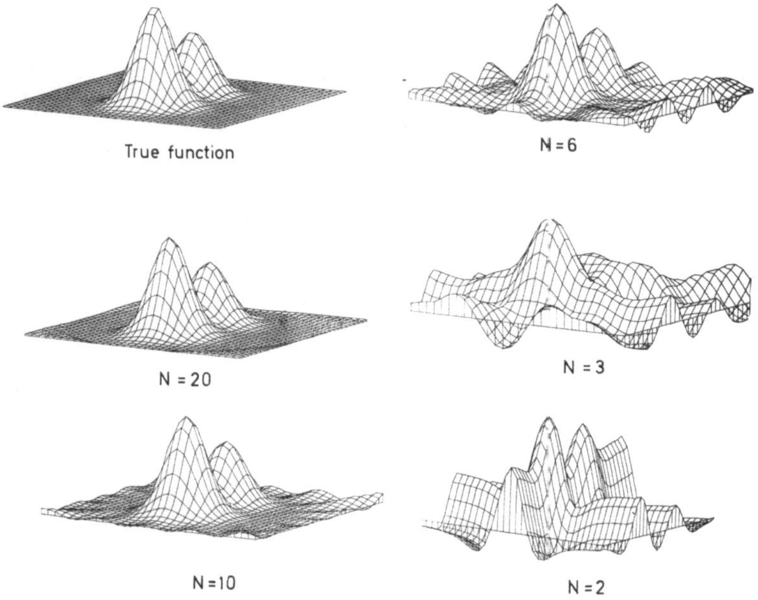

Fig.6.5. Reconstruction of two Gaussians F_2 using N viewing angles

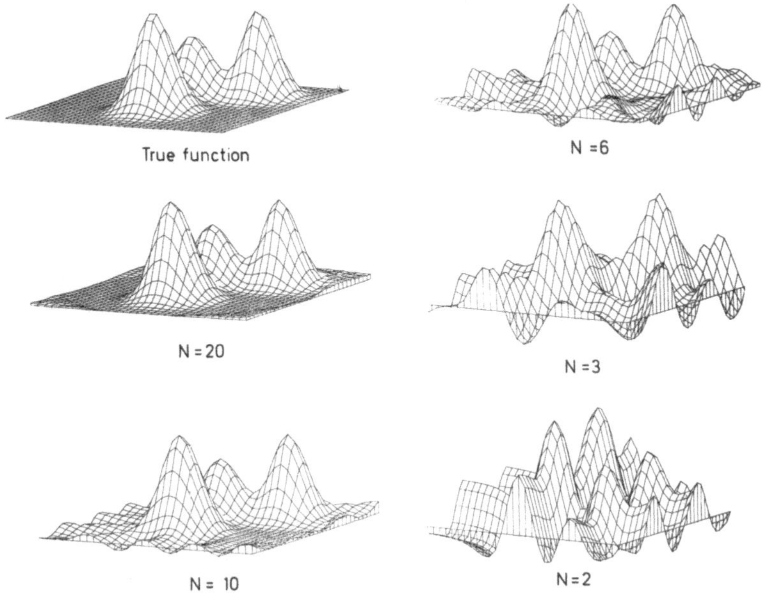

Fig.6.6. Reconstruction of three Gaussians F_3 using N viewing angles

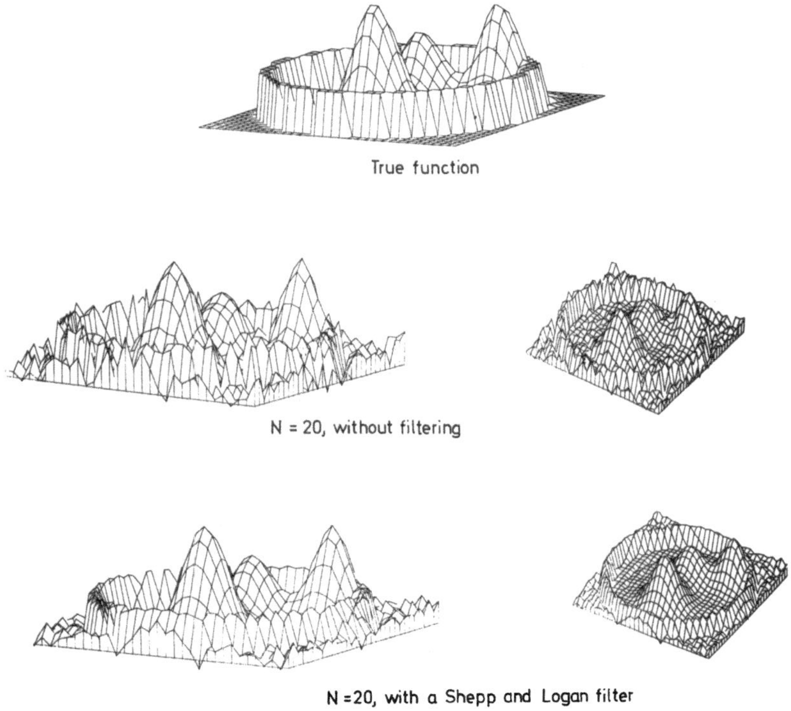

Fig.6.7. Reconstruction of three Gaussians F_3 surrounded by a circular "top hat" function using 20 viewing angles

ed function to converge to approximately the true function even when the true function has no absolute bandlimit, and with a relatively small number of scans, a summation of Gaussian distributions can be accurately reconstructed. In the axisymmetric example of F_1, it is obviously possible to reconstruct this Gaussian function with only a single noiseless projection since any other projection data would be redundant. This a priori knowledge of symmetry was not utilized in the reconstruction of F_1. In examples where axial symmetry is expected, several projections can be taken, then averaged to increase the S/N ratio of the system. The averaged projection data can then be used repeatedly in the algorithm [$N = (\pi/2)M$ would be an optimal choice]. It is interesting to note the similarity between using a small number of scans and truncating a Fourier series or Fourier transform at too low a frequency. The same type of ringing and Fourier transform equivalent of Gibbs phenomenon occur.

The selection of filter function had little effect on either rms error or image quality for F_1, F_2, and F_3. The reconstruction of F_4 was greatly improved by the use of the modified filter function. This is due to the greater approximate bandlimit of F_4 compared to F_3. Only the reconstruction of F_4 with 20 projections is illustrated here since the results from 10 or fewer projections do not yield meaningful images. The difference between the accuracy of reconstruction for the continuous Gaussian functions (F_1, F_2, and F_3) and the discontinuous function F_4 is due to the greater approximate bandlimit of F_4. It is anticipated that the choice of filter function will greatly effect the accuracy of reconstruction with finite S/N ratios.

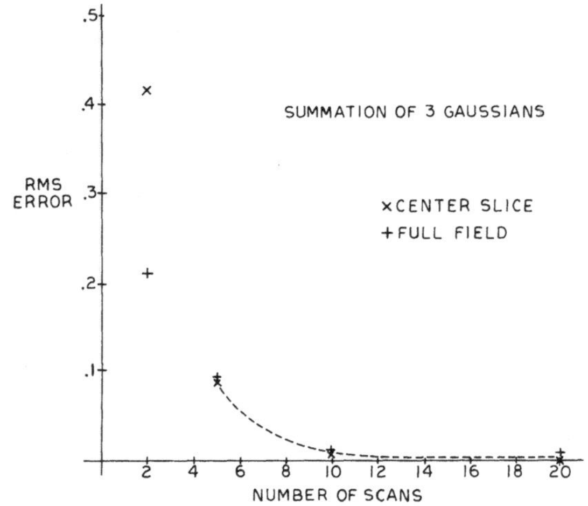

Fig.6.8. Effect of the number of viewing angles N on reconstruction accuracy

Figure 6.8 illustrates the functional dependence of rms error on N (angular scanning) for F_3. Two rms error terms are generated. One indicates the deviation from the true function over an equally spaced (10×10) array covering the function space. The other error term specifies the error produced by a row of 10 points along the x axis of this array approximately at $y = 0$. This plot of rms error as a function of N for F_3 is representative of the error curves generated for the other functions. They all exhibit rather chaotic rms error magnitude at $N < 4$ and all approach rms error $= 0$ at large N.

In generating these figures, M (lateral scanning at a fixed angle) was chosen to be 100 and the question of why to allow M to be so large with all the effort to reduce N is certainly a valid one. What the approximate bandlimit of combustion distributions is can only be hypothesized without experimentation. If a large number of samples can be obtained along a single axis without too great a penalty in measurement acquisition time, several important gains can be made. By taking the transform of the projection, the approximate bandlimit of the function at least in this direction can be ascertained, if the S/N ratio is large and if the projection is not undersampled. Knowing the bandlimit of the function is essential in the determination of the filtering and data processing requirements in reconstruction. Having a large number of samples can also improve the system's S/N ratio especially if the sampling frequency is much larger than the function bandlimit, since these samples could then be filtered before the convolution algorithm is implemented.

6.3.6 Algebraic Reconstruction

Algebraic reconstruction technique (ART) is another commonly used algorithm from which functions can be reconstructed from their projections. There are many excellent papers which thoroughly describe the existing theory and numerical implementation of ART [6.13,23]. A review of this literature yields the conclusion that the algebraic algorithms may be very useful in reconstruction from a small number of projections [6.24,32]. The major advantage that ART has in comparison to the Fourier based algorithms is its ability to perform nonlinear operations. The ability to constrain the intermediate function values to lower and upper level bounds reduces greatly the ringing which may be present in Fourier based reconstructions. One major penalty involved using the algebraic reconstruction algorithms is the requirement of significant computer time to process the data. Direct comparisons on a function basis with anticipated S/N conditions are necessary to prove the advantage of one algorithm over the other.

6.3.7 Benefits of Additional Digital Signal Processing

Use of the convolution algorithm does not preclude the implementation of additional processing before and after the reconstruction. Each projection could be filtered before its use in the convolution algorithm. The knowledge that negative absorption coefficients are nonexistent might be utilized to zero these points and linearly interpolate to a contiguous positive valued point. The reconstructed matrix can also be integrated along the same ray geometries and these terms compared with original projection values to determine a measure of reconstruction accuracy.

6.3.8 Conclusion

The accurate reconstruction of sums of Gaussian distributions of bandlimit equivalent to those tested here, can be accomplished with a physically manageable number of uniformly spaced projections $N \geq 5$, using the convolution algorithm. The accuracy in a noiseless system is limited by the ringing caused by the undersampling in angle of the function space (for $M \gg N$). This produces ringing effects similar to those generated by the truncation of the Fourier transform of the function at a frequency below its approximate bandlimit. As would be anticipated from the previous statement, discontinuous functions are more difficult to reconstruct than continuous functions because of the larger spread in their power spectrum. Regardless of the anticipated function bandlimit, it is advantageous to sample each projection at as high a rate as possible. This would allow the determination of the function's bandlimit to facilitate the selection of filter and other processing requirements.

6.4 Experimental Aspects

The first three sections of this chapter have examined the potential and limitations of optical tomography. The foremost attraction of absorption methods is their potential for low concentration threshold (1 ppm in situ). Another remarkable feature is the large number of source photons produced per unit time in the spectral frequency range of interest: even a low power laser beam (10^{-2} Watt) contains 10^{11} photons/μs. Thus, a resolution requirement of a 100×100 element grid, given a 20 kHz flame fluctuation (50 μs), corresponds to a beam intensity $I_0 = 10^{11} \times 10^{-4} \times 50 = 5 \times 10^8$ photons, well above a statistically meaningful signal. Hence, the ability to scan rapidly and to measure flame processes *in real time*.

Section 6.3 showed also that five viewing angles yield sufficient accuracy, especially near the center of the flame. The experimental configuration shown in Fig. 6.9 should provide the means to observe a flame from five directions simultaneously. Such sets of beam splitters are used successfully in holographic applications [6.22]. An alternative would be to use five or more inexpensive CW sources as was done for NO and CO in one dimension [6.3].

Figure 6.10 illustrates the source-sensor configuration for one beam only. It includes a differential amplifier to eliminate source fluctuations [6.1,33]. It also uses a raster, which could be of the rotating gold-coated (IR) mirror variety [6.34]. Since scanning is used, a single photomultiplier can be used rather than a sensor array (or camera): the time-dependent signal is stored in digitized or analog form and processed later.

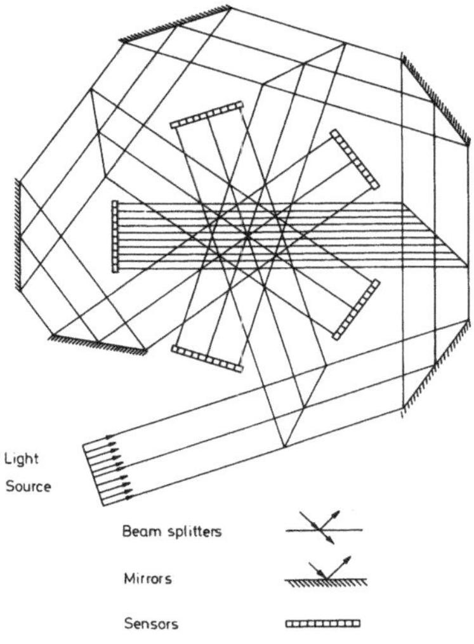

Fig.6.9. Instantaneous multiangular scanning (N = 5, M = 100)

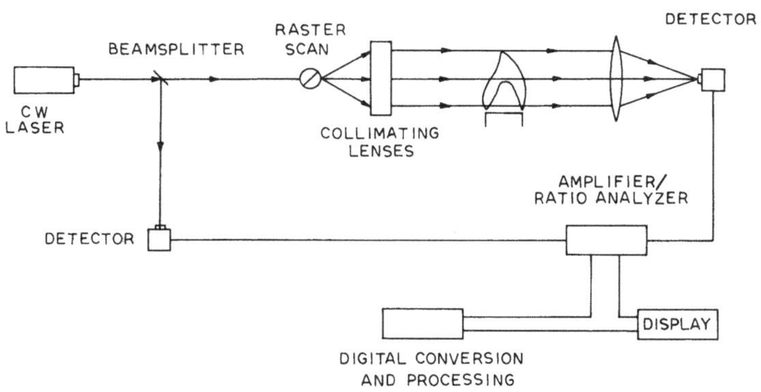

Fig.6.10. Optical scanning of a flame (1 beam only)

As can be seen from [6.1] there are a number of laser sources whose frequency corresponds to combustion radicals and pollutants. However, not all are continuous sources (CW) and most are in the visible range where very few molecules absorb. One is then led to conceive of several types of experiments:

I) a CW dye laser (visible range) for investigating some of the molecules which absorb in the near UV-visible region: OH, C_2, NO_x, CH (see [6.11]).
II) a CW dye laser (visible range) to be used in conjunction with tracers (Na, for instance) or particulates.
III) a tunable or "discretely" tunable laser [6.1] which could pick a line from an IR rotational-vibrational band. HANST [6.5] listed a series of discrete source-pollutant combinations (SO_2, CO, CH_4, ...).
IV) a traditional noncoherent diagnostics system (NO, CO) as mentioned earlier.

If the knowledge of the temperature T is required, then two measurements are necessary (see Sect.6.1). If the emitted beam includes two frequencies sufficiently different to make the use of (6.2) unambiguous, a system of filters and a chopper will make continuous recording possible. Another usable (if expensive) source would be obtained by the merging of the beams of two tunable lasers.

If only concentrations are of interest, there exists in each vibrational-rotational band one rotational line ($J \simeq 10$) whose strength is practically independent of temperature [6.4,35]. In such a case, a single tunable laser would be adequate for concentration measurements.

The two-dimensional (sectional) results discussed in Sect.6.3 were obtained on a PDP-11. If one wishes to measure three-dimensional grids (i.e., scan in two dimensions, TV style) the reconstruction time and memory requirements are likely to be more important. There exist also numerical programs which display graphic perspectives of three-dimensional maps obtained by tomography [6.25]. The possibility to study visually the progress of three-dimensional combustion processes in a quantitative slow motion manner could be very useful.

This outline of experimental possibilities shows a combination of techniques which have been already demonstrated separately. However, some are not entirely troublefree and it is not planned to attempt a full experiment at this point. Rather, it is proposed to test first a single scanning beam in both slicing and TV modes (2D and 3D), as illustrated in Fig.6.10. An important output would be the projections $P(r,\theta_0)$ of a number of typical flames. Their frequency analysis (see Sect.6.3) would yield a reliable estimate of the number of viewing angles required in the full experiment.

Acknowledgements. The authors wish to acknowledge the support received from the Air Force Office of Scientific Research (AF 77-3439) and from the National Engineering Laboratory of the National Bureau of Standards (NBS G8-9013). Among the many scientists who contributed their ideas and time to this project, special mention should be made of Drs. R.S. Ledley, W.K. McGregor, R.J. Santoro and W.G. Mallard.

References

6.1 C.K.N. Patel: Science *202*, 157-173 (1978)
6.2 E.D. Hinkley (ed.): *Laser Monitoring of the Atmosphere*, Topics in Applied Physics, Vol. 14 (Springer, Berlin, Heidelberg, New York 1976) p. 274
6.3 M.G. Davis, W.K. McGregor, J.D. Few: "Spectral Simulation of Resonance Band Transmission Profiles for Species Concentration Measurements: NO Bands as an Example"; Rpt. AECD-TR-74-124, Arnold Engineering Development Center, AAFS Tennessee 37389 (Jan. 1975)
6.4 D.E. Burch, D.A. Grayvnak: "Infrared Gas Filter Correlation Instrument for in Situ Measurement of Gaseous Pollutants"; Rpt. EPA-65012-74-094, Philco-Ford Corp., Newport Beach, CA (1974)
6.5 P.L. Hanst: Appl. Spectrosc. *24*, 161-174 (1970)
6.6 J. Reid, J. Shewchun, B.K. Garside, B.A. Ballit: Laser Focus, October 1977, pp. 45
6.7 F.B. Dunning: Laser Focus, May 1978, pp. 72-77
6.8 D.Q. Wark, H.E. Flemming: Month. Weather Rev. *94*, 351-377 (1966)
6.9 M.T. Chahine: J. Atmosph. Sci. *27*, 960-967 (1970)
6.10 J.T. Jeffries: *Spectral Line Formation* (Blaisdell, New York 1968)
6.11 C.M. Chao, R. Goulard: "Nonlinear Inversion Techniques in Flame Temperature Measurements", in *Heat Transfer in Flames*, ed. by N.H. Afgan, J.M. Beer (Scripta Book, Washington D.C. 1974) pp. 295-337
6.12 D.W. Sweeney, C.M. Vest: Int. J. Heat and Mass Transfer *17*, 1443-1454 (1974)
6.13 R. Gordon, G.T. Herman, S.A. Johnson: Sci. Amer., October 1975, pp. 56-68
6.14 W. Swindell, H.H. Barrett: Phys. Today, December 1977, pp. 34-41

6.15 F.P. Chen, R. Goulard: J. Quant. Spectrosc. Radiat. Transfer *16*, 819-827 (1976)
6.16 R.A. Brooks, G. di Chiro: Phys. Med. Biol. *21*, 689-732 (1976)
6.17 P.T. Radulovich, C.M. Vest: "Determination of Three-Dimensional Temperature Fields by Holographic Interferometry"; Rpt. INTFL-7601, Department of Mechanical Engineering, The University of Michigan, Ann Arbor, Mich. (1976)
6.18 J.S. Bendat, A.G. Piersol: *Random Data: Analysis and Measurement Procedures* (Wiley Interscience, New York 1971)
6.19 J.A.L. Thomson, A.R. Davis, K.G.P. Sulzmann: "Conceptual Design of an Airborne Laser Doppler Velocimeter"; Rpt. PD-B-76-118, Physics Dynamics Inc., Berkley, Calif. (1976)
6.20 G.N. Ramachandran, A.V. Lakshmenarayanan: Proc. Natl. Acad. Sci. U.S. *68*, 2236-2240 (1971)
6.21 R.S. Ledley: Comput. Biol. Med. *6*, 239-246 (1976)
6.22 R.D. Matulka, D.J. Collins: J. Appl. Phys. *12*, 1109-1119 (1971)
6.23 R. Gordon, R. Bender, G.T. Herman: J. Theor. Biol. *29*, 471-481 (1970)
6.24 G.N. Minerbo, J.G. Sanderson: "Reconstruction of a Source from a Few (2 or 3) Projections"; Rpt. LA-6747-MS, Los Alamos Scientific Laboratory (1977)
6.25 R.P. Kruger, T.M. Cannon: Material Evaluation, April 1978, pp. 75-80
6.26 E.L. Ritman, R.E. Sturm, E.H. Wood: "Needs, Performance Requirements, and Proposed Design of Special Cardiopulmonary and Circulatory Dynamics", in Proc. 1975 Workshop on Reconstruction Tomography in Diagnostic Radiology and Nuclear Medicine, ed. by M.M. Ter-Pogossian (University Park Press, Baltimore 1977) pp. 431-451
6.27 B.E. Openheim: "Reconstruction Tomography from Incomplete Projections", in Proc. 1975 Workshop on Reconstruction Tomography in Diagnostic Radiology and Nuclear Medicine, ed. by M.M. Ter-Pogossian (University Park Press, Baltimore 1977) pp. 155-183
6.28 G.K. Kowalski: IEEE Trans. NS-*24*, 2006-2016 (1975)
6.29 L.A. Shepp, J.A. Stein: "Simulated Reconstruction Artifacts in Computerized X-Ray Tomography" in Proc. 1975 Workshop on Reconstruction Tomography in Diagnostic Radiology and Nuclear Medicine, ed. by M.M. Ter-Pogossian (University Park Press, Baltimore 1977) pp. 33-48
6.30 L.A. Shepp, B.F. Logan: IEEE Trans. NS-*21*, 21-43 (1974)
6.31 Y.S. Kwoh, I.S. Reed, T.K. Truong: IEEE Trans. NS-*24*, 1990-1998 (1977)
6.32 M. Schlindwein: IEEE Trans. NS-*25*, 1135-1143 (1978)
6.33 K.G.P. Sulzmann, J.E.L. Lowder, S.P. Penner: Combust. Flame *20*, 177-191 (1973)
6.34 C.E. Baker: "Laser Displays", in *Development in Laser Technology*, ed. by J.N. Forsyth, SPIE Seminar Proceedings, Vol. 20 (SPIE, Palos Verdes Estates, Calif. 1970) pp. 141-145
6.35 F.P. Chen, R. Goulard: "Some Aspects of Optical Pollutant Measurement Systems for Jet Engine Exhaust Flows", AIAA Paper No. 76-108, AIAA, New York (1975)
6.36 R.N. Bracewell, A.C. Riddle: Astrophys. J. *150*, 427-434 (1967)
6.37 R.N. Bracewell, J.R. Cos, Jr.: "An Overview of Reconstruction Tomography and Limitations Imposed by a Finite Number of Projections", in Proc. 1975 Workshop on Reconstruction Tomography in Diagnostic Radiology and Nuclear Medicine, ed. by M.M. Ter-Pogossian (University Park Press, Baltimore 1977) pp. 3-32

7. Polarization Utilization in Electromagnetic Inverse Scattering

W.-M. Boerner

With 11 Figures

It is the objective of this chapter to review the progress in electromagnetic inverse scattering over the past three decades by placing major emphasis on the radar target imaging problem. The complete description of electromagnetic scattering processes implies polarization and to recover the descriptive parameters of a scatterer from the measured field given the incident field requires the knowledge of the target scattering matrices and their particular properties. In radar target discrimination, identification and imaging use of the entire spatial frequency domain of the radar cross section must be made leading to various approximate approaches whose measurement inputs differ greatly from those used for the inverse problems in optics. We begin with an introductory section (Sect.7.1) providing general definitions of the electromagnetic inverse problems and their relation to other chapters in this volume. Next, we introduce in Sect.7.2 vector diffraction integrals and formulas both in frequency and time domains which are required to derive all components of the scattering matrices, show their relations, and briefly outline novel computer-assisted numerical methods. In Sect.7.3 properties of the radar cross section, its related scattering matrices and various radar target operators are derived placing major importance on optimal polarization descriptors, i.e., the cross polarization (maximum) and co-polarization (minimum) null pairs which in the radar case are of great use in the unique description of target and clutter. We then review pertinent inverse scattering theories in the various electromagnetic spatial frequency regimes (of the radar cross section) in Sect.7.4 and we will identify their limitations and polarization sensitivity. In Sect.7.5 we show how vector holography is intrinsically related to the electromagnetic inverse problem and how polarization plays a dominant role in electromagnetic imaging. Finally, in Sect.7.6 we summarize the state of the art, define still existing open problems, and we suggest new approaches by using methods developed in other physical fields.

7.1 Scope

Although the inverse problem of electromagnetic scattering encompasses that of optical scattering, we shall be mainly concerned about macroscopic wave interaction phenomena and microscopic material properties will not be considered here. Therefore, the content of this chapter will differ in nature somewhat from that of the other chapters, which are essentially dealing with the recovery of characteristic parameters of the microscopic structure of materials in the optical region. We will not be concerned with the polarizability of matter and/or optical devices as reviewed most recently by BENNETT and BENNETT [7.40], but rather with properties of the coherent radar scattering matrix in relation to the problem of vector diffraction from radar targets. Subsequently, we shall place increased emphasis on the peculiar properties of the radar cross section scattering matrix, its associated optimal polarizations, i.e., the use of the cross/co-polarization null pairs in electromagnetic inverse problems, rather than investigating properties of the time-averaged incoherent Stoke's scattering or 4×4 Mueller matrix as applied to remote sensing of material properties of radar targets. It is the objective to introduce novel methods developed during the past three decades in radar imaging which should also be of considerable use in advancing the state of the art of inverse scattering in optics.

Here we introduce definitions of the inverse problem in electromagnetics. Next, we discuss its relation to the inverse optical problem providing relevant limitations.

7.1.1 Definitions of the Electromagnetic Inverse Problem

Inverse techniques have been developed in many seemingly unrelated fields of physical sciences where the characteristic descriptors of a medium are estimated from experimental data, obtained from measurements made usually at a distance from the medium, utilizing the laws that relate these characteristic parameters to the experimental data in a given situation. Various descriptions of the nature of inverse problems are given in the literature (see, e.g. [7.1-6]) and solutions to inverse problems have become of increasing importance in aeronomy, geophysical exploration, radar target imaging, remote sensing, material and medical diagnostics, etc. An elaborate survey of inversion techniques as applied to a variety of disciplines is compiled in a NASA Technical Memorandum [7.1] and more recently a comprehensive state of the art review of electromagnetic inverse scattering was attempted in [7.5] containing an extensive list of references. It would be far beyond the scope of this chapter on polarization utilization in electromagnetic inverse scattering to provide a complete treatment of the topic. Instead, it will be one of our main objectives to scrutinize those methods and techniques which look promising to be applied or generalized to the three-dimensional electromagnetic vector case, i.e., the case in which polarization effects must be considered.

Whereas, on the one hand, in the direct problem of electromagnetic diffraction, total a priori information on the size, shape, and material constituents of a scatterer, the incident field vector and its orientation with respect to the fixed scatterer coordinate system are given, and the scattered field is to be calculated everywhere over the total frequency domain; on the other hand, in the inverse problem we need to recover the size, shape, and constitutive characteristics of an a priori unknown scattering target with the knowledge of the incident field and the resulting scattered field data.

7.1.2 Definitions of Exact, Unique, and Approximate Methods

Exact solutions of the direct problem of electromagnetic scattering satisfying Maxwell's equations are possible for a limited number of canonical, perfectly conducting shapes [7.7]; and for more complex conducting and/or nonconducting shapes approximate methods need to be used [7.8] which are based on approximate boundary conditions, e.g., such as the Kirchhoff high frequency or the Leontovich scalar impedance approximations [7.7]. The exact solution to the electromagnetic inverse problem would require the inversion of Maxwell's equations or of the vector diffraction integral [7.9], which do not exist and other methods must be sought. Several limited attempts of deriving exact electromagnetic inverse scattering theories have been made and we refer here to the concept of electromagnetic inverse boundary conditions introduced in [7.10,11] or an approach of generalizing the Gel'fand-Levitan approach of quantum mechanics [7.4,12] to the electromagnetic case [7.13-15]. In either case the question of exact versus unique solution must be raised, i.e., an approximate method may provide a self-consistent unique and approximate solution (see, e.g. [7.16]), whereas an exact method may result in a necessary but not sufficient exact solution (see, e.g. [7.11]).

Since measurement data in practice are given only over a limited frequency band, various approximate inverse theories applicable to the specific spatial frequency region of the scatterer's cross section, for which scattered field data are available, must be sought which provide unique, self-consistent solutions. In Sect.7.4 various inverse techniques applicable only within such limited regions of the spatial frequency domain will be discussed and limitations on accuracy and well-posedness of solution in dependence of the format and completeness of the available data will be clarified.

7.1.3 Incompleteness and A Priori Knowledge, Data Limitedness and Self-Consistency

In practice, the measured field data are limited in format (phase, amplitude, doppler, polarization), aspect (monostatic and/or bistatic over a limited aperture; discrete and sparse) and frequency (discrete multifrequency, band-limited) and various limited approximate techniques may have to be applied depending upon the degree of completeness of data as well as on the amount of a priori knowledge available on the nature of the target.

For example, in the case of radar target identification and/or imaging, it can be assumed that usually a rather extensive amount of a priori knowledge on both target and clutter properties is available so that a trade-off on the amount of measured data is possible. If only the monostatic radar cross section for one aspect over a wide (multifrequency) band is available, the target ramp response method of KENNAUGH and co-workers [7.17,18] may be applied (see Sect.7.4.3) given the a priori knowledge that the target is rotationally symmetric, perfectly conducting, and incidence is along the invariant target axis so that the average shape in the physical optics limit excluding any fine structure can be reconstructed.

Should the measured data be too limited in frequency so that the target ramp response method does not apply, target signature comparison or techniques using advanced principles of estimation theory [7.19] and pattern recognition [7.20] need to be used. Using a nearest neighbor discrimination technique [7.21] in n-dimensional feature space, LIN and KSIENSKI [7.22] showed how specific radar targets can be identified with very low false alarm rate given a sufficiently complete library on target signatures. Given in addition polarization information [7.23], this nearest neighbor separation method becomes increasingly more reliable.

Furthermore, another highly approximate engineering approach developed by YOUNG [7.24] may be used instead of the Bojarski-Lewis physical optics inverse theory [7.25,16] in case monostatic scattering data is not given for a sufficiently high number of discrete aspect directions covering the total unit sphere of directions. In this highly approximate target portrayal method developed from a stereo three-look angle concept, the overall approximate target shape is portrayed using the target ramp response method (see Sect.7.4.3) for a very limited number of aspect view angles spread uniformly over the unit sphere of directions.

In case data are band and aperture limited, applicable inverse theories may become ill-conditioned and the method of regularization needs to be applied [7.26] to obtain a self-consistent solution as was discussed in [7.27, 28] in relation to the physical optics far-field inverse scattering theory [7.29]. We need to emphasize here that questions of data incompleteness, and limitedness, of available a priori knowledge on the nature of a scatterer, and of self-consistency of a solution for the two electromagnetic physical optics approximations of KENNAUGH and co-workers in the time domain [7.17,18], and of BOJARKSI and LEWIS in the frequency domain [7.25,16] are best treated in a mixed coordinate representation of RADON's projection space [7.30]. We have shown [7.31,32] that the two formulations represent a Fourier-Radon transform pair and that a discussion of general data limitedness versus given a priori knowledge is best accomplished in the projection space [7.33-36]. Furthermore, we should emphasize the fact that the solution to an inverse problem for the general data limited, highly sparse aspect case for which sampling theorems [7.36] do no longer apply may become nonlinear [7.13] and also in this particular case a treatment of the nonlinear inverse problems in Radon's projection space utilizing BACKUS-GILBERT's method [7.37] and solitary wave theory [7.38,39] should prove highly useful.

In essence, we find that whenever a self-consistent, unique solution to an inverse problem is to be found given overall limited measurement data, an equal amount of complementary a priori information on the nature of the target equivalent to the unavailable information is required. We are safe to state that a very significant amount of a priori information, explicitly contained in polarization information, had been discarded in the past, and it is the purpose of this review to show how access to this important polarization information can be obtained and incorporated into the general noncooperative target, limited data case of electromagnetic inverse scattering.

7.2 The Vector Diffraction Integral, Its Far-Field Approximations, and Some Tauberian Relations

It is the purpose of this section to present a few basic concepts of electromagnetic vector diffraction theory in order to establish some theoretical background in terms of which the approximations introduced in later sections can be understood with greater insight. The discussion will be confined to a purely formal presentation of final results with the intent to provide some mathematical and physical motivation for the approximations and procedures described in this chapter and the reader is referred to [7.7-9,41] for more details.

7.2.1 Basic Scattering Phenomena, Nomenclature, and Radar Definitions

The inverse scattering phenomena that are the subject of this chapter are governed by Maxwell's equations which, using the MKSI system of units and $\exp(i\omega t)$ time dependence with ω denoting the radiant temporal frequency and t time, may be written as

$$\begin{aligned} &\nabla \times \underline{E} = -i\omega\underline{B} \qquad \nabla \cdot \underline{B} = 0 \qquad \underline{B} = \mu\underline{H} \\ &\nabla \times \underline{H} = i\omega\underline{D} + \underline{J} \quad \nabla \cdot \underline{D} = \rho \qquad \underline{D} = \varepsilon\underline{E} \quad , \end{aligned} \tag{7.1}$$

where $\underline{E}$ is the electric field strength, $\underline{H}$ is the magnetic field strength, $\underline{D}$ is the dielectric displacement, $\underline{B}$ is the magnetic induction, $\underline{J}$ is the current density, and ρ is the charge density, μ and ε are the permeability and permittivity of the medium, respectively. We assume throughout this presentation that the propagation medium is homogeneous, isotropic, and, except where otherwise stated, source free

$$\underline{J} = 0 \quad , \quad \rho = 0 \quad . \tag{7.2}$$

The boundary conditions separating two dielectric media with local normal $\hat{a}_n$ to the interface directed from medium 1 to 2 become

$$\hat{a}_n \times (\underline{E}_1 - \underline{E}_2) = 0 \qquad \hat{a}_n \cdot (\underline{D}_1 - \underline{D}_2) = 0$$
$$\hat{a}_n \times (\underline{H}_1 - \underline{H}_2) = 0 \qquad \hat{a}_n \cdot (\underline{B}_1 - \underline{B}_2) = 0 \quad , \tag{7.3}$$

whereas the boundary conditions at the surface of a perfect conductor of outward local normal $\hat{a}_n$ are

$$\hat{a}_n \times \underline{E} = 0 \qquad \hat{a}_n \cdot \underline{B} = 0 \quad , \tag{7.4}$$

and the approximate impedance boundary condition or LEONTOVICH condition [7.42] applied to the total fields $\underline{E}$, $\underline{H}$ becomes

$$\hat{a}_n \times (\underline{E} \times \hat{a}_n) = \eta Z_0 \hat{a}_n \times \underline{H} \quad , \quad \underline{E} = \underline{E}_i + \underline{E}_s \quad , \quad \underline{H} = \underline{H}_i + \underline{H}_s \quad , \tag{7.5}$$

where the subscripts s and i denote scattered and incident field, respectively, $Z_0 = (\mu_0/\varepsilon_0)^{\frac{1}{2}} = (120\pi)\Omega$ is the free space intrinsic impedance, and η is the relative averaged local surface impedance [7.7].

Because the major purpose of this chapter is to deal with electromagnetic vector inverse theories applicable to the radar case in the mm-to-m-wavelength regions, the quantity of primary interest is the distribution of electromagnetic energy far from the scattering body which can be described in terms of a normalized quantity, the radar cross section abbreviated commonly as RCS [7.43-45], that characterizes the scattering properties of a target. The radar cross section σ can be defined as

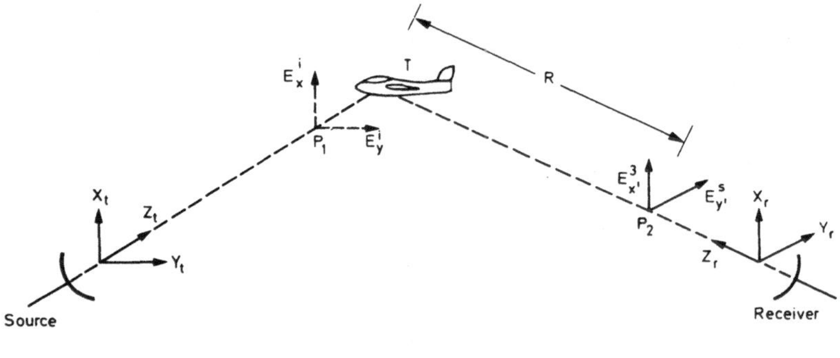

Fig. 7.1. Illustration of the radar range equation

$$\sigma = \lim_{R \to \infty} 4\pi R^2 |\underline{H}_s(\theta_s,\phi_s)/\underline{H}_i(\theta_i,\phi_i)|^2 \quad \text{or}$$

$$\sigma = \lim_{R \to \infty} 4\pi R^2 |\underline{E}_s(\theta_s,\phi_s)/\underline{E}_i(\theta_i,\phi_i)|^2 \quad , \tag{7.6}$$

where R denotes the range and is the distance from the observation point to the origin of a coordinate system centered in or near the scattering body as illustrated in Fig.7.1. The cross section σ is a function of the direction of the observation point (θ_s,ϕ_s) relative to the propagation direction of the incident field (θ_i,ϕ_i), in case those are co-linear $(\theta_s=\theta_i,\phi_s=\phi_i)$ σ defines the monostatic cross section, otherwise the bistatic cross section. The extension of these expressions to the general polarization-dependent case is presented in Sect.7.3.

The monostatic or back scattering cross sections of a perfectly conducting sphere of radius a and electric dimension ka (with $k=2\pi/\lambda=\omega/c$ and c denoting speed of electromagnetic propagation), plotted in Fig.7.2, serve as a standard for comparison and definition of the various spatial frequency regimes within which different asymptotic approximations are valid [7.8,7,46] as indicated in the figure. The inverse scattering theories associated with the various spatial frequency domains of σ and their polarization-dependent properties are discussed in Sect.7.4.

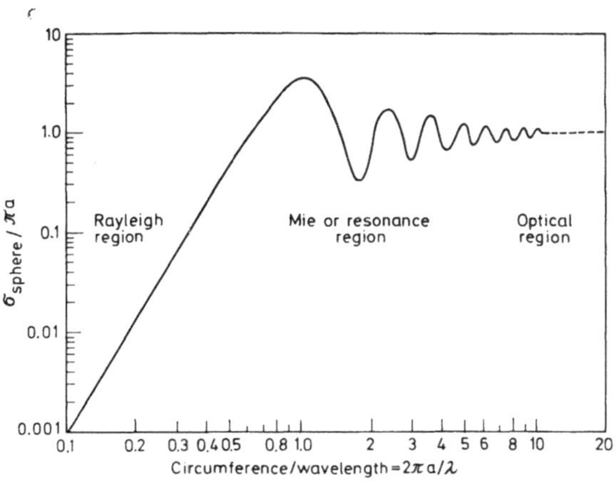

Fig. 7.2. Radar cross section of a sphere of radius a at wavelength λ

In certain inverse scattering theories developed in Sects.7.4 and 5, we require to reconstruct the exact field from the far field, where the scattered far fields $\underline{E}_s$, $\underline{H}_s$ are defined in terms of the incident fields $\underline{E}_i$, $\underline{H}_i$ and the diffracted fields $\underline{E}$, $\underline{H}$ by

$$\underline{E}_s = \lim_{R\to\infty} (\underline{E}-\underline{E}_i) \quad , \qquad \underline{H}_s = \lim_{R\to\infty} (\underline{H}-\underline{H}_i) \quad ,$$

$$\underline{E}_s(\underline{r}) \simeq \underline{U}(\underline{k},\hat{a}_r)\exp(-ikr)/r \quad , \quad \underline{H}_s(\underline{r}) \simeq \underline{V}(\underline{k},\hat{a}_r)\exp(-ikr)/r \quad , \qquad (7.7)$$

where $\underline{U}$, $\underline{V}$ are the complex electric and magnetic scattering vectors and depend only on the propagation vector $\underline{k}$ and the direction $\hat{a}_r = \underline{r}/|\underline{r}|$. Thus, for a given incident field $\underline{E}_i = \underline{U}_i \exp(-i\underline{k}\cdot\underline{r})$, $\underline{H}_i = \underline{V}_i \exp(-i\underline{k}\cdot\underline{r})$, $\underline{U}$ and $\underline{V}$ are functions of θ and ϕ alone in a spherical coordinate system (r,θ,ϕ) with origin in the near-field region [7.10,11].

In order to facilitate comparisons amongst various theories discussed in the remainder of this chapter, symbols to be used throughout are defined in Table 7.1.

Symbol	Meaning	Symbol	Meaning
$\underline{E}$, $\underline{H}$	electric, magnetic vector fields	[J],[K],[L]	Pauli spin matrices
$\underline{D}$, $\underline{B}$	dielectric, magnetic flux densities	ψ	target orientation angle
$\underline{J}$, ρ	vector current, scalar charge density	ν	target skip angle
$\omega = 2\pi f$	radiant frequency	γ	characteristic angle
$k = 2\pi/\lambda$	propagation constant	[U]	unitary target transformation matrix
$c = (\varepsilon\mu)^{-\frac{1}{2}}$	speed of em propagation	$[\Gamma]$	target characteristic operator
$Z = (\mu/\varepsilon)^{\frac{1}{2}}$	intrinsic em impedance	$A(r')$	target silhouette function
$\hat{a}_n$	local outward unit normal	$\rho(\underline{\kappa})$	monostatic complex field cross section
R, r	range	$\Gamma(\underline{\kappa})$	augmented complex field cross section
(r,θ,ϕ)	spherical coordinates	$\gamma(x)$	characteristic target shape function
dS	surface element	f_I, f_U, f_R	impulse, step, ramp response
σ	scattering cross section	$\Delta(\underline{x},\hat{p})$	directional derivative
$\underline{h}$	polarization vector	$\hat{\gamma}(\xi,\hat{p})$	Radon transform of $\gamma(\underline{x})$
$\underline{g}$	Stoke's vector	$\tilde{\gamma}(\underline{\kappa})$	Fourier transform of $\gamma(\underline{x})$
α	absolute phase	$R_{\parallel,\perp}$	Fresnel coefficients
τ	ellipticity angle	K	curvature
ϕ	polarization orientation angle	$\rho(s)$	radius of curvature
[S]	target scattering matrix	$M = \underline{r}\cdot\hat{a}_n$	Minkowski support function
[M]	Mueller matrix	$\underline{A}$, $\underline{B}$	surface vectors

7.2.2 The Stratton-Chu Vector Diffraction Integral and the Vector-Current Integral Equations

Because in most inverse scattering theories applicable to the radar case, where the characteristic target parameters must be deduced from the far-field data, it is necessary to resort to asymptotic methods of calculating the far field in terms of parameters more immediately associated with the near field – for example, sources, surface currents, and surface charges. The inversion of the resulting asymptotic expressions to recover the surface characteristic parameters is then sought and in the treatment of this chapter emphasis is placed on polarization utilization. Therefore, a basic vector integral representation for an electromagnetic field in terms of induced currents and charges on the scatterer surface is required which will assist us in deriving and/or generalizing existing polarization insensitive scalar inverse theories to polarization sensitive vector inverse theories. STRATTON and CHU [7.47] arrived at such a representation in which the total fields $\underline{E}$, $\underline{H}$ can be expressed in terms of the incident fields $\underline{E}_i$, $\underline{H}_i$ and of integrals over the surface of a scattering body (see Fig.7.3) which may be dielectric, finitely or perfectly conducting in nature. The electric and magnetic field integral equations are given by

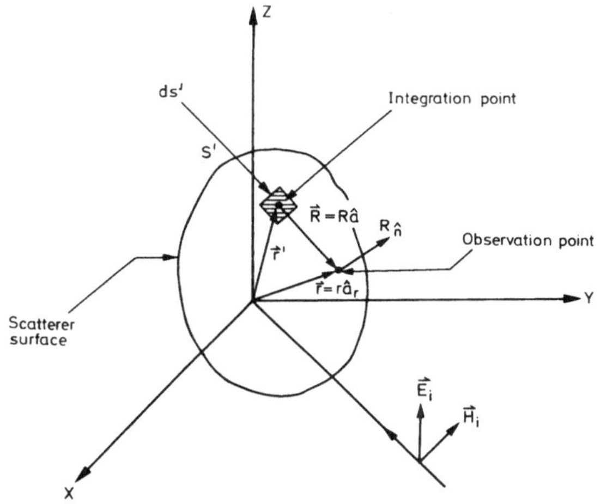

Fig. 7.3. Target geometry for diffraction integral

$$\underline{E}(\underline{r}) = \underline{E}_i + 1/4\pi \iint_{\Sigma} \left[-i\omega\mu(\hat{a}_n \times \underline{H})G + (\hat{a}_n \times \underline{E}) \times \nabla G + (\hat{a}_n \cdot \underline{E})\nabla G\right] dS'$$

$$\underline{H}(\underline{r}) = \underline{H}_i + 1/4\pi \iint_{\Sigma} \left[(\hat{a}_n \times \underline{H}) \times \nabla G + (\hat{a}_n \cdot \underline{H})\nabla G + i\omega\varepsilon(\hat{a}_n \times \underline{E})G\right] dS' \quad . \qquad (7.8)$$

The terms $(\hat{a}_n \times \underline{H})$, $(\hat{a}_n \times \underline{E})$, $(\hat{a}_n \cdot \underline{E})$ and $(\hat{a}_n \cdot \underline{H})$ express the induced electric surface charge and magnetic surface charge densities, respectively.

In the radar case the characteristic dimension ℓ of the scattering body is usually of the order of or several times larger than the radar operating wavelengths λ_0, in which case the high frequency asymptotic techniques are suited best [7.5]. We require the high frequency asymptotic far-field integral expressions for the purpose of deriving the corresponding high frequency approximate inverse scattering theories in Sect.7.4.2.

7.2.3 Far Scattered Fields in the Physical Optics Limit and Their Vector Corrections

Using the far-field asymptotic expansion of the Green's function G(R),

$$\lim_{\sigma\to\infty} G(|\underline{r}-\underline{r}'|) \simeq \exp(-ikr)/r \cdot \exp(ik\hat{a}_r\cdot\underline{r}') \quad , \tag{7.9}$$

the electric and magnetic field integral equations can be reduced to the normalized expressions for $\underline{U}$ and $\underline{V}$, where

$$\begin{aligned} \underline{U}(k,\hat{a}_r) &= ik/4\pi \iint_\Sigma \left[Z_0(\hat{a}_n\times\underline{H})-(\hat{a}_n\times\underline{E})\times\hat{a}_r-(\hat{a}_n\cdot\underline{E})\hat{a}_r\right]\exp(ik\hat{a}_r\cdot\underline{r}')dS' \\ \underline{V}(k,\hat{a}_r) &= ik/4\pi \iint_\Sigma \left[Z_0^{-1}(\hat{a}_n\times\underline{E})+(\hat{a}_n\times\underline{H})\times\hat{a}_r+(\hat{a}_n\cdot\underline{H})\hat{a}_r\right]\exp(ik\hat{a}_r\cdot\underline{r}')dS' \end{aligned} \tag{7.10}$$

may be regarded as exact relations for the far-field vector amplitudes even though the far field is an approximation [7.48].

The appearance of magnetic surface charge and current densities in (7.10) is meaningful for a dielectric material of finite conductivity. These terms will disappear if the scattering body is a perfect conductor in which case

$$\begin{aligned} &\hat{a}_n\times\underline{E} = \hat{a}_n\times(\underline{E}_i+\underline{E}_s) = 0 \quad , &&\hat{a}_n\times\underline{H} = \underline{J}_\Sigma \quad , \\ &\hat{a}_n\cdot\underline{E} = \rho_\Sigma \quad , &&\hat{a}_n\cdot\underline{H} = 0 \quad , \end{aligned} \tag{7.11}$$

and ρ_Σ, $\underline{J}_\Sigma$ denote the induced electric surface charge and current densities on the scattering surface Σ, respectively. We note here that in the geometrical optics limit the condition $\hat{a}_n\cdot\underline{E}=0$ is a necessary and sufficient inverse boundary condition for the exact recovery of Σ in the illuminated region of a scatterer [7.11].

Using (7.11) we are able to rewrite (7.10) in terms of its far-field asymptotic formulations, where in the frequency domain we obtain

$$\begin{aligned} \hat{a}_n\times\underline{E}_i(\underline{r}) &= 1/(2\pi i\omega)\iint_\Sigma \hat{a}_n\times\left\{-\omega\varepsilon\mu\underline{J}_\Sigma(\underline{r})+[\nabla\cdot\underline{J}_\Sigma(\underline{r})]\nabla'G\right\}dS' \\ \underline{J}_\Sigma(\underline{r}) &= \underline{J}_{p0}(r) + \underline{J}_c(x) = 2\hat{a}_n\times\underline{H}_i(\underline{r}) + 1/(2\pi)\iint_\Sigma \hat{a}_n\times[\underline{J}_\Sigma(\underline{r})\times\nabla'G]dS' \end{aligned} \tag{7.12}$$

and similarly in the time domain we obtain

$$\begin{aligned} \hat{a}_n\times\underline{E}_i(r,t) &= 1/(2\pi)\iint_\Sigma \hat{a}_n\times\Big[(\mu/\partial)/\partial t\underline{J}_\Sigma(r',t)-\rho_\Sigma(r',t)/\varepsilon\underline{r}/r^3 \\ &\qquad -1/\varepsilon\,\partial/\partial t\,\rho_\Sigma(\underline{r}',t)\underline{r}/cr^2\Big]dS' \\ \underline{J}_\Sigma(\underline{r},t) &= \underline{J}_{p0}(\underline{r},t) + \underline{J}_c(\underline{r},t) = 2\hat{a}_n\times\underline{H}_i(\underline{r},t) + 1/(2\pi)\iint_\Sigma \hat{a}_n \\ &\qquad \times[1/c\,\partial/\partial t\,\underline{J}_\Sigma(\underline{r}',t)+\underline{J}_\Sigma(\underline{r}',t)/r]\times\underline{r}/r^3dS' \quad . \end{aligned} \tag{7.13}$$

Whereas (7.12) was used mainly in analytical studies of classical boundary value solutions applied to targets with surfaces coinciding with the coordinate surface of separable coordinate systems [7.7,8,46], (7.13) has become of increasing interest with the advent of high-speed digital computers using a time-stepping procedure as reviewed most recently in detail by BENNETT [7.6] and presented by FELSEN [7.48].

It should be noted that the surface current density expressions approximate $\underline{J}_\Sigma$ in (7.12,13) and contain the physical optics current densities $\underline{J}_{p0}$ which according to Kirchhoff [7.49] may be defined as

$$\underline{J}_{p0} = \begin{cases} 2\hat{a}_n \times \underline{H}_i & \text{illuminated region } \Sigma_+ \text{ of } \Sigma \\ 0 & \text{shadow region } \Sigma_- \text{ of } \Sigma \quad . \end{cases} \tag{7.14}$$

The remainder $\underline{J}_c$ in (7.12,13) represents the polarization-dependent correction term of the exact vector diffraction formulation.

The magnetic far-field integral equation for the perfectly conducting case then becomes in the frequency domain

$$\underline{H}_s(\underline{r}) = (-ik/4\pi R) \iint_{\Sigma_+} (\underline{J}_\Sigma \times \hat{a}_r)\exp(-ik|\underline{r}-\underline{r}'|)dS' \tag{7.15}$$

with $\underline{J}_\Sigma = \underline{J}_{p0} + \underline{J}_c$ defined in (7.12). Similarly, the time domain magnetic far-field integral equation becomes with $\underline{J}_s = \underline{J}_{p0} + \underline{J}_c$ defined in (7.13)

$$\underline{H}_s(\underline{r},t) = (1/4\pi cR) \iint_{\Sigma_+} \left\{\partial/\partial t[\underline{J}_\Sigma(\underline{r}',\tau)]\times\hat{a}_r\right\}\Big|_{\tau=t-r/c} dS' \quad . \tag{7.16}$$

These far-field integral equations provide the basis for the derivation of the physical optics far-field inverse scattering techniques given in Sect. 7.4.2.

It should be noted that the time-domain solution is of particular importance to the electromagnetic inverse problem because all the electromagnetic information about a radar target is contained in the impulse response resulting from an incident ideal impulse. The radar cross section or equivalently the entire frequency response can be obtained from the impulse response by a Fourier transform. The response of the target due to any radar waveform can be obtained from the impulse response by a simple convolution procedure.

7.2.4 Time-Domain Target Modeling: Utilization of Some Tauberian Theorems

For the purpose of target classification and target shape estimation the impulse response $f_I(t)$, the step response $f_u(t)$, and the ramp response $f_R(t)$ provide a particularly useful characterization since they are closely related to the actual target geometry. Since time-domain modeling can provide deeper insight into electromagnetic transient phenomena relevant to the low as well as high frequency asymptotic inverse scattering theories, some Tauberian theorems [7.50,51] are revisited in the following.

In time-domain modeling, the scattering process is modeled by a passive linear two-port with time-invariant parameters, where for an input e(t) and a two-port impulse response $f_I(t)$, the output is f(t) such that

$$f(t') = \int_{t_1}^{\infty} f_I(t)e(t'-t)dt \quad , \tag{7.17}$$

where $t' = (t-R/c)$ has been introduced here to remove the time delay between scatterer and observer (Fig.7.4). The impulse response waveform $f_I(t')$ is the response for an impulsive Dirac delta input, i.e., $e(t') = \delta(t')$, where the lower limit in (7.17) indicates the initial value for which $f_I(t')$ departs from zero. The conceptual two-port has a frequency-dependent phasor response $F(i\omega)$ which in the radar target case is related to the co-polarized monostatic cross section as

$$\sigma(\omega) = |F(i\omega)|^2 \tag{7.18}$$

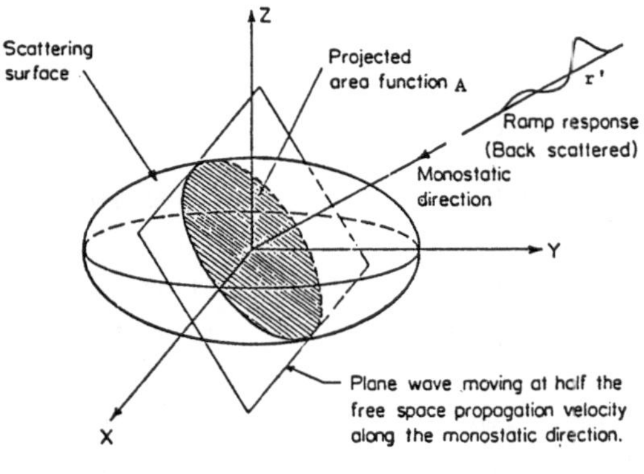

Fig. 7.4. Projected area function $A(r')$ along the direction of incidence r'

so that $F(i\omega)$ and $f_I(t')$ form a Laplace transform pair which for $s = i\omega$ becomes

$$F(s) = c\int_{0}^{\infty} f_I(t')e^{-st'}dt' \tag{7.19}$$

and

$$f_I(t') = (1/2\pi c)\int_{-j\infty}^{j\infty} F(s)e^{t's}ds \quad . \tag{7.20}$$

The impulse response $f_I(t')$ is the time-dependent electromagnetic field strength produced at the output terminals when $e(t) = \delta(t')$; and once $f_I(t')$ is known, the response waveform for any incident waveform can be determined by (7.17). In the remainder of this chapter repeated use of the step response $f_u(t')$ and the ramp response $f_R(t')$ will also be made, where

$$f_u(t') = \int_{t_1}^{\infty} f_I(\tau)u(t'-\tau)d\tau \tag{7.21}$$

and

$$f_R(t') = \int_{t_1}^{\infty} f_I(\tau)(t'-\tau)u(t'-\tau)d\tau \quad . \tag{7.22}$$

At sufficiently low frequencies ($ka \ll 1$), the phasor response $F(s)$ of a scattering object may be expanded into a Taylor series about the origin $s = 0$ [7.52]

$$\lim_{s\to 0} F(s) \simeq \sum_{n=0}^{\infty} a_n s^n \quad , \tag{7.23}$$

where, by employing a Tauberian expansion [7.50] on (7.20), we obtain

$$\int_0^{\infty} (t')^n f_I(t')dt' = (-1)^n n! a_n/c \tag{7.24}$$

and the terms a_n are related to the terms appearing in the Rayleigh-Gans low frequency expansion of the radar cross section as will be discussed in Sect. 7.4.1.

Similarly, it can be shown according to some other Tauberian theorem [7.50,51] that if

$$\lim_{s\to\infty} F(s) \simeq \sum_{n=0}^{\infty} A_n s^{-n} e^{-st} \quad , \tag{7.25}$$

then the individual terms in (7.25) are related for $n = 0$ to $f_I(t)$, for $n = 1$ to $f_u(t)$, for $n = 2$ to $f_R(t)$, and lower order discontinuities in $f_I(t)$ as defined in (7.21,22). Thus, if the high frequency asymptotic expansion of $F(s)$ in (7.19) contains a term $A_n s^{-n} e^{-st}$, the impulse response waveform undergoes a discontinuous jump of the form $A_n \exp(-s\tau)$ in its (n-1)th order derivative at $t = \tau$. In particular, terms of the form $A_n \exp(-st)$ in the asymptotic expansion of (7.25) are contributions predicted by geometric optics resulting for $n = 0$ into an impulse of magnitude A_0 at $t = \tau$ in $f_I(t)$ as will be discussed in Sects.7.4.3 and 4.

Before vector inverse scattering theories are introduced, we shall next consider effects of polarization on the scattered field, and, in turn, on the radar cross section.

7.3 The Radar Scattering and Target Polarization Matrices

Recent advances in broadband antenna theory and technology have made possible polarization radar systems with sufficient isolation between the co- and cross-polarized channels for both linear and circular polarization measurement facility over a wide band of harmonic frequencies. Therefore, we are now able to measure accurately amplitude, relative (and also absolute) phase and Doppler information of the four components of the RCS polarization scattering matrix [S]. Given this total set of information over a wide frequency band and a wide range of aspect directions, it is possible for each aspect to map the characteristic radar target properties in terms of its associated co- and cross-polarization nulls onto the Poincaré polarization sphere or its associated polarization maps. The resulting two pairs of optimal polarizations map onto one great circle path on the polarization sphere, and for each target as well as each type of natural or man-made clutter, the co-polarization nulls possess very particular distributions limited to specific polarization discriminant regions on that sphere. This peculiar property provides a very powerful tool in radar target imaging and it will assist us in extracting the useful target signal from clutter perturbed measurement data. A systematic development of the basic underlying principles not found in treatments of polarization in optics [7.53,54] and only in a rather incomplete presentation in those for radio waves [7.8,45,55-58] is given.

7.3.1 Basic Electromagnetic Polarization Descriptors

In consistency with the notation introduced in Table 7.1, we define the incident plane time harmonic electromagnetic wave with the harmonic magnetic field as shown in Fig.7.5a given by

$$\underline{H}_i(\underline{r},t) = \underline{a}_H H_0 \exp[i(\omega t-\underline{k}_i\cdot\underline{r})] \tag{7.26}$$

so that

$$\underline{E}_i(\underline{r},t) = \underline{a}_H \times \underline{k}_i(Z_0/k)H_0 \exp[i(\omega t-\underline{k}_i\cdot\underline{r})] = \underline{h}_i \exp[i(\omega t-\underline{k}_i\cdot\underline{r})] \quad ,$$

where $\underline{h}_i$ describes the complex polarization vector of the incident field. In radar propagation, it is common usage to decompose any complex polarization vector into its horizontal (parallel to earth's surface) and vertical components

$$\underline{h} = h\hat{h} = \begin{pmatrix} E_h \\ E_v \end{pmatrix} = (h_h\hat{h}_h + h_v\hat{h}_v)\exp(i\alpha) = \begin{pmatrix} a_h \\ a_v \exp(i\delta) \end{pmatrix}\exp(i\alpha) \quad ,$$

$$a^2 = a_h^2 + a_v^2 \quad , \tag{7.27}$$

where δ is the phase difference between the h and v channels of the antenna; and α is the "absolute phase" of the antenna; it determines the phase reference of the antenna at time $t = 0$. Introducing the ellipticity angle τ, given in terms of the ratio of the minor-to-major axes of the polarization ellipse, $\tan\tau = (a \sin\tau)/(a \cos\tau)$, and the polarization orientation angle ψ as defined in Fig.7.5b, the polarization vector defined in (7.27) may be rewritten as

$$\underline{h}(a,\alpha,\tau,\psi) = a\begin{pmatrix}\cos\psi & -\sin\psi\\ \sin\psi & \cos\psi\end{pmatrix}\begin{pmatrix}\cos\tau\\ i\sin\tau\end{pmatrix}\exp(i\alpha) \qquad (7.28)$$

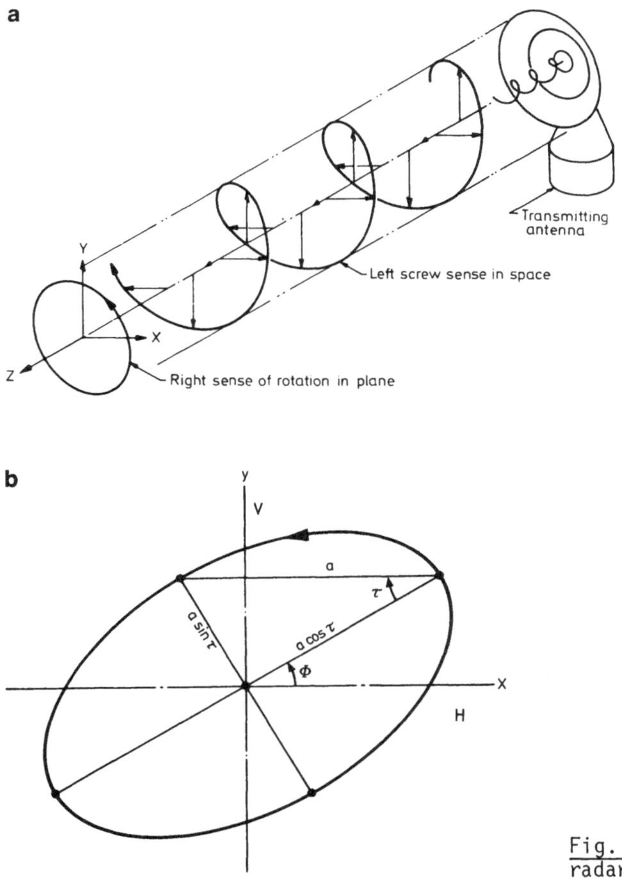

Fig. 7.5a,b. Definition of radar polarization

which for negative values of τ describes a left sensed and for positive values a right sensed polarization ellipse.

Using the above definition of $\underline{h}$, the four components of the associated Stokes vector $\underline{g} = (g_0, g_1, g_2, g_3)$, which describes the averaged power properties of $\underline{h}$, become

$$g_0 = a^2 = a_h^2 + a_v^2 = E_h^2 + E_v^2 = I$$

$$g_1 = a^2 \sin 2\tau = 2a_h a_v \sin\delta = 2\,\mathrm{Im}\{E_h E_v\} = V$$

$$g_2 = a^2 \cos 2\tau \cos 2\psi = a_h^2 - a_v^2 = |E_h|^2 - |E_v|^2 = Q$$

$$g_3 = a^2 \cos 2\tau \sin 2\psi = 2a_h a_v \cos\delta = 2\,\mathrm{Re}\{E_h E_v\} = U \quad , \tag{7.29}$$

where

$$g_0^2 = g_1^2 + g_2^2 + g_3^2 = I^2 = (Q^2+U^2+V^2) \quad .$$

It should be noted here that frequently the modified Stokes parameters $I_h = (I+Q)/2$ and $I_v = (I-Q)/2$ are used [7.69].

If we now introduce the auxiliary parameter

$$u = (h_h - ih_v)/(h_h + ih_v) \quad ,$$

the properties of the polarization vector $\underline{h}$ and/or the Stokes vector $\underline{g}$ can be described on the Poincaré polarization sphere [7.59] as illustrated in Fig.7.6, where with conventional spherical coordinates (r,θ,ϕ)

$$r = a^2 \quad , \quad \theta = \pi/2 - 2\tau = \mathrm{arc}\cos[(1-|u|^2)/(1+|u|^2)] \quad ,$$

$$\phi = 2\psi = \mathrm{phase}\{u\} \quad , \tag{7.30}$$

and a, ψ, τ denote the magnitude, the tilt or polarization orientation angle and the ellipticity angle, respectively, as defined in (7.27). It should be noted here that the absolute phase α is not accommodated in this description. On the Poincaré sphere various states of general elliptical polarization are mapped uniquely, as illustrated in Fig.7.5:

i) all linear polarizations $\theta = \pi/2 - 2\tau = \pi/2$ lie on the equator of zero ellipticity (τ=0) with horizontal polarization being at zero latitude $\phi = 2\psi = 0$ and vertical polarization at $\phi = 2\psi = \pi$.

ii) the zenith (θ=0, τ=π/4) describes right sensed circular, and the nadir (θ=π, τ=$-\pi$/4) left sensed circular polarization so that for $(\theta-\pi/2) \gtreqless 0$ or $\tau \gtreqless 0$ polarization is right sensed elliptically, linear, and left sensed elliptically polarized, respectively.

iii) the Poincaré sphere is a very useful tool for mapping statistical polarization distributions; and for random polarization such as produced by thermal radiation the states are uniformly distributed over the polarization sphere. (7.31)

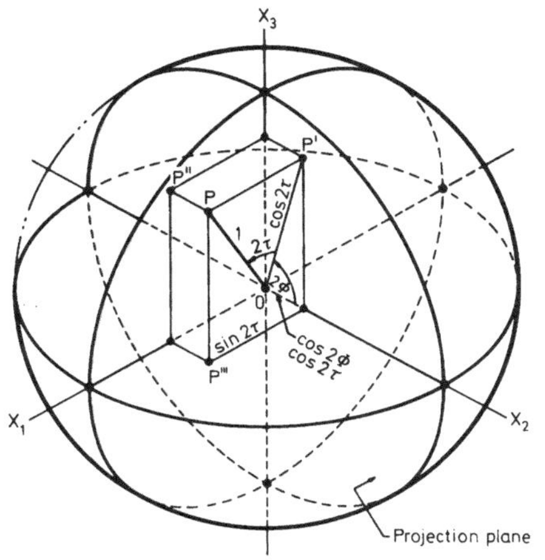

Fig. 7.6. The Poincaré polarization sphere

In electromagnetic inverse problems the properties of a remote medium or target are determined from the scattered or received (j) vector field given the incident or transmitted (i) vector field and thus a polarization pair $(\underline{h}_j, \underline{h}_i)$ describes the interaction phenomenon. There exist four canonical electromagnetic (radar) polarization pairs which we need to introduce here to facilitate comprehension of the remainder.

The canonical polarization pairs may be defined in terms of the power $P(\underline{h}_j, \underline{h}_i)$ received by $\underline{h}_j = (a_j, \alpha_j, \tau_j, \psi_j)$ due to that transmitted by $\underline{h}_i = (a_i, \alpha_i, \tau_i, \psi_i)$

$$P(\underline{h}_j, \underline{h}_i) = \frac{1}{2} a_j^2 a_i^2 [1 + \sin 2\tau_j \ \sin 2\tau_i + \cos 2(\psi_j + \psi_i) \cos 2\tau_j \ \cos 2\tau_i] \qquad (7.32)$$

and specifically we may define as illustrated in Fig.7.7 and shown in [7.61-63]:

Orthogonal: $\underline{h}_j = (\underline{h}_i)^*_\perp \ : \ P(\underline{h}_i^*, \underline{h}_i) = 0$

$\underline{h}_0 = \underline{h}_j(a_j = \text{arb}, \alpha = \text{arb}, \tau_j = -\tau_i, \phi_j = \phi_i + \pi/2)$

antipodal, but undetermined in a and α

Cannot be employed as a unique descriptor

Transverse: $\underline{h}_j = \underline{h}_i^* \ : \ P(\underline{h}_i, \underline{h}_i^*) = P_{max} = \frac{1}{2} a_i^2 a_j^2$

$\underline{h}_i = \underline{h}_j(a_j = a_i, \alpha_j = \alpha_i, \tau_j = \tau_i, \phi_j = -\phi_i)$

Optimal reception: Antenna matching

Symmetric: $\underline{h}_j = \underline{h}_i(-\phi_i, -\tau_i) \ : \ P(\underline{h}_i, \underline{h}_i) = a_i^2 a_j^2 \cos^2 2\tau_i$

$\underline{h}_s = \underline{h}_j(a_j = a_i, \alpha_j = \alpha_i, \tau_j = -\tau_i, \phi_j = -\phi_i)$

Most frequent radar target polarization

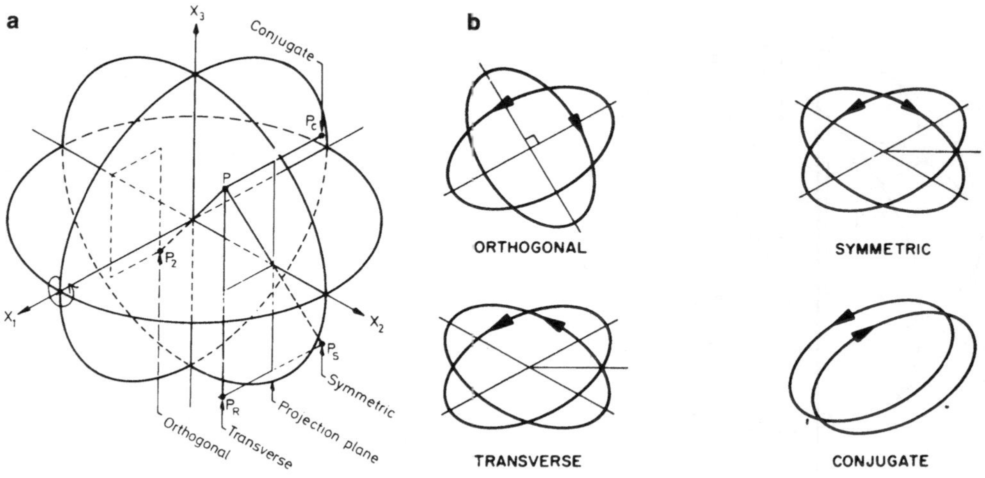

Fig. 7.7a,b. Basic polarization pairs

Conjugate: $\underline{h}_j = \underline{h}_i(-\tau_i)$

$$h_C = \underline{h}_j(a_j=a_i, \alpha_j=-\alpha_i, \tau_j=-\tau_i, \phi_j=\phi_i)$$

Precipitation (circular pol.) clutter .

7.3.2 Radar Scattering Matrices and Radar Measurables

The limitations of the conventional radar range equation expressed in (7.6) in terms of the radar cross section $\sigma(\theta_s, \phi_s; \theta_i, \phi_i)$ were recognized when precipitation clutter [7.64] severely limited radar target detection. To overcome these limitations, SINCLAIR [7.65] initiated a systematic approach to the analysis of radar target detection in clutter using polarization. Using linear horizontal (parallel to earth's surface) and vertical polarization as base polarizations

$$\underline{h} = a_h \hat{h}_h + a_v \hat{h}_v = h\hat{h} \quad , \quad h^2 = |E_h|^2 + |E_v|^2 \quad , \tag{7.33}$$

we may define the bistatic "linear polarization-restricted" scattering matrix $[\sqrt{\sigma}]$ with absolute phase in terms of the incident polarization $\underline{h}_i(\theta_i, \phi_i)$ and the received (scattered field) polarization $\underline{h}_s(\theta_s, \phi_s)$ as

$$\underline{h}_s = [\sqrt{\sigma}(\theta_s, \phi_s; \theta_i, \phi_i)]\underline{h}_i \quad , \tag{7.34}$$

where with the absolute phase expressed in terms of α_{hv}

$$[\sqrt{\sigma}(\theta_s,\phi_s;\theta_i,\phi_i)] = \frac{\exp(i\alpha_{hv})}{4\pi R^2}\begin{bmatrix}\sqrt{\sigma}_{hh}\exp[i(\alpha_{hh}-\alpha_{hv})] & \sqrt{\sigma}_{hv} \\ \sqrt{\sigma}_{vh}\exp[i(\alpha_{vh}-\alpha_{hv})] & \sqrt{\sigma}_{vv}\exp[i(\alpha_{vv}-\alpha_{vh})]\end{bmatrix} . \quad (7.35)$$

Since in practice the absolute phase cannot easily be measured without interferometric (holographic) techniques [7.65], the bistatic scattering matrix with relative phase σ_{hh}, σ_{vv}, σ_{hv}, $(\alpha_{hh}-\alpha_{hv})$, $(\alpha_{vv}-\alpha_{hv})$, and $(\alpha_{vh}-\alpha_{hv})$ is used, where the latter three are either positive or negative. For the relative phase case we may assign to the absolute value $|\alpha_{vh}-\alpha_{hv}|$ an arbitrary sign so that in total nine positive real quantities must be measured to describe $[\sqrt{\sigma}(\theta_s,\phi_s;\theta_i,\phi_i)]$ in the general bistatic relative phase case [7.67].

In the following we shall limit ourselves exclusively to the monostatic relative phase case for which $(\theta_s=\theta_i,\phi_s=\phi_i)$ and we may define the "normalized monostatic scattering matrix [S] with relative phase" in terms of two arbitrary but orthogonal base polarization vectors $\hat{h}_A$ and $\hat{h}_B$ so that with $\underline{h} = h_A\hat{h}_A + h_B\hat{h}_B$

$$\underline{h}_s = [S]\underline{h}_i \quad , \quad [S] = \begin{bmatrix} S_{AA} & |S_{AB}| \\ |S_{AB}| & S_{BB}\end{bmatrix} \quad , \quad \begin{matrix} S_{AB} = S_{BA} \\ \alpha_{AB} = \alpha_{BA}\end{matrix} . \quad (7.36)$$

Thus, we require to measure five real or seven positive real quantities to determine [S] completely since reciprocity $S_{AB} = S_{BA}$ and conservation of energy must be satisfied in the monostatic relative phase case as is shown in [7.67].

There exist an infinite number of general elliptical orthogonal polarization base vectors $\hat{h}_A$, $\hat{h}_B$ and an infinite number of possible invariant transformations which as we shall show in Sect.7.3.3 can best be performed using the Poincaré polarization sphere [7.59] expressing any general polarization $\underline{h}$ as a linear combination of some horizontal $\hat{h}_h$ and vertical $\hat{h}_v$ polarizations.

Linear and circular polarization are two distinct special cases of elliptical polarization used widely in radar as reference polarizations [7.43-45, 57] so that their basic relationships need to be given here. If we define linear polarization by $\underline{h}^L = (E_h,E_v)^T$ and circular polarization by $\underline{h}^C = (E_\ell,E_r)^T$, where E_h, E_v and E_ℓ, E_r denote the electric fields of the horizontal, vertical and left, right sensed circular components, respectively, so that $\underline{h}_s^L = [L]\underline{h}_i^L$, $\underline{h}_s^C = [C]\underline{h}_i^C$

$$[L] = \begin{bmatrix} L_{hh} & L_{hv} \\ L_{vh} & L_{vv}\end{bmatrix} \quad , \quad [C] = \begin{bmatrix} C_{\ell\ell} & C_{\ell r} \\ C_{r\ell} & C_{rr}\end{bmatrix} \quad , \quad (7.37)$$

then it can be shown [7.57,64] that

$$E_\ell = (E_h+iE_v)/\sqrt{2} \quad , \quad E_r = (E_h-iE_v)/\sqrt{2}$$
$$C_{\ell\ell} = |(L_{hh}-L_{vv})/2+iL_{hv}| \quad , \quad L_{hv} = L_{vh}$$
$$C_{\ell r} = |(L_{hh}+L_{vv})/2| \quad , \quad C_{\ell r} = C_{r\ell}$$
$$C_{rr} = |(L_{hh}-L_{vv})/2-iL_{hv}| \quad . \tag{7.38}$$

For a perfectly conducting surface of radii of curvature large compared to the wavelength, one may assume that $L_{hh} = L_{vv}$ so that we observe relations among the radar cross section matrix components $\sigma_{AB} = k|S_{AB}|^2$ of the kind

$$\sigma_{\ell\ell} = \sigma_{rr} = \sigma_{hv} = \sigma_{\ell r} = \sigma_{r\ell} = \sigma_{hh} = \sigma_{vv} \quad . \tag{7.39}$$

Similarly, the σ_{AB} for any combination of transmitted and received polarizations can be expressed in terms of a mixed representation of the S_{AB} so that for a radar transmitting circular (ℓ: left, r: right) and receiving linear (h: horizontal, v: vertical), it can be shown [7.57] that

$$\sigma_{\ell h} = (k/2)|L_{hh}+iL_{vv}|^2 \quad , \quad \sigma_{\ell v} = (k/2)|L_{hv}+iL_{vv}|^2 \quad ,$$
$$\sigma_{rh} = (k/2)|L_{hh}+iL_{hv}|^2 \quad , \quad \sigma_{rv} = (k/2)|L_{hv}-iL_{vv}|^2 \quad . \tag{7.40}$$

In general, none of the components of the radar cross section matrix $[\sigma]$ as defined, e.g., in [7.68] are equal on an instantaneous basis [7.57], because the phase and the amplitude of each of the S_{AB} are fluctuating functions of time. We thus require to introduce another scattering matrix description whose components are expressed in terms of statistically averaged measured quantities.

The *Stokes reflection or Mueller scattering matrix* [M] is obtained by relating the 4-component Stokes vectors $\underline{g}_s(I_h, I_v, U, V)$, associated with the 2-component polarization vector $\underline{h}_s^L$ and $\underline{h}_i^L$ such that for $\underline{h}_s^L = [S(h,v)]\underline{h}_i^L$ the modified Mueller matrix $[M_m]$ may be expressed in terms of the components S_{AB} as shown in [7.69] as $\underline{g}_s = [M_m]\underline{g}_i$ so that

$$[M_m] = \begin{bmatrix} |S_{hh}|^2 & |S_{hv}|^2 & \mathrm{Re}\{S_{hh}S_{hv}^*\} & -\mathrm{Im}\{S_{hh}S_{hv}^*\} \\ |S_{vh}|^2 & |S_{vv}|^2 & \mathrm{Re}\{S_{vh}S_{vv}^*\} & -\mathrm{Im}\{S_{vh}S_{vv}^*\} \\ 2\mathrm{Re}\{S_{hh}S_{vh}\} & 2\mathrm{Re}\{S_{hv}S_{vv}^*\} & \mathrm{Re}\{S_{hh}S_{vv}^*+S_{hv}S_{vh}^*\} & -\mathrm{Im}\{S_{hh}S_{vv}^*-S_{hv}S_{vh}^*\} \\ 2\mathrm{Re}\{S_{hh}S_{vh}\} & 2\mathrm{Im}\{S_{hv}S_{vv}^*\} & \mathrm{Im}\{S_{hh}S_{vv}^*+S_{hv}S_{vh}^*\} & \mathrm{Re}\{S_{hh}S_{vv}^*-S_{hv}S_{vh}^*\} \end{bmatrix} , \tag{7.41}$$

where all of the 4×4 elements are real. The Stokes "reflection matrix" [M] represents the target in terms of measurements of power in a similar way as Sinclair's scattering matrix [S] represents the target in terms of field strength measurements. It should be noted here that in the monostatic radar case very frequently the normalized presentation of [S] is used, i.e., the S_{AB} in [S] are normalized with respect to the invariant $p = \sum_i^A \sum_j^B |S_{ij}|^2 = \left(|S_{AA}|^2 + |S_{BB}|^2 + 2|S_{AB}|^2\right)$ as will be discussed later on (7.45).

For the case of random time-dependent clutter variations, <[M(t)]> is required for which in the incoherent case, the Stokes parameters of independent waves are additive similar to the coherent case in which the components of [S] are additive. Using the common definition of the Stokes parameters defined in (7.29) by $\underline{g} = \underline{g}(I,Q,U,V)$, the elements m_{ij} of the 4×4 time-averaged Mueller matrix become

$$
\begin{aligned}
m_{11} &= \tfrac{1}{2}\left\langle |S_{hh}|^2 + |S_{vv}|^2 + 2|S_{hv}|^2 \right\rangle , & m_{22} &= \tfrac{1}{2}\left\langle |S_{hh}|^2 + |S_{vv}|^2 - 2|S_{hv}|^2 \right\rangle \\
m_{33} &= \left\langle \mathrm{Re}\{S_{hh}S_{vv}^*\} + |S_{hv}|^2 \right\rangle , & m_{44} &= \left\langle |S_{hv}|^2 - \mathrm{Re}\{S_{hh}S_{vv}^*\} \right\rangle \\
m_{12} = m_{21} &= \tfrac{1}{2}\left\langle |S_{hh}|^2 - |S_{vv}|^2 \right\rangle , & m_{34} = m_{43} &= -\left\langle \mathrm{Im}\{S_{hh} - S_{vv}^*\} \right\rangle \\
m_{13} = m_{31} &= \left\langle \mathrm{Re}\{(S_{hh} + S_{vv})S_{hv}^*\} \right\rangle , & m_{23} = m_{32} &= \left\langle \mathrm{Re}\{(S_{hh} - S_{vv})S_{hv}^*\} \right\rangle \\
m_{14} = m_{41} &= -\left\langle \mathrm{Im}\{(S_{hh} - S_{vv})S_{hv}^*\} \right\rangle , & m_{24} = m_{42} &= -\left\langle \mathrm{Im}\{(S_{hh} + S_{vv})S_{hv}^*\} \right\rangle .
\end{aligned}
$$

We have presented here the conversions from [S] to both $[M_m]$ and [M] for clarity as one finds inconsistent usage of the Mueller matrix elements leading to misinterpretation of correlations existing between electromagnetic measurement and "clutter truth" when using the co-variance matrix test [7.70].

Before introducing the target and clutter characteristic operators (Sect. 7.3.4), we shall first discuss the properties of optimal polarizations from which additional target/clutter characteristic operators can be extracted.

7.3.3 Kennaugh's Optimum Polarization Pairs

Numerically, the transformation properties of [S(A,B)] defined in (7.36) from any one orthogonal polarization pair $\underline{h} = h_A\hat{h}_A + h_B\hat{h}_B$ to another orthogonal pair $\underline{h}' = h_{A'}\hat{h}_{A'} + h_{B'}\hat{h}_{B'}$ on the same polarization sphere of radius $p = |S_{AA}|^2 + |S_{BB}|^2 + 2|S_{AB}|^2 = |S_{A'A'}|^2 + |S_{B'B'}|^2 + 2|S_{A'B'}|^2 = p'$ can be expressed in terms of a single complex transformation parameter ρ and its complex conjugate ρ^* so that

$$[S'(A',B')] = [T]^T[S(A,B)][T] \quad , \tag{7.42}$$

and for a normalized representation of

$$[T] = (1+\rho\rho^*)^{-1}\begin{bmatrix} 1 & -\rho^* \\ \rho & 1 \end{bmatrix} \tag{7.43}$$

the transformed matrix elements $S'_{A'B'}$ become

$$\begin{aligned}
S'_{A'A'} &= (1+\rho\rho^*)^{-1}\left[S_{AA}+\rho^2 S_{BB}+\rho(S_{AB}+S_{BA})\right] \\
S'_{A'B'} &= (1+\rho\rho^*)^{-1}\left[-\rho^* S_{AA}+\rho S_{BB}+S_{AB}-\rho\rho^* S_{BA}\right] \\
S'_{B'A'} &= (1+\rho\rho^*)^{-1}\left[-\rho^* S_{AA}+\rho S_{BB}+S_{BA}-\rho\rho^* S_{AB}\right] \\
S'_{B'B'} &= (1+\rho\rho^*)^{-1}\left[\rho^{*2} S_{AA}+S_{BB}-\rho^*(S_{AB}+S_{BA})\right] \quad .
\end{aligned} \tag{7.44}$$

We note that if $S_{AB} = S_{BA}$ then $S'_{A'B'} = S'_{B'A'}$ for all ρ, i.e., if reciprocity is satisfied for any one pair of orthogonal polarizations, it is satisfied for all such pairs. Furthermore, we must emphasize the important property that for any one given aspect and at one fixed frequency, the transformation is invariant so that for assumed reciprocity and conservation of energy [7.67]

$$\begin{aligned}
\det\{[M]\} &= \det\{[M']\} = (S_{AA}-S_{BB}-|S_{AB}|^2) = (S'_{A'A'}S'_{B'B'}-|S'_{A'B'}|^2) \quad , \\
\mathrm{sp}\{[M]\} &= \mathrm{sp}\{[M']\} = \sum_i^A \sum_j^B |S_{ij}|^2 = \sum_{i'}^{A'} \sum_{j'}^{B'} |S'_{i'j'}| \quad , \\
p &= \left(|S_{AA}|^2+|S_{BB}|^2+2|S_{AB}|^2\right) = \left(|S'_{A'B'}|^2+|S'_{B'B'}|^2+2|S'_{A'B'}|^2\right) \quad .
\end{aligned} \tag{7.45}$$

There exist two pairs of optimal polarization which can be associated with (7.44) and are useful to express the five independent real components of [S] in the monostatic relative phase case on the polarization sphere of radius $p = \sum_i^A \sum_j^B |S_{ij}|^2$ subject to (7.45) which were first discovered and defined by KENNAUGH [7.71,72].

The *co-polarization null* (minimal polarization) *pair* is obtained from (7.44) by setting $S'_{A'A'}$ and $S'_{B'B'}$ to zero, where the minimal polarization pair is defined by the two solutions of

$$0 = S_{AA} + \rho^2 S_{BB} + \rho^2 S_{AB} \quad ; \tag{7.46}$$

the *cross-polarization* (maximal polarization) *pair* is obtained from (7.44) by setting $S'_{A'B'} = S'_{B'A'} = 0$ so that the two solutions for maximal polarizations become

$$0 = -\rho^* S_{AA} + \rho S_{BB} - \rho\rho^* S_{AB} + S_{BA} \quad ;$$

and the four roots of these equations lie on one main circle path of the associated Poincaré sphere of radius $p = \sum_{i}^{A}\sum_{j}^{B}|S_{ij}|^2$.

By definition of (7.30,44), we let $q = (1-i\rho)/(1+i\rho)$ with ρ being solutions of the optimal polarization so that the coordinates $(\theta;\phi)$ of the nulls on the Poincaré sphere of radius p are given by

$$\theta = \pi/2 - 2\tau = \text{arc cot}\left\{2\text{Im}\{q\}/(1-\left[(\text{Re}\{q\})^2+(\text{Im}\{q\})^2\right]\right\}$$

$$\phi = \text{arc tan}\left\{(\text{Re}\{q\}) / (\text{Im}\{q\})\right\} \quad .$$

Thus given [S] in terms of any orthogonal polarization pair $\underline{h} = h_A\hat{h}_A + h_B\hat{h}_B$, we can uniquely determine the location of the two "CO-POL" and the two "X-POL" nulls, which lie on one great circle path [7.72,62,63] so that the X-POL nulls are orthogonal, i.e., lie at antipodal points on the polarization sphere of radius p, whereas the CO-POL nulls are bisected by the line joining the X-POL nulls as shown in Fig.7.8. It should be noted here that under certain conditions of target symmetry the two distinct CO-POL nulls may degenerate into one double solution being identical to one of the X-POL nulls.

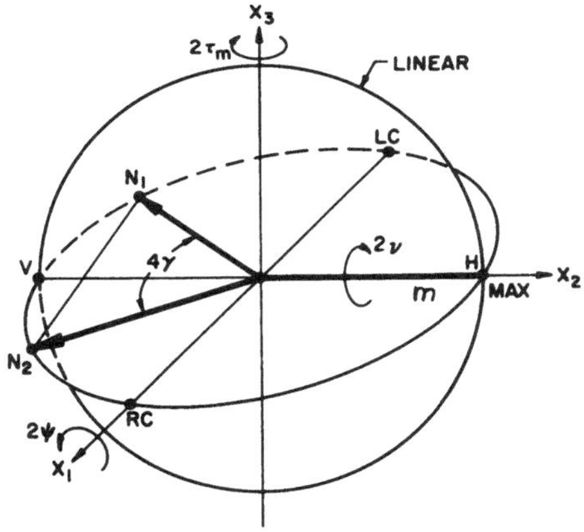

Fig. 7.8. Definition of Huynen's polarization fork

We need to emphasize here that each specific type of target and/or clutter produces its own characteristic CO-POL null distribution within bounded regions on the Poincaré sphere (see Fig.7.9a,b). In fact, it has been demonstrated that the co-polarization null distribution of clutter is strongly affected in the presence of a target and thus, the co-polarization null concept should serve as a useful target versus clutter detection as well as imaging discriminant.

7.3.4 Radar Target and Clutter Characteristic Operators

In this subsection we consider radar polarization characterization operators for three distinct cases: i) the single target developing the theory fundamental to all of the three cases; ii) the distributed target case of clutter providing a succinct summary of the state of the art; and iii) synthetic aperture imagery developing the concept of the complete $p = \sum_{ij}^{AB} |S_{ij}|^2$ image.

Single Radar Target Classification

Based on the inspiration of SINCLAIR [7.65], on a series of papers on radar antenna polarization [7.73], and on the important pioneering studies of KENNAUGH [7.71,72] and GENT [7.74], an early attempt at single radar target classification was made by COPELAND [7.75] using the received complex voltage with rotating linear polarization illumination. This concept was further advanced by HUYNEN [7.60-63], who expanded on the measurement conceptual studies on [S], by GRAVES [7.68], and CRISPIN [7.77]. A significant summary of the early state-of-the-art of radar measurements was presented at the Radar Reflectivity Measurement Symposium 1964 [7.78-80], in a subsequent Special IEEE Proceedings issue on Radar Reflectivity Studies 1965 [7.61,81,82], and further developed in the Russian literature [7.58,83-87]. Of particular interest is the theory of radar target phenomenology developed by HUYNEN [7.62] utilizing the deterministic polarization-agil characteristics of isolated single targets contained in [S].

HUYNEN [7.63] introduced a set of five independent canonical target discriminant parameters: *the target orientation angle* ψ, *the ellipticity angle* τ_m *of maximum target polarization, the relative phase expressed in terms of the target skip angle* ν, *the target characteristic identification angle* γ, *and the target characteristic magnitude* m *(see Fig.7.8), and in addition the absolute phase* α *may be defined*. Introducing the rotational group or Pauli-spin matrices [J], [K], and [L] related to the unit matrix [I] by

$$[J]^2 = -[I] \quad , \quad [K]^2 = -[I] \quad , \quad [L]^2 = [I] = [J] \cdot [K] = [K] \cdot [J] \quad ,$$

where

$$[I] = \begin{bmatrix} 1 & 0 \\ 0 & 1 \end{bmatrix} \quad , \quad [J] = \begin{bmatrix} 0 & -1 \\ 1 & 0 \end{bmatrix} \quad , \quad [K] = \begin{bmatrix} 0 & i \\ i & 0 \end{bmatrix} \quad , \quad [L] = \begin{bmatrix} -i & 0 \\ 0 & i \end{bmatrix} \quad , \qquad (7.47)$$

it can be shown for $\alpha = 0$ [7.60] that [S] defined in (7.36) becomes

$$[S] = [U^*(\psi, \tau_m, \nu)][\Gamma(m, \gamma)][U^*(\psi, \tau_m, \nu)]'$$

where [U]' denotes the transpose of [U], and

$$[U(\psi,\tau_m,\nu)] = \exp(\psi[J])\ \exp(\tau_m[K])\ \exp(\nu[L])$$

$$[\Gamma(m,\gamma)] = m\begin{bmatrix}1 & 0\\ 0 & \tan\gamma\end{bmatrix} = m/2\ \cos^2\gamma\begin{bmatrix}1+\cos 2\gamma & 0\\ 0 & 1-\cos 2\gamma\end{bmatrix}$$

$$sp\{[S]\} = p = \frac{1}{4}\ sp\{[\Gamma]\} = m^2/4(\tan^2\gamma) \geqq 0$$

$$[U'(\psi,\tau_m,\nu)] \cdot [U^*(\psi,\tau_m,\nu)] = [I] \quad . \qquad (7.48)$$

The definition of Huynen's target scattering operator defined in (7.48) shows that [S] can be diagonalized by a unitary transformation matrix $[U(\psi,\tau_m,\nu)]$, where ψ, τ_m, and ν are rotation angles of the sphere about three orthogonal axes as shown in Fig.7.8, and the remaining target characteristic operator $[\Gamma(m,\gamma)]$ expresses the properties of the co-polarization null locations in terms of the target characteristic identification angle γ and the target characteristic magnitude m. It has been shown that propagation losses and other relative constant multipliers of σ can be absorbed in m [7.58] so that target polarization effects can be determined in terms of the normalized target characteristic operator

$$\Gamma(m=1,\gamma) = \begin{bmatrix}1 & 0\\ 0 & \tan^2\gamma\end{bmatrix} \quad , \qquad (7.49)$$

which can be mapped onto the gamma target map [7.63]. The peculiar properties of the normalized target characteristic operator $[\Gamma(m=1,\gamma)]$ and of the unitary transformation matrix $[U(\psi,\tau_m,\nu)]$ enables Huynen to establish the *polarization fork* concept for single targets, relating the co-polarization null locations (Fig.7.8:N_1,N_2) with an angle of (4γ) between the nulls and the center of the normalized (m=1) sphere. From inspection of Fig.7.8, we observe that the location of the cross-polarization nulls C_1 = Max and C_2 = V, bisect the angles spanned by N_1 and N_2, and for symmetric target shapes C_1 and C_2 lie on the equator of the polarization sphere which specifies purely linear polarization. HUYNEN [7.78,62,63] provided examples and interpretations for simple canonical target shapes including values for (m,ψ,τ_m,ν). We note here that the quantity $\alpha = \alpha_{hv}$ defined in (7.35) disappears with power measurements. It may be altered arbitrarily by moving the target along the radar-line-of-sight leaving the target attitude otherwise unchanged. The absolute phase is a mixed target parameter which characterizes target surface geometry and composition as well as a target's spatial composition and can be recovered only by using holographic techniques [7.66,88]. Summarizing, we note that the *polarization fork concept* establishes a powerful *single target discriminator* which provides deep insight into a target's polarization properties [7.58].

The Time-Varying Distributed Target

The subject of time-varying distributed targets was developed mostly independent of the above cited literature on single targets. The early work of GENT

[7.74] is exceptional because he also discussed distribution of single targets. Statistical models for terrain are given by SPETNER and KATZ [7.89]. The question of whether reciprocity is valid for rough surface scattering was studied by AMENT [7.90], and KO [7.91] presented an introduction with application to partially polarized scattering. A classical treatise on scattering from rough surfaces, treated mostly by scalar theory, was published by BECKMANN and SPIZZICHINO [7.55] and updated by BECKMANN [7.56], where the work of FUNG [7.92-94] on vector scattering theory is analyzed considering effects of depolarization of electromagnetic waves. The older literature is referenced by BORN and WOLF [7.53] including aspects of the theory of partial coherence which have been more recently analyzed in the excellent textbooks by ISHIMARU [7.69] and in STROHBEHN [7.95]. We will record here also some of the earlier noteworthy contributions to wave scattering phenomena from rough surfaces and various kinds of clutter such as the high frequency asymptotic theories based on Kirchhoff's approximation by HAGFORS [7.96] and STOGRYN [7.97]. Overall reviews are given in [7.96,98] on radar astronomy, in [7.7,8,46] on distributed radar scatter, in [7.99] on earlier work in light scattering, more recently in [7.64,100] on precipitation scatter, in the excellent textbook [7.57] on radar reflectivity of land and sea, and in [7.101-104] on sea scatter.

In the reviews given during the recent workshop on radar back scatter from terrain [7.105], it became apparent that very little has been accomplished in the western literature except for the extensive studies by LEADER [7.106] and recent studies by GOUGH and BOERNER [7.107] to make full use of Mueller matrix component characteristics. The unique information content of co-polarization null characteristics of clutter, long since recognized in the Russian literature as summarized in the early sixties by KANAREYKIN et al. [7.58], has not been utilized except in some isolated most recent studies by DALEY [7.108], IOANNIDIS and HAMMER [7.109], and MORGAN [7.110]. There it is shown that the co-polarization nulls of various kinds of clutter map into very limited distinct regions of the polarization sphere of less than 5% aerial extent over the total surface of the unit sphere. For example, precipitation clutter maps into small circular patches centered about the poles of circular polarization on the Poincaré sphere as was predicted by KENNAUGH [7.72]; whereas rough sea clutter maps into semi-lunar domains centered about horizontal linear polarization (see Fig.7.9a). A similar result [7.105] should also be expected for the rough ground terrain case. We emphasize here that recently it was particularly POELMAN who in a series of papers [7.241-244] demonstrated the great importance of this particular phenomenon of co-pol null distributions to radar target versus clutter discrimination.

In summary, we must emphasize here that the utilization of the peculiar co/cross-polarization null properties in the characterization of clutter is far from completed. In particular, coherency tests applied to various spectral components of the clutter nulls should assist us in separating the useful coherent target signal from incoherent clutter.

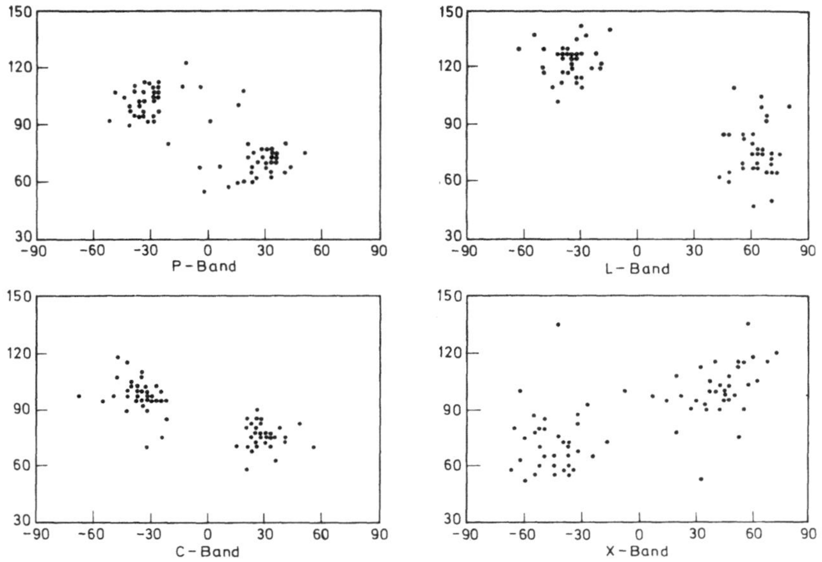

Fig. 7.9a. Co-polarization null locations of sea clutter [7.110,245]

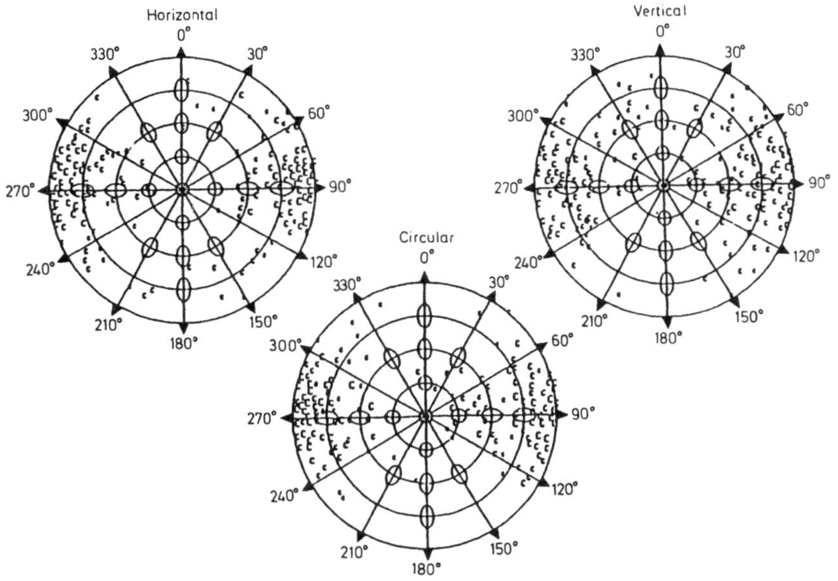

Fig. 7.9b. Co-polarization null location of chaff [7.246] (plotted on POELMAN's extended polarization chart [7.244])

Synthetic Aperture Imagery

Finally, we must note that use of polarization should also increase the quality of SLAR (side looking aperture radar) and SAR (synthetic aperture radar) [7.111] as was reviewed by MARDER [7.112] and BOERNER [7.5]. In traditional SLAR, SAR, or microwave imaging as applied to remote sensing of rough surfaces, use was made of antenna configurations such that polarization was chosen horizontally with respect to the mean surface contour. The radar image so produced with horizontal electric field is referred to as being like-polarized, for example is denoted as an HH image. However, the heterogeneous nature of most earth and sea surfaces causes the signal to depolarize [7.56]. Factors affecting the polarization of the return signal include surface roughness and geometry of the target, moisture content of soil [7.113,114] and material surface properties, the overburden haze, shadowing and multiple reflection, Bragg diffraction as well as the incidence angle and radar frequency [7.57,105]. By using a radar or imaging system with dual orthogonal polarization facility it is possible to irradiate a target "completely" in terms of polarization and to receive simultaneously the returning echoes with both polarizations, whether linear, circular, or general elliptical. Such radar systems are described in [7.115-117] and a rather illuminating report on its applications to geophysics is given by McCAULEY [7.117]. It should be noted that in most of these systems only $|S_{HH}|^2$, $|S_{HV}|^2$ or $|S_{VV}|^2$, $|S_{VH}|^2$ had been recorded simultaneously but not the total information on the relative phase scattering matrix [S].

We must emphasize here that it is essential and *absolutely* necessary to record total information on [S] and to store $p = \sum_{i}^{A}\sum_{j}^{B}|S_{ij}|^2 = \left(|S_{AA}|^2+|S_{BB}|^2+2|S_{AB}|^2\right)$ which is polarization invariant and contains complete target information. Recording of either $|S_{AA}|^2$, $|S_{AB}|^2$, and/or $|S_{BB}|^2$ is incomplete and will result in the loss of information. (7.50)

Intuitively making use of this concept, it was shown by GNISS et al. [7.118, 119] how the information content, the resolution, the image fidelity and feature enhancement of microwave images can be increased if the three independent scatter matrix components $|S_{HH}|^2$, $|S_{VV}|^2$ and $|S_{HV}|^2$ are superimposed onto one image in an incoherent manner, thus satisfying the condition that $p = \sum_{i}^{A}\sum_{j}^{B}|S_{ij}|^2$ is invariant as is shown in Fig.7.10. Finally, we note that any two orthogonal pairs of general elliptical polarization can be employed and the result of producing the image according to (7.50) will always provide the same quality.

Fig. 7.10 a) Polarization effects on the microwave images of a simple text object (left: flat plate, center and right: dihedral corner reflectors, f = 36 GHz, 3 dB resolution = 2, 4λ, [7.119] b) Image reconstruction of complex object using total polarization information [7.119] ►

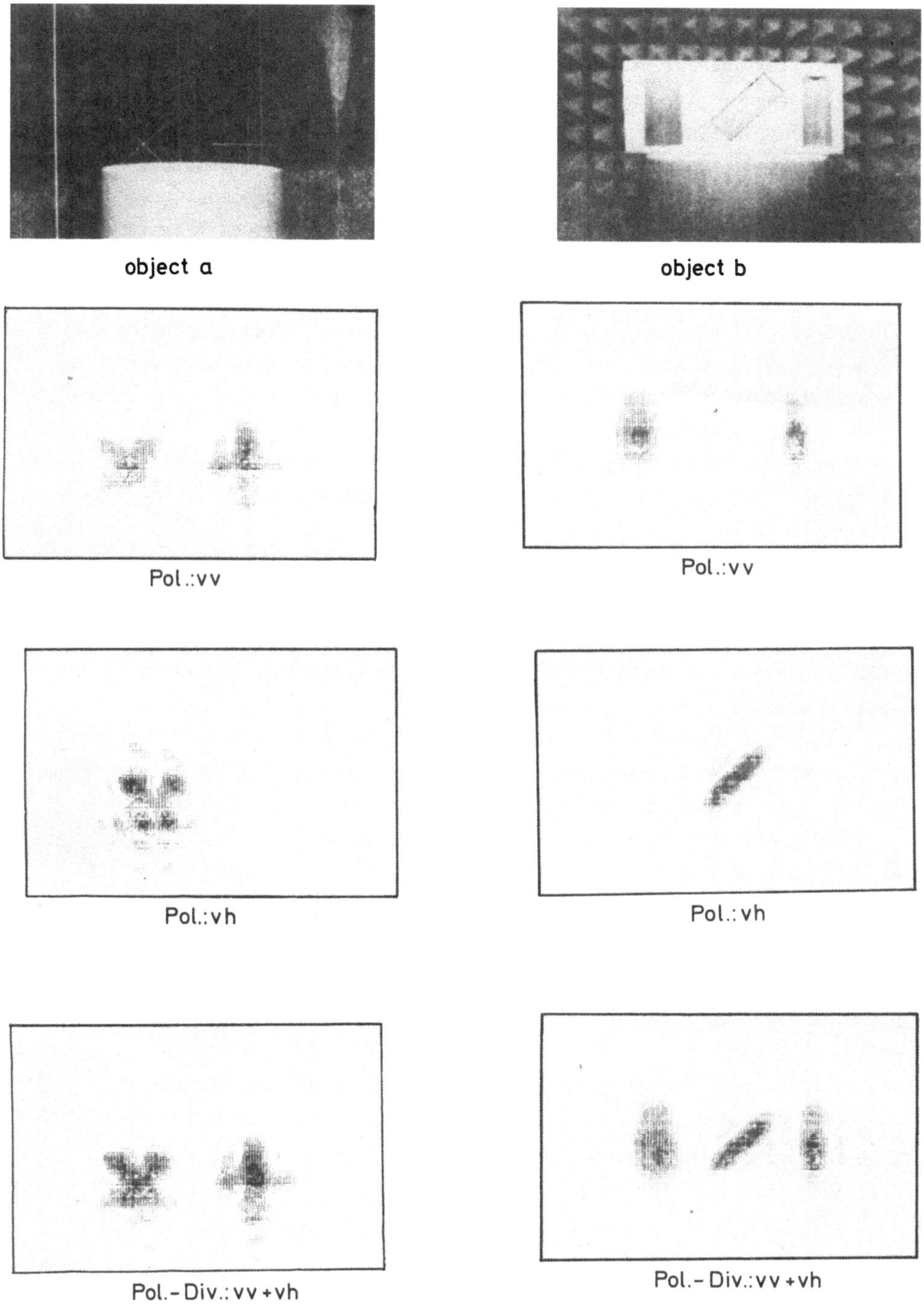

Fig. 7.10a,b. Figure caption see opposite page

7.4 Inverse Scattering Theories in Various Electromagnetic Frequency Regimes

Although a considerable amount of scientific effort has been expanded on theoretical approaches to electromagnetic direct/inverse scattering, the prospects of expressing scattering characteristics of complex shapes successfully and simply are still inadequate. Since no radar system can make complete measurements in practice to define a target uniquely, the data processing procedure must possess an approximate rather than exact basis developed from inverse scattering approaches restricted to specific RCS frequency regimes such as: a) What is the size and the characteristic length of a target? b) What is the silhouette/profile of the target? c) What are the main features in the shape of the target? d) What is the radiating current distribution on the target surface? e) What are the relative positions and sizes of the main scattering centers of the target?

Furthermore, in order to obtain a proper perspective of the electromagnetic inverse scattering problem, it is necessary to appreciate the following aspects:

1) There does not appear to exist any unequivocal derivation of the uniqueness conditions (or necessary and sufficient conditions [7.11]) which apply in general to inverse scattering problems for arbitrary scatterers.

2) There seems to be little hope of applying radio interferometric techniques developed in radio astronomy [7.98]. It appears that the incoherency and statistical constituency of astronomical objects are necessary for the efficacy of the astronomer's technique which yet may be explored to hold true also in radar target imaging [7.19,120].

3) There still is no truly applicable technique that has been derived from the vast body of theory on quantum inverse scattering [7.4,15].

4) Little or no efforts have been made to incorporate the powerful polarization characteristic operators summarized in Sect.7.3 into inverse scattering theories.

The most reasonable approach to electromagnetic inverse scattering is to use *physical optics* (PO) as the central basis which gives remarkably accurate results for the more significant parts of the radar scattering patterns of many types of radar targets [7.7-9,46]. However, it is absolutely necessary to be able to construct corrections to physical optics to account for the following effects:

Polarization: back scattering from simple shapes (excluding distributed targets), as calculated from PO, displays no polarization dependence although measurement data do.

Target Conductivity: PO only applies to targets with metallic surfaces [7.7-9].

Bistatic Scattering: as was shown in [7.121] PO is in disagreement with the reciprocity theorem and disagreement tends to become more pronounced as the bistatic angle increases [7.122].

Multiple Reflection: the corrections to PO for multiple reflection effects are of the same order as the physical optics term over long portions of the target.

Creeping Wave Contributions: PO does not account for diffraction in the penumbra and umbra regions.

Therefore, it will be necessary to use partial solutions of frequency-regime-restricted inverse theories that can be recovered from measurement data made available by wide (multifrequency) band polarization radar systems over the total RCS spatial frequency regime of practical interest (see Fig. 7.2) for the purpose of improving on the PO solution.

7.4.1 The Low Frequency Regime: Rayleigh-Gans Theory

In the Rayleigh or low frequency region [7.123] the three-dimensional objects provide interesting and useful conditions on the impulse and transient wave forms [7.123,124]. At sufficiently low frequencies ($ka \ll 1$), the phasor response $F(s)$ defined in (7.23) can be expanded in a Taylor series resulting in (7.24), where according to the Rayleigh-Gans theory [7.99]

$$a_0 = a_1 = 0 \quad , \quad a_2 \simeq V \quad , \quad \lim_{s\to 0} \sigma = As^4V^2 \quad , \tag{7.51}$$

$$A = (4/\pi c^4)(1+e^{-\tau}/\tau) \quad , \quad \tau = (\text{length/width}) \quad ,$$

V denotes the volume of the scatterer.

It is to be noted that a_2 depends upon the shape, orientation and constitutive parameters of the scatterer *as well as* on polarization [7.124]. Although radar targets do not usually fall within the Rayleigh regime (target dimension ℓ smaller than wavelength: $k\ell \ll 1$), other scatterers such as precipitation particles, i.e., hydrometeors, do, and (7.51) was one of the first and is one of the most frequently used inverse scattering formulas [7.125]. We shall make further use of this condition when dealing with the Kennaugh-Cosgriff-Moffatt transient response method in Sect.7.4.3.

7.4.2 The Resonant Frequency Regime: Natural Frequency Expansion

In the *Mie* or *resonant region* either the low or the high frequency approximations fail and one has to resort to lengthy computational methods. The first solution within this spatial frequency regime was given for dielectric spheres [7.126], however, there does not exist any analytical asymptotic approximatión providing explicitly any profile characteristic parameters applicable to developing an inverse theory. The most direct and effective methods for predicting scattering properties of an object in the resonance region have been the *modal expansion* [7.127] and the *singularity expansion method* (SEM) developed in electromagnetic pulse (EMP) scattering as discussed in detail in [7.128]. Although SEM is not exclusively limited to the analysis of scattering problems in the resonance region, we will consider here its potential use for developing a powerful inverse scattering theory.

The basic idea of SEM is to express the electromagnetic behavior in terms of the singularities in the complex frequency plane which is a well-known problem in acoustics, seismic shock wave analysis and electrical network theory. For finite conducting bodies in free space, various monographs recently were given in [7.245]. A broad band pulse excites the corresponding poles which are referred to as the *natural frequencies* of the object. Following the fundamental contributions of TESCHE [7.129], MARIN [7.130] showed that an isolated, closed shape, perfectly conducting object possesses a resonant frequency singularity expansion of the currents induced on its surface, where the integral equation for the time-harmonic induced current derived from the magnetic-field formulation [7.131] can be casted into operator form

$$[A(\omega)] \cdot \underline{j} = \left(\frac{1}{2}\,[I]-[L]\right) \cdot \underline{j} = \underline{j}^{inc} = \hat{a}_n \times \underline{H}_i \quad . \tag{7.52}$$

The natural current distributions are obtained by evaluating $[A(\omega)] \cdot \underline{j} = 0$ and $[A(\omega)]^{-1}$ is bounded except for simple poles ω_n and $[A(\omega{=}0)]^{-1}$. There exists for $\omega \neq \omega_n$

$$[A(\omega)]^{-1} = \sum\left[(\omega-\omega_n)^{-1}+\omega_n^{-1}\right]\left[<[B_n]\underline{j}_n,\underline{h}^*>\right]^{-1}\underline{j}_n\underline{h}_n + [A(0)]^{-1} \quad , \tag{7.53}$$

where the coupling coefficient $C_{\ell n}$ can be defined [7.130] by

$$C_{\ell n} = \left\langle \underline{j}_\ell^{inc}, \underline{h}_{\ell n}\right\rangle [<[B_{\ell n}]\cdot\underline{f}_{\ell n},\underline{h}_{\ell n}>]^{-1} \quad , \tag{7.54}$$

with $\underline{h}$ denoting here the interior surface magnetic field.

MARIN [7.130] showed that whereas a target's set of natural frequencies is independent of aspect, polarization and excitation, the coupling coefficients $C_{\ell n}$ are not, i.e., the residues associated with the natural frequencies are polarization dependent. We should note here that the actual transfer function of a closed target possesses an infinite number of resonances, and it remains to be determined if a finite number of dominant resonances can adequately describe the transfer function for purposes of target identification [7.132-134]. In particular, the following questions in relation to recovering the natural frequencies from measured data [7.135] need to be addressed: i) Does the hypothesis of a finite number of dominant eigenfrequencies hold up for complicated geometrical objects such as aircraft? ii) Will the resonances for similar but different target types differ appreciably in some region of the complex s plane? iii) Can resonances in the higher frequency region of the RCS spectrum be used for discrimination? iv) Will polarization properties enter the estimation procedure?

In most of the relevant publications the eigenfrequencies are recovered from measurement data using like-polarization, i.e., S_{AA} or S_{BB} which according to

$$p(\omega) = \sum_{i}^{A} \sum_{j}^{B} S_{ij} = \mathit{invariant}$$

does not contain complete target information. It is well known that scattering in the resonance region is highly polarization sensitive [7.7-9] and the relation between the coupling coefficient of (7.54) and the unitary transformation matrix $[U(\psi,\tau_m,\nu)]$ defined in (7.48) needs to be further investigated in relation to a target's invariant properties, i.e., the set of eigenfrequencies and the properties of a target's characteristic scattering operator $[\Gamma(m,\gamma)]$ may be related, and the residues associated with the natural frequencies are polarization dependent.

7.4.3 Physical Optics Far-Field Inverse Scattering Theories: Broad-Band Approach

When the wavelength λ is small compared with the characteristic dimensions ℓ of the scattering body, asymptotic high-frequency approximations are sought because at high frequencies the convergence properties of eigenfunction series solutions are very poor [7.7-9]. Propagation of waves for $k\ell \gg 1$ is described by means of geometric optics (GO) approximations and related inverse theories are of narrow-band nature and will be considered in Sect.7.4.4. Phenomena which represent deviations from the GO laws, for example, the penetration of waves into the shadow regions, are known as diffraction [7.49-53]. Various wideband HF-inverse diffraction theories are reviewed in detail in [7.5] and in the following only a very brief summary emphasizing polarization dependence of various PO inverse scattering theories is presented.

Fourier Transform Method of Physical Optics

For the case of perfectly conducting shapes satisfying Kirchhoff's PO current approximation defined in (7.14), it is possible to relate the far scattered magnetic field (7.15) and the surface current (7.12) in terms of a Fourier transform pair [7.136].

If we consider a wave traveling along the z direction and incident on a scatterer of surface $S(\xi,\eta,\zeta)$ producing a surface current $\underline{J}(\xi,\eta,\zeta) = a_\xi \hat{a}_\eta + a_\eta \hat{a}_\eta$, the transverse magnetic fields of $\underline{H}_s(x,y,z)$ using the far-field approximation of (7.9) become with

$$R = R_0 + \xi - z - (\xi x+\eta y+\zeta z)/R_0 + (x^2+y^2+z^2)/2R_0 \quad ,$$

$$H_x(x,y,z) \simeq ik \exp(i\omega t)/4\pi R_0 c \exp\left\{ik\left[z-(x^2+y^2+z^2)/2R_0\right]\right\}$$

$$\cdot \iiint a_\eta(\xi,\eta,\zeta) \exp(-ik\zeta) \exp[ik(\xi x+\eta y+\zeta z)/R_0]d\xi d\eta d\zeta \qquad (7.55)$$

and H_y is identical in form except that a_η is replaced by $-a_\xi$ [7.137]. Note, since H_z is negligible, a_ζ cannot be recovered (only possible for near-field condition). The ζ dependence $\exp[-ik\zeta(1-z/R_0)]$ may be further reduced by assuming that the receiving system cannot be made to focus on the target at different ranges, i.e., we require $z=0$ in which case [7.138] with $X = (kx)/(2\pi R_0)$, $Y = (ky)/(2\pi R_0)$, we may define

$$U_{xy}(X,Y) = H_{xy}(x,y,0)i2\lambda R_0 c \exp\left\{-i\left[\omega t-k(x^2+y^2)/2R_0\right]\right\} \qquad (7.56)$$

and

$$b_{\xi,\eta}(\xi,\eta) = \int a_{\xi,\eta}(\xi,\eta,\zeta) \exp(-ik\zeta)d\xi \quad .$$

If now, measurements are made over a limited domain D only, and if it can be shown that $\underline{U}(x,y)$ is negligible outside D, then D is the extension required of the radar receiving system for it to be possible to reconstruct at least parts of $\underline{J}$ in terms of b_η and b_ξ so that

$$b_{\eta,\xi}(\xi,\eta) = \iint\limits_D U_{x,y}(X,Y) \exp[-i2\pi(\xi X+\eta Y)]dXdY \quad , \qquad (7.57)$$

which establishes the Fourier transform relation of (7.56) over the limited aperture D. Note, if the radar waves are all singly reflected, then wide-band radar transmission can be used to resolve the target in range along ξ direction provided the bandwidth is large enough to isolate all distinct scattering centers on the target [7.136]. Thus, instead of using (7.56) and defining

$$U_x(X,Y,t) = i\lambda R_0 cH_x(X,Y,t) \exp[i2\pi(\xi X+\eta Y)] \quad , \qquad (7.58)$$

it follows that $a_\eta(\xi,\eta,\zeta) \exp(-ik\zeta)$ and $U_x(X,Y,t)$ form a Fourier transform pair, where k represents the wave number at the center frequency and it must be assumed that the bandwidth is a sufficiently small fraction of the outer frequency in order for the variation of a_η to be negligible over the band. The numerical solution for recovering a_η and a_ξ is discussed in [7.5], and we will conclude with the remark that POFFIS is essentially the basis of most wide-band, short-pulse inverse scattering techniques which became attractive with the advent of wide-band radars having resolution of the order of decimeters [7.25]. Furthermore, with the most recent availability of wide (multi-frequency) band radar systems possessing dual orthogonal polarization measurement facilities, it is possible to improve on image quality by incoherently superimposing the three independent components reconstructed from σ_{AA}, σ_{BB}, σ_{AB} according to

$$p = \sum_i^A \sum_j^B |S_{ij}|^2 \; \textit{invariant}$$

as defined in (7.45).

POFFIS in Time, Frequency, and Projection Domain

Having established the necessary relations for the recovery of the reconstructable part of the surface current distribution, it is pertinent to inquire how the physical parameters can be related to the form of the reradiating current distribution using physical optics far-field inverse scattering (POFFIS). There exist two major POFFIS theories: i) *the time-domain approach* [7.17,18] relating the target ramp response $f_R(t)$, defined in (7.22), with the target silhouette area function $A(z{=}ct'/2)$ as

$$f_R(t') = -1/\pi c^2 A(z)\big|_{z=ct'/2} \tag{7.59}$$

and, ii) *the* $\underline{\kappa}$ *frequency-space formulation* [7.25] (see [7.16]), in which the monostatic field cross section ρ ($\underline{\kappa}{=}2k\hat{a}_i$) is related via the augmented field cross section $\Gamma(\kappa) = 2\pi[\rho(\kappa)+\rho^*(-\kappa)]/|\underline{\kappa}|^2$ to the characteristic target shape function $\gamma(\underline{x}) = 1$ or 0 for $\underline{x}$ within or outside the body D via the $\underline{\kappa}$ to $\underline{x}$ space Fourier transform identity

$$\begin{aligned} \gamma(\underline{x}) &= (2\pi)^{-3} \iiint \Gamma(\underline{\kappa}) \exp(i\underline{\kappa}\cdot\underline{x}) d^3\underline{\kappa} \quad , \\ \rho(\underline{\kappa}) &= i/2\pi \iint_{\underline{\kappa}\cdot\hat{a}_n>0} \underline{\kappa} \cdot \hat{a} \exp(-i\underline{\kappa}\cdot\underline{x}) dS \quad . \end{aligned} \tag{7.60}$$

The ramp response identity, applied as formulated in (7.59), can only be used to recover the weighted shape of axially symmetric targets within the illuminated region for incidence along the invariant axis of a rotationally symmetric target, which may be restated in the broad-band formulation which was defined by KELLER [7.139] as

$$A(z) = cz/2 \int_{-\infty}^{\infty} F(\omega)/\omega^2 \exp[i\omega(2z/c)] d\omega \quad , \tag{7.61}$$

indicating that input data need to be given over the total frequency range. On the other hand the $\underline{\kappa}$ to $\underline{x}$ space Fourier transform identity of (7.60) states that if the back scattered field could be measured in amplitude and relative phase at all frequencies $\omega = kc$ and at all aspects $\underline{\kappa}/|\underline{\kappa}|$, then $\Gamma(\underline{\kappa})$ would be known for all $\underline{\kappa}$, and (7.60) would yield a self-consistent solution of the inverse diffraction problem for a perfectly conducting scatterer in the PO limit [7.16]. It should be noted that, when scattering data at all frequencies but for a single aspect angle are used [7.31] equivalent to a line in $\underline{\kappa}$ space, say $\underline{\kappa} = (0,0,\kappa_3) = -(2\omega)/c$, then (7.60) yields only $A(x_3=z)$. The cross sectional area or target silhouette function normal to the radar line of sight $(x_3=z)$ is

$$A(x_3=z) = 2/\sqrt{\pi}\,\mathrm{Re}\left\{\int_{-\infty}^{\infty} \rho(0,0,\kappa_3)\kappa_3^{-2}\exp(i\kappa_3 x_z)d\kappa_3\right\} , \qquad (7.62)$$

which bears a direct relationship with the broad-band formulation of Keller given in (7.61). The close relationship existing between (7.61,62) and the formal derivation of (7.60) given in [7.5] suggests that a strict transform relation between the formulation of (7.59) and (7.60) exists. Such a relationship was established in [7.140] with the aid of the projection transform theory of RADON [7.30] as is shown in more detail in [7.38,35,140]. Introducing the augmented silhouette area function or complete projection as illustrated in Fig.7.4

$$A_c(r') = A(r')\Big|_0^{r_0} + A(-r')\Big|_\ell^{r_0} , \qquad (7.63)$$

which is obtained from $A(r')\big|_0^{r_0'}$, extracted from $f_R(t',\underline{\kappa}/|\underline{\kappa}|)$, and from $A(-r')\big|_\ell^{r_0'}$, extracted from $f_R(t',\underline{\kappa}/|\underline{\kappa}|)$, and smoothened across the shadow boundary at r_0' so that $A_c(r')$ extends across the length of the intersection from $r' = 0$ across the shadow boundary at $r' = r_0'$ to the antipodal specular point in the shadow region at $r' = \ell$. We note here that the truncation of $f_R(t')$ at $r' = r_0'$ and its extension into the umbra region or, equivalently, curve fitting of two complementary truncated silhouette area functions across the shadow boundary is not resolved [7.35]. It should be noted that for the rotationally symmetrical target case with nose-on incidence MOFFATT [7.141] and YOUNG [7.145] obtained rather satisfactory results by extending $A(r')$ from r_0' to ℓ, i.e., utilizing $f_R(t')$ up to its first zero-crossing which requires more subtle analyses taking into account various Tauberian theorems referred to in Sect.7.2.3.

In Radon projection [7.5] or mixed representation [7.142] space ($\underline{\xi}$: unit vector in direction of incidence $\underline{\kappa}/|\underline{\kappa}$; q: Euclidean distance along projection line r' from origin), the Radon transform $\hat{\gamma}(\underline{\xi},q)$ of the characteristic silhouette function $\gamma(\underline{x})$ in $\underline{x}$-image space as defined in (7.60) is given in [Ref.7.5, Chap.5.3] as

$$\hat{\gamma}(\underline{\xi},q) = \iiint_{-\infty}^{\infty} \gamma(\underline{x})\delta[q-(\underline{\xi}\cdot\underline{x})]d^3\underline{x} \quad , \tag{7.64}$$

where $\delta(q)$ denotes Dirac's delta function, and $\hat{\gamma}(\underline{\xi},q)$ relates to its Fourier transform $\tilde{\gamma}(\alpha\underline{\xi})$ in α space [Ref.7.5, Chap.8.2.3] as

$$\hat{\gamma}(\underline{\xi},q) = (2\pi)^{\frac{1}{2}} \int_{-\infty}^{\infty} \tilde{\gamma}(\alpha\underline{\xi}) \exp(i\alpha q)dx \quad , \tag{7.65}$$

where α is a single variable denoting reciprocal space radius. For the particular aspect direction $\underline{\kappa} = (0,0,\kappa')$, i.e., projection for a particular $(\underline{\xi},q)$ along r', $\alpha = \kappa'$, the relation between $\hat{\gamma}(\underline{\xi},q)$ and $A_c(r',\underline{\kappa}/|\underline{\kappa}|)$ as well as between $\tilde{\gamma}(\kappa'\underline{\xi})$ and $\Gamma(\underline{\kappa})$ needs to be established. For targets with rotational symmetry the relation was established in [7.31] and for the general nonsymmetrical case in [7.32], where it is shown [7.35] that

$$\hat{\gamma}(\underline{\xi},q) = A_c(r',\underline{\kappa}/|\underline{\kappa}|) \quad \text{and} \quad \tilde{\gamma}(\kappa'\underline{\xi}) = \Gamma(\underline{\kappa}) \quad \text{for} \quad \underline{\kappa} = (0,0,\kappa') \quad . \tag{7.66}$$

Thus, we have established that the identities (7.59) and (7.60) establish a Radon-Fourier transform pair in mixed $(\underline{\xi},q)$ representation space. This relation requires further exhaustive analysis particularly with respect to the complementation and curve-fitting procedures adopted in reconstructing $A_c(r')$ and $\Gamma(\underline{\kappa})$ in (7.63) and in (7.60), respectively, as is shown in [7.140].

The Limited Aperture Problem

Since in practice data are given or usually known only over a subset of $\underline{\kappa}$ space, the Fourier transform of (7.60) can no longer be used to obtain $\Gamma(\underline{\kappa})$ and in turn $\gamma(\underline{x})$ directly. In [7.25] it was seemingly first shown that if D is the limited region in which $\Gamma(\underline{\kappa})$ is known, one may choose $K(\underline{\kappa})$ being zero outside Q and nonzero inside so that $K(\underline{\kappa}) \cdot \Gamma(\underline{K}) \equiv G(\underline{\kappa})$ and

$$g(\underline{x}) = (2\pi)^{-1} \iiint_{-\infty}^{\infty} G(\underline{\kappa}) \exp(i\underline{\kappa}\cdot\underline{x})d^3\underline{\kappa} \quad . \tag{7.67}$$

Then if (7.67) can be solved, the size ka and the shape of the target can be recovered at least within D [7.16] (see also [7.5,35]). A unique solution for the band and aspect-limited $(\underline{\kappa}<|\kappa|<\kappa_+, \Omega<\underline{\kappa}|\underline{\kappa}|)$, ill-posed case of (7.67) was given in [7.28,29] making use of the directional derivative approach [7.143]. The Fourier transform $\tilde{\Delta}(\underline{x},\hat{p})$ of the directional derivative $\Delta(\underline{x},\hat{p})$ of (7.60) across the bounding surface Σ of B in direction $\hat{p}$ with $\hat{a}_n$ being the local outward unit normal to Σ is given by

$$\tilde{\Delta}(\underline{x},\hat{p}) = \iint_{\Sigma} \hat{p} \cdot \hat{a}_n \exp(-i\underline{\kappa}\cdot\underline{\xi})dS(\xi) \quad . \tag{7.68}$$

Using multi-dimensional stationary phase evaluation [7.28,29,144], the reconstructed directional derivative $\Delta_K(\underline{x},\hat{p})$ over the limited $K(\underline{\kappa})$ aperture defined in (7.67) becomes

$$\mathrm{Re}\{\Delta_K[\underline{x},\hat{p},(\underline{\xi})]\} \simeq \left[\hat{a}_n(\underline{\xi})\cdot\hat{p}(\underline{\xi})/2\pi D^{\frac{1}{2}}\right] \sin[|\kappa(\underline{x}-\underline{\xi})|]\Big|_{\kappa_-}^{\kappa_+} \quad . \tag{7.69}$$

This important identity [7.28] behaves as a band-limited delta function with a central lobe peaking on the target surface Σ in regions with surface normals $\hat{a}_n$ falling within the family of aperture directions, with the height of the central lobe being proportional to (κ_+,κ_-) and the width to $2\pi/\kappa_+$ with $\kappa_+ >> \kappa_-$. Thus, the specular target regions are identifiable with a resolution that depends directly on the high frequency character of the data and on the bandwidth of information available. With the use of the directional derivative concept, the ill-posedness of (7.67) can be avoided to obtain a solution for the band- and aspect-limited case which is shown in detail in [7.140].

Polarizational Correction

The POFFIS method suffers one critical deficiency in that it is polarization insensitive, as explained in detail in the introduction to Sect.7.4.

Since the PO approximation of (7.14) is polarization independent, depolarization effects [7.56] of nonrotationally symmetric bodies with nonidentical principal radii of curvature will strongly affect the accuracy of POFFIS shape reconstruction [7.5,35]. It will be necessary to develop a vector extension of scalar POFFIS utilizing available data of co/cross-polarized components of the polarization scattering matrix defined in (7.36) and of the first-order vector corrections of the temporal PO approximation which were given in (7.12). Using the notation introduced in Sects.7.2.2 and 3, where (7.12) may be defined with respect to (7.14) as $\underline{J}_\Sigma = \underline{J}_{PO} + \underline{J}_c$ so that (7.15) results for $\underline{J}_{PO}$ in

$$\underline{H}_{PO} = (2\pi)^{-1}\partial^2 A(r')/\partial r'^2(\hat{a}_1 H_{i_1} + \hat{a}_2 H_{i_2})/r \quad , \tag{7.70}$$

and the corresponding solution for $\underline{J}_c$ becomes

$$H_{PO_c} = (2\pi)^{-1}\partial^2 A(r')/\partial r'^2(\hat{a}_1 H_{i_1} - \hat{a}_2 H_{i_2})(K_1-K_2)/r \quad , \tag{7.71}$$

where $\hat{a}_1$, $\hat{a}_2$ are orthogonal unit vectors along the incident magnetic field and K_1, K_2 denote the associated principal curvatures along $\hat{a}_1$ and $\hat{a}_2$. Whereas (7.70) results by integration into the ramp response identity (7.59), (7.71) becomes a step response identity [7.6,140]

$$\underline{f}_u(t') = \frac{1}{2}\pi A(r')(\hat{a}_1 H_{i_1} - \hat{a}_2 H_{i_2})(K_1 - K_2)/r \quad . \tag{7.72}$$

From inspection of (7.70-72), we find that the correction terms (7.72) are certainly nonnegligible in the use of (7.59) if the principal curvatures differ appreciably in which case the POFFIS identities (7.59,60,66) become highly inadequate. The anti-symmetric nature of (7.71,72) also shows that even if the cross-polarized components are neglected in applying POFFIS, different results in shape reconstruction are obtained if incidence is chosen purely along $\hat{a}_2$ for nonsymmetrical target shapes as was demonstrated beyond doubt in [7.18,140,141,145,146].

Thus, instead of simplistically generalizing a scalar wave approach to a vector wave solution [7.147], we need to revisit the highly complex diffraction processes of skew incidence on nonsymmetrical targets using properties of the radar scattering matrices introduced in Sect.7.3. In particular, the important relation

$$p = \sum_i^A \sum_j^B |S_{ij}|^2 = |S_{AA}|^2 + |S_{BB}|^2 + 2|S_{AB}|^2$$

defined in (7.45) must be given appropriate consideration in future vector POFFIS investigations [7.140].

7.4.4 Geometrical Optics Inverse Scattering Asymptotic Theories

In geometrical optics (GO: in [7.148,149,52]), the propagation of energy between two points $P(s_0)$ and $Q(s_0+s)$ occurs according to Fermat's principle along the geodesic where the variation in intensity is dictated by energy conservation, i.e., the energy flux in a tube of rays must be the same at all points along the tube.

$$|E(P)| = |E(Q)|(dS_Q/dS_P)^{\frac{1}{2}} = |E(Q)|\{(\rho_1\rho_2)/[(\rho_1+s)(\rho_2+s)]\}^{\frac{1}{2}} \exp(-iks) \quad , \tag{7.73}$$

with dS_Q and dS_P denoting the curvilinear cross sections of elementary ray tubes at P and Q, respectively, which are inverse proportional to the Gaussian curvatures $[(\rho_1+S)(\rho_2+S)]^{-\frac{1}{2}}$ and $(\rho_1\rho_2)^{-\frac{1}{2}} = (K_1K_2)^{\frac{1}{2}}$, respectively. The properties of local reflection of fields at material surfaces satisfy Snell's law and Fresnel's coefficients $R_{\parallel,\perp}$ [7.52,53]. The GO result yields an infinite value on the caustics $A(\rho_1=-s)$ and $B(\rho_2=-s)$ and the fields must be evaluated by alternate asymptotic evaluations [7.150]. Useful extensions to GO as well as PO are KELLER's geometrical theory of diffraction (GTD) [7.151], UFIMTSEV's physical theory of diffraction (PTD) [7.152], and the equivalent current method (ECM) [7.153,154] for predicting HF scattering of bodies with edges, corners, vertices, tips and rough surfaces. An excellent exposition of these techniques has recently been given in [7.155] otherwise we refer to the special issues on diffraction [7.246] for a succinct summary of these methods.

In contrast to POFFIS methods which are mainly wide (multifrequency) band techniques, GO inverse scattering (GOIS) has become the basis for narrow-band techniques including the scattering center discrimination technique based on Keller's GTD. In the following only the more important narrow-band GOIS methods will be introduced placing major emphasis on their polarization corrections.

GOIS and the Minkowski Problem

Based on the GO postulate of (7.73), KELLER [7.156] introduced the *GO equivalent curvature method* applicable to the profile inversion of rotationally symmetric targets, where the range-normalized radar cross section σ of (7.6) becomes in the GO limit

$$\sigma(\theta,\phi) = R_{\parallel,\perp}(\theta/2)/4K(\theta/2,\phi) \quad , \tag{7.74}$$

where $R_{\parallel,\perp}(\theta/2)$ represents the respective (parallel, normal) energy reflection coefficient and $K(\theta/2,\phi)$ is the Gaussian curvature at the specular point for a perfectly conducting scatterer with symmetry of revolution. For the monostatic case, (7.74) may be rewritten as shown in [7.156-159] with $r(z)$ representing the radial surface coordinate of a body of rotational symmetry about the z axis as

$$\begin{aligned}
\sigma(u) &= \pi K^{-1}(u) = \pi r[1+(dr/dz)^2]/(d^2r/dz^2)\\
r(u) &= \left\{1/\pi \int_0^u \sigma(u)\sin^2 u\, du\right\}^{\frac{1}{2}}\\
z(u) &= 1/\pi \int_0^u [\sigma(u)/r(u)]\sin^2 u\, du \quad ,
\end{aligned} \tag{7.75}$$

where u describes the direction of incidence as defined in [7.159]. Numerous examples for bodies of specific rotational symmetry are worked out in [7.156-159] and a numerical procedure was provided by WEISS [7.159]. It should be noted that a rather similar approach was given for the acoustical case by KINBER [7.160], and that for singly curved bodies [7.157] the BLASBERG approximation [7.161] must be used.

In [7.156], the general case of arbitrary smooth and convex closed shape surfaces in three-dimensional space is considered and reduced to Minkowski's inverse problem [7.162,163,146] in differential geometry, in which the topological image of a closed surface $\hat{X}(x,y,z)$ is mapped onto the unit sphere of directions $\hat{a}_n(\xi,\eta)$ in terms of i) the Gaussian curvature $K = (K_1K_2)^{\frac{1}{2}} = (\rho_1\rho_2)^{-\frac{1}{2}}$

defining the Minkowski problem, or ii) the sum of the principal radii of curvature $\rho_1+\rho_2$ defining the Christoffel-Hurwitz problem, where use is made of the Minkowski support function

$$M = \underline{r} \cdot \hat{a}_n \quad , \quad dM(\xi,\eta,\zeta) = \underline{r} \cdot \underline{\underline{dn}} \quad , \quad x = M_\xi \quad , \quad y = M_\eta \quad , \quad z = M_\zeta \quad (7.76)$$

and the subscripts denote differentiation with respect to ξ,η,ζ. The two characteristic equations are then given by

$$\text{i)} \quad K(\xi,\eta,\zeta)^{-1} = \rho_1\rho_2 = (M_{\xi\xi}M_{\eta\eta}-M_{\xi\eta}) + (M_{\xi\xi}M_{\zeta\zeta}-M_{\xi\zeta}) + (M_{\eta\eta}M_{\zeta\zeta}-M_{\eta\zeta}) \quad ;$$

$$\text{ii)} \quad (\rho_1+\rho_2) = -(M_{\xi\xi}+M_{\eta\eta}+M_{\xi\eta}) \quad ; \qquad (7.77)$$

subject to the condition [7.164]

$$\text{iii)} \quad \int_\Omega K^{-1}(\theta,\phi)\hat{a}_n(\theta,\phi)d\Omega = 0 \quad ;$$

and the Minkowski inverse problem is to recover $x = M_\xi$, $y = M_\eta$, and $z = M_\zeta$ from $K = (K_1K_2)^{\frac{1}{2}}$, $\rho_1 = K_1^{-1}$, and $\rho_2 = K_2^{-1}$.

In particular, KELLER [7.156] showed that the data provided by the GO RCS as defined in (7.74) do not suffice to determine the shape of the scatterer nor any part of it. But, if two functions $\sigma_+(\theta,\phi)$ and $\sigma_-(\theta,\phi)$ are given corresponding to two incident waves coming from opposite directions and if the reflection coefficient is also known, then $\sigma_c = \sigma_+ + \sigma_-$ determines $K(\theta,\phi)$ over the complete Minkowski sphere, and the scalar inverse problem in the GO limit has a unique solution. Yet, a formal reconstruction was not carried out. Keller's approach was extended by WATERMAN and WEISS [7.165], who proposed a global numerical solution for general bodies which was based on a vector-differential equation

$$\underline{\gamma}_\theta \times \underline{\gamma}_\phi = \hat{a}_n \sin\theta \; K^{-1}(\theta,\phi) \quad , \qquad (7.78)$$

where $\gamma(\theta,\phi)$ describes the closed shape, convex surface in spherical coordinates (θ,ϕ). A method of numerical solution was outlined for the case of a general prolate spheroid but inversion was not carried out. Another extension of the approach given in [7.156] was attempted by PAYNE [7.166] using a comparison polyhedron, i.e., a general spheroidal scatterer of prescribed major axes which recover the equivalent local curvature at the specular point.

Independently and unaware of results due to PAYNE [7.166], a similar method employing the concept of equivalent curvatures was introduced for the two-dimensional scalar case [7.167] and for the three-dimensional vector case [7.146].

Vector Extension of GO Equivalent Curvature Inverse Method

As was shown in [7.156-161] the leading term in the GO asymptotic expansion of the scattered field depends primarily on the local radius of curvature at the specular point of the illuminated target region (7.74). This behavior in back scattering direction is the foundation of the "Equivalent Curvature Approach" introduced in [7.166] for the 2-D, and in [7.146] for the 3-D case using an iterative averaging method which compares the averaged magnitude of the back scattered field, scattered by the unknown, with that resulting from a known comparison model scatterer. It should be noted that this method is applicable only when the local radii of curvature are large enough so that creeping wave effects are negligible in determining the field distribution near the back scattering angle.

In the general three-dimensional case the recovery of the major axes of the comparison spheroid requires the use of formulations provided by the Minkowski inverse problem, briefly summarized in [7.76,77] and discussed in more detail in [7.5,146]. In particular, it is shown that a formal solution of recovering the coordinates of the unknown scattering surface requires at least three radar measurables which can be used to recover K, K_1, and K_2. In turn, this requirement can be satisfied only by using complete polarization information employing the first-order correction to the PO ramp response method defined in (7.70-72). Various numerical results are presented in [7.146] proving how the co/cross-polarized components of [S] defined in (7.36) can be used most efficiently to recover the shape and the size of perfectly conducting smooth, convex, closed shape scatterers, where use was made of computational results given in [7.6,18,145,168,169].

Scattering Center Discrimination: Kell's Monostatic-Bistatic Equivalence Theorem

The scattering center discrimination technique [7.121,170-172] draws heavily from the monostatic-bistatic equivalence theorem derived by KELL [7.122] and extended to the general polarization case by BICKEL [7.172]. Starting from the Stratton-Chu vector diffraction integral defined in (7.8), KELL [7.122] derived the "*bistatic scattering integral*" in cylindrical coordinates (ρ,θ,z) as

$$\sigma(\theta_s,\theta_i) = (\pi/\lambda^2)\left|\int \underline{J}(z)\,\exp[-i2k_0 z\cos(\beta/2)]dz\right|^2 \quad , \tag{7.79}$$

where $k_0 = 2\pi/\lambda_0$ with λ_0 being the operating wavelength, and where $\underline{J}(z)$, which is not precisely known, is a composite expression of surface geometry, of illuminating and observing ray geometry, and of surface plus creeping wave propagation effects. A cylindrical coordinate system (ρ,θ,z) is used to relate monostatic and bistatic properties, such that its z axis is coincident with the bisector of the bistatic angle β formed by the transmitter, target centroid and receiver [7.170]. It is found that the above integral can be subdivided into a sum of integrals, each of which is taken over the range of z, within which its integrand is continuous and contributions only come at the end points of each subregion integral which define the scattering centers. The contribution of each scattering center to the total bistatic integral corresponds to the neighborhood of the scattering center, over which the phase of the net phase of the integrand in (7.79) remains within $\pi/2$ of its value and which applies to both monostatic and bistatic scattering. Phase changes are indicated by the factor $k_0 \cos(\beta/2)$ and it follows that small bistatic angles (less than 10°) have little effect on wavelength, whereas at larger bistatic angles the reradiation characteristics of the scattering centers are more important. Evaluating (7.9) over various separated centers or "*co-phased patches*" yields the total RCS

$$\sigma = \left|\sum_{m=1}^{m} \sqrt{\sigma_m} \exp(i\theta_m)\right|^2 \quad , \quad \phi_m = 2k_0 z_m \cos(\beta/2) + \xi_m \quad , \qquad (7.80)$$

where σ_m, ϕ_m, z_m and ξ_m denote the RCS individual phase, distance to the reference (first) phase center, and residual phase contributions of the mth center including creeping wave path length phase corrections, where $\sqrt{\sigma_m}$, z_m and ξ_m are insensitive to the bistatic angle β over the range of β considered in (7.79). The monostatic-bistatic equivalence theorem then follows:

The bistatic cross section of aspect angle α and bistatic angle β is equivalent to the monostatic radar cross section measured on the bisector at a frequency lower by a factor of $\cos(\beta/2)$. (7.81)

Using narrow-band reconstruction methods, it was shown mainly at Cornell Aeronautical Laboratories [7.170,171] how the scattering centers can be determined and allocated from back scattered data requiring considerably increased efforts in data processing as compared to the wide-band short-pulse POFFIS approaches. A vector generalization is given in [7.173]. It should be noted here, that narrow band interferometric methods based on (7.81) have been developed also in other fields [7.174], such as, aperture synthesis [7.171], speckle interferometry [7.175], and in synthetic interferometer radar [7.176].

7.5 Vector Holography and Polarization Utilization

Another approach to radar target shape reconstruction deals with electromagnetic interferometry [7.177] and wavefront reconstruction or holography

[7.138]. In most treatments of holography (see, e.g. [7.178-181]) the vector nature of the field was neglected since the assumption was made, that at optical wavelengths, any flat surface will scatter – in part – a coherent plane wave diffusely in all directions with statistically uniform depolarization [7.181]. That these assumptions do not even hold in the optical region has been shown recently in [7.182,183] and in the previous sections it was demonstrated that polarization effects cannot be neglected in the low frequency to high frequency regions [7.55-58]. Therefore, in the microwave to millimeter wave regions, targets under consideration require a vector treatment of holography [7.184] taking into consideration total polarization information which here is named vector holography. In Sects.7.5.1 and 2 recent results on vector holography and related problems pertinent to microwave imaging and radar target mapping are summarized.

Very closely related to vector holography are inverse scattering theories using extended boundary conditions which also involve the time-averaged "irradiances" of two interfering coherent waves [7.10,11] and are used to allocate the target surface, the material surface properties of a target as well as its size. It should be noted here that *holography* was defined by GABOR [7.185,186] as "*total recording*", which actually can only apply to vector holography in which amplitude and phase of not only one scalar but of both polarization components are recorded [7.187] and reconstructed accordingly [7.188]. The concept of such inverse boundary conditions and the concept of vector holography are reviewed in Sect.7.5.3; and in Sect.7.5.4 a new unified near-field theory is described, which is of immediate relevance to establishing exact vector inverse scattering theories.

7.5.1 Vector Wavefront Reconstruction and Interferometry

Wavefront reconstruction has been aptly named holography [7.185,186] which in the definition of Gabor only implied true recording of amplitude and phase. An electromagnetic wave is completely described only, if amplitude and phase for both polarization components are known in a plane normal to the direction of propagation (the third component is redundant for a charge-free region). Hence, we must generalize Gabor's definition of holography and consider the problem of *vectorial reconstruction* or *vector holography*. One of the first methods of recording the entire vectorial information was given by LOHMANN [7.187] and verified experimentally in [7.188-191].

In the technique proposed in [7.187] two orthogonally polarized reference beams along the orthogonal unit base vectors $\hat{a}_p$ and $\hat{a}_c$ are expressed as $\underline{R}(\underline{r}) = \hat{a}_p R_p + \hat{a}_c R_c$ which interfere with the object wave $\underline{O}(r)$ decomposed along the same orthogonal base vectors $\hat{a}_p$ and $\hat{a}_c$ as $\underline{O}(r) = \hat{a}_p O_p(\underline{r}) + \hat{a}_c O_c(\underline{r})$ so that the irradiance of the two interfering waves becomes with κ representing some constant multiplier

$$I(\underline{r}) = \kappa[\underline{R}(\underline{r})+\underline{O}(\underline{r})] \cdot [\underline{R}(\underline{r})+\underline{O}(\underline{r})]^* \quad . \tag{7.82}$$

The recording and reconstruction geometries of the experimental setup were so arranged that false images in reconstruction due to diffraction of one of the reference beams from the fringes found by interference between the orthogonally polarized reference beam and the object wave, do not overlap the desired image. Variations of these methods are given in [7.192] and by KURTZ [7.188] who encoded some of the polarization information in a single linearly polarized reference beam, whose polarization angle is rotated sequentially making use of the concept of holographic multiplexing first introduced by LEITH and UPATNIEKS [7.193]. These methods of sequential multiplexed recording of phase, amplitude and polarization information have been used in polarizing microholography [7.192] and in crystallography [7.188]. Another variation of Lohmann's principal method was given by SOM and LASSARD [7.194], who showed how the principle of angular selectivity of volume holograms could be used to suppress spurious images [7.195] of the depolarized components, or how to retain complete polarization information in vectorial image reconstruction. It should be noted that volume holograms produced with two orthogonally polarized reference waves of rough surface targets [7.196] showed remarkable image improvements in fidelity and image quality of the reconstructed waveforms [7.197].

Another important aspect, namely that of depolarization, caused either at the surface of the object or by the diffuser, has been grossly neglected in many studies, particularly those related to image reconstruction of targets embedded in random clutter or speckle [7.182,175]. It is well known that the co-polarized component of a linearly polarized wave back scattered from an extended object is determined mainly by segments of the object having a small curvature, whereas the cross-polarized component is caused mainly by scattering at segments of the surface with large curvatures, edges and point discontinuities. Therefore, the information of the latter waves containing very essential information about the detailed structure of a target is lost in the process of scalar holography if the reference wave is orthogonal to the cross-polarized component, since two waves which are polarized in mutually perpendicular directions cannot interfere. Experimental studies in speckle holography [7.175,182,198] have clearly demonstrated that both, contrast ratios and intensity distributions, are strongly influenced by depolarization effects which have been analyzed theoretically in [7.199]. More recently, we have shown in [7.183,200] how depolarization effects can appreciably degrade image quality of reconstructed holographic images leading to i) a decrease in fringe visibility, ii) a loss of information of object segments having extreme curvature, and iii) an extra distortion factor due to the detector's nonlinearity (film gamma). We have shown theoretically and verified experimentally that these effects can easily be eliminated if a "rotated reference" beam [7.183] is used on the condition that the angle of rotation would distribute the reference power between the two orthogonal scattered components of the object wave proportional to their strength.

Use of these studies has also been made in studies on holographic imaging of trans-illuminated particle fields generating a local optically processed reference wave [7.197], where we have shown how the depolarized component, if appropriately filtered, transformed in the spatial frequency domain, and polarization rotated, can be used as a local reference wave similar to the low frequency components of the co-polarized wave [7.201]. Another important approach to generate a reference wave from data supplied by the object wave is the "point-reference holographic technique" proposed by PORTER and co-workers [7.184,202-204]. In this method, the object itself acts as a point reference if the reference frequency is sufficiently smaller than the imaging frequency and a three-dimensional vector treatment taking polarization into account is given in [7.203].

7.5.2 Polarization Dependence in Millimeter and Microwave Holography [7.247]

Methods of wavefront reconstruction have found applications far beyond the *optical region* [7.138] and of particular interest to radar target mapping and image reconstruction are recent advances, made in acoustical holography [7.205] and in microwave holography [7.206] which provided useful input to the development of *true* and *quasi-holographic* microwave imaging techniques. In "*true holography*", image coordinates are two orthogonal distances normal to the radial coordinates; whereas in "*quasi-holography*", one of the distances is range and it provides the basis of one-dimensional holograms obtained in *synthetic aperture radar* [7.111,112,45]. At millimeter to meter wave frequencies both amplitude and phase can be recorded either by a discrete distribution of antennas [7.207], on continuous recording materials such as thermoplasts [7.208] or liquid crystals [7.209] that change under microwave illumination. A recent state-of-the-art review of the subject matter was given in [7.206] which, however, deals only with the scalar treatment of microwave holography, with emphasis placed on planar hologram recordings. Extensive reviews of European investigations are given in [7.210-213] and other holographic broad-band imaging systems have been analyzed in depth in [7.116,214], where also the Doppler frequency spectra of rotating targets are analyzed in [7.211,212,214].

In most of the wide-band coherent microwave radars, similar to *SAR* used in terrain mapping [7.57,105], the complex signal and spectral densities are used to perform the image synthesis based on the Fraunhofer holographic principle [7.138] or equivalently on the antenna synthesis method [7.215-217] which, in principle, are closely related to PO Fourier transform inverse scattering discussed in Sect.7.4.2. Recently, TRICOLES and YUE [7.218] and HOWARD [7.219] carried out in-depth feasibility studies of microwave imaging systems applicable to target imaging; however, as in most of the systems discussed and reviewed [Ref.7.5, Chap.7, 249,250] little emphasis was given to polarization effects which are not negligible [7.220].

As was shown in Sect.7.3.4, only if the target is ideally conducting, smooth and of small curvature compared to wavelengths, may depolarization effects be neglected. However, at microwave frequencies the scatterers most likely under consideration do possess polarization-sensitive sections [7.118] and scattering is described in terms of the polarization scattering matrix defined in (7.35,36). Again, we emphasize the importance of the polarization transformation invariance of (7.45) in vector holography, and in concluding this section, we refer to Sect.7.3.4 and Fig.7.10 which clearly demonstrates the importance of $p = \sum_{ij}^{AB}|S_{ij}|^2$ in microwave vector holography.

7.5.3 The Postulate of Inverse Boundary Conditions

In problems of scattering and diffraction, the shape and the material constituents of the scatterer all of which are a priori known completely, together with the prespecified incident field, may be incorporated into the boundary conditions. On the other hand, in the inverse problem, in general, no specific information about the scatterer may be assumed. Therefore, in this inverse case such "inverse boundary conditions" must be sought, which depend neither on the shape nor the material properties of the scattering body, but allow to specify those characteristic parameters uniquely from the near field which need to be recovered from far-field measurements.

The problem of recovering the near field from the far-field measurements is still an open problem [7.221,222] although remarkable progress has been made recently as is documented in other chapters in this volume. Thus, assuming that the near field can be recovered from measured far-field data, the question arises as to how many and which characteristic parameters must be defined uniquely to determine the shape, the size, and the material constituents of an unknown scatterer, given the fields everywhere within and in its neighborhood.

As it has been discussed in breadth, the PO current approximation of (7.14) represents an "inverse boundary condition" according to (7.11) for perfectly conducting shapes which may be rewritten here as

$$|\underline{E}_i| - |\underline{E}_s| = 0 \text{ on a perfectly conducting surface in the HF limit.} \quad (7.83)$$

This inverse boundary condition which is necessary and sufficient to allocate the surface locus (see [7.10]) approximates the true conducting surface of a scatterer in the HF limit. However, in practice, it must be assumed that a scatterer is not perfectly conducting and that measurement data can be made available also, in the low frequency and resonant regions commonly requiring less measurements. The objective then was to investigate whether other inverse boundary conditions exist, which enable unique specification of shape and averaged material surface parameters (see [7.223]) of a scatterer. One such extended boundary condition was introduced in [7.224] for the case of a perfectly electric conducting scatterer

$$\underline{E} \times \underline{E}^* = 0 \quad , \quad \underline{E} = \underline{E}_i + \underline{E}_s \tag{7.84}$$

as was verified in [7.223,10], where in addition the inverse condition $\underline{H} \times \underline{H}^* = 0$ on a perfectly magnetic conducting body was analyzed. A generalization of these necessary but not sufficient conditions was then sought in [7.223] utilizing the properties of the Leontovich scalar impedance boundary condition defined in (7.5).

Assuming that (7.5) represents the *direct boundary condition*, WESTON and BOERNER [7.223] showed that associated with it are two surface vectors

$$\begin{aligned} \underline{A} &= \underline{E} \times \underline{E}^* - |\eta|^2 \underline{H} \times \underline{H}^* \\ \underline{B} &= \eta \underline{E}^* \times \underline{H} - \eta^* \underline{E} \times \underline{H}^* \end{aligned} \tag{7.85}$$

which satisfy the following *inverse boundary conditions*

$$\underline{A} \cdot \underline{B}^* = 0 \quad , \quad A^2 - B^2 = 0 \quad , \quad \hat{a}_n \cdot \underline{A} = \hat{a}_n \cdot \underline{B} = 0 \tag{7.86}$$

which are necessary but not sufficient. An analysis of the properties of these inverse boundary conditions as well as its computational applicability was carried out in [7.10,11,225,226]. Although the results obtained are excellent as is illustrated in Fig.7.11, it was found that the above inverse conditions are not complete and it is not possible to determine simultaneously the size, the shape, and the material surface properties (η) of an unknown scatterer unless some a priori information on one of these three characteristic parameters is given.

It is interesting to note that the surface vectors $\underline{A}$, $\underline{B}$ defined in (7.85) could be expressed in units of W/m^2 and can be related to the generalized time-averaged Poynting vector, which when expanded contains terms like

$$\underline{E} \cdot \underline{E}^* = (\underline{E}_i + \underline{E}_s) \cdot (\underline{E}_i + \underline{E}_s)^* \quad , \quad \underline{H} \cdot \underline{H}^* = (\underline{H}_i + \underline{H}_s) \cdot (\underline{H}_i + \underline{H}_s)^* \quad . \tag{7.87}$$

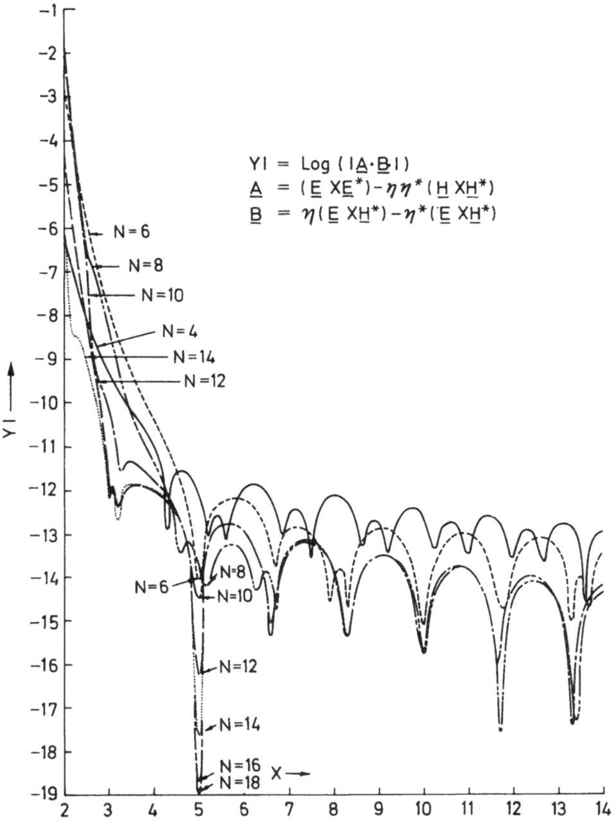

Fig. 7.11a. Plot of Y_1 and Y_2 versus radiant vector X

If we then assume that the incident field represents a reference field and the scattered field the object field in the definition of (7.82), it is evident that the inverse boundary conditions defined in (7.86) are closely related to vector holography. This fact is demonstrated in Fig.7.11, and from inspection of the pseudo-surfaces generated in addition to the locus by the inversion conditions (7.86), as was analyzed in great detail by AHLUWALIA [7.227], where great care was also given to the polarization properties of the fields. It should be emphasized here that in optical holography only the term $\underline{E} \cdot \underline{E}^*$ is of interest and not $\underline{H} \cdot \underline{H}^*$, since *photographic material* responds only to electric but not magnetic time-averaged energies [7.40,53,228]. Thus, in optical holography only shape information can be retained, whereas in vector holography we should be able to recover information on both shape and material constituent parameters. The concept of necessary but not sufficient *inverse boundary conditions* will require further studies in connection with

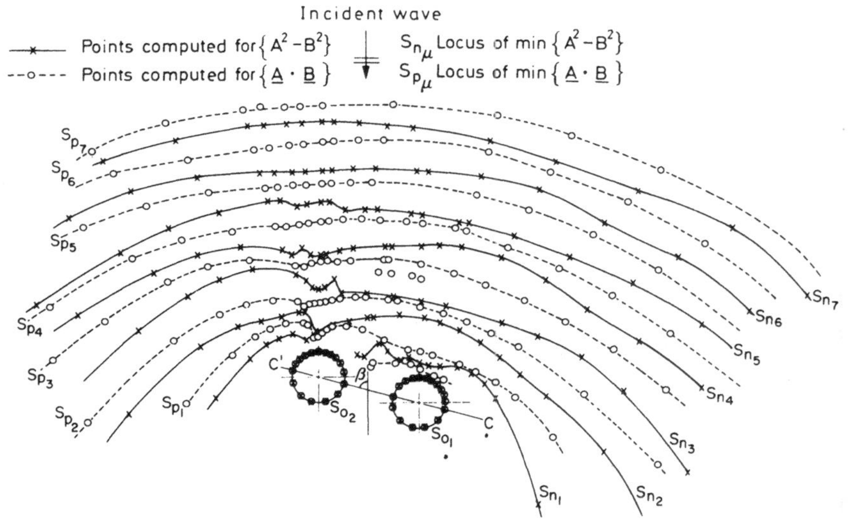

Fig. 7.11b. Plot of the loci $S_{n\mu}$ and $S_{p\mu}$ of successive minima Y_1 and Y_2 of Fig.7.11a [7.225]

systematic methods of discriminating the proper from the pseudo loci of either surface location or surface impedance.

It should be mentioned here that a rather similar approach was used in [7.229] to synthesize the impedance boundary condition on an opaque scatterer from the knowledge of far-field data and the auxiliary condition that as R tends to infinity, one has $\mathrm{Re}\{\iint(\underline{E}\times\underline{H}^*)\cdot \underline{dS}\}=0$. This approach of recovering the surface impedance, though different, also emphasizes the relations which may exist between vectorial impedance boundary conditions and various time-averaged electromagnetic stress tensors [7.230].

7.5.4 Near-Field Approach to Vector Inverse Scattering

More recently a new unified theory of near-field analysis and measurement relevant to this study was developed by WACKER [7.231] and it has become instrumental for developing the excellent near-field measurement facility of the Electromagnetics Division, Boulder [7.232]. Although this powerful near-field measurement technique, which marks a definite breakthrough and advance on the first useful approach developed by BROWN and JULL [7.233], was developed primarily for the purpose of determining far-field properties of antennas and scatterers from near-field measurements, it will also become of great interest to verifying vector inverse theories. The unified theory developed by WACKER [7.234] is based upon generalized scattering matrix theory and utilizes symmetries arising from relativistic and gauge invariances. This general theory incorporates natural orthogonalities of TE and TM modes on planar, circular, cylindrical and spherical measurement surfaces which enable rather simplified yet extremely accurate near-field measurements in terms of modal expansions of the vector fields. It is thus possible to reduce the otherwise incomprehensible vector problem to a set of orthonormal scalar problems which implicitly contain the principles of optimal polarization. The measurement facility will also enable exact interferometric vector near-field measurements relevant to various inverse methods discussed in this and other chapters of this volume. For details on the underlying basic theory containing many novel expansions of vector translation theorems and efficient computer programs, we refer to [7.231-234] and none of the lengthy detailed formalism required to describe this new unified vector near-field approach is given here. In concluding this section, we emphasize that a very powerful near-field measurement range utilizing complete polarization information also useful for the verification of various vector inverse theories has been made available to us. With the aid of this superb facility and further advanced measurement systems, it should in fact be possible to recover completely all of the strongly polarization-dependent vector quantities for reconstructing the surface vectors $\underline{A} = \underline{E} \times \underline{E}^* - |\eta|^2 \underline{H} \times \underline{H}^*$ and $\underline{B} = \eta \underline{E}^* \times \underline{H} - \eta^* \underline{E} \times \underline{H}^*$ defined in (7.85) and thus we should be able to verify general properties of vector holography.

7.6 Conclusions

This chapter emphasizes the fact that in electromagnetic inverse problems we are dealing strictly with a vector problem. It is the common trend to shy away from vector problems and rather treat the simplified scalar cases which more often cannot be implemented in practice. Thus it was considered essential to bridge the gap separating real measurement observations and inverse theories resulting from more or less crude approximations to Maxwell's equations. Keeping this important aspect in mind, in the following, some very brief essential comments are made in the summary of this state-of-the-art review, on defining some of the still open questions which should be analyzed now, and on suggesting new approaches to electromagnetic vector inverse theory by introducing new inverse concepts recently developed in other fields of physical sciences.

7.6.1 Summary

In this chapter we have addressed the vector nature of the electromagnetic scattering phenomena which implies polarization. Although the phenomenon of polarization seems to be well understood in optics, the rather important optimal polarization properties of the radar scattering matrix have been overlooked. Therefore, the very useful concepts of co/cross-polarization nulls have been stressed, since they will undoubtedly become instrumental, in radar signal versus clutter discrimination and in recovering the useful coherent target signal from clutter-perturbed data. Similarly, the important properties of the invariant

$$p = \sum_{i}^{A} \sum_{j}^{B} |S_{ij}|^2$$

which states that the total scattering matrix must be recovered, otherwise polarization information essential to completely describe the properties of a scatterer is lost, which is of particular importance to radar target mapping and holographic imaging in the mm-to-m-wave range. However, with the remarkable progress made in the theory and design of adaptive broad-band antennas with dual polarization facility [7.235], we are now able to measure with sufficient accuracy amplitude, relative and absolute phase, Doppler information, and polarization of radar signals and thus we are ready to implement vector inverse scattering theories.

7.6.2 Unresolved Vector Inverse Problems

Starting with Maxwell's equations and the Stratton-Chu vector diffraction integral, the basis was laid in Sect.7.2 for developing the necessary scattering matrix and vector scattering theories of Sects.7.3-5 essential for deriving vector inverse theories.

To this date there does not exist a self-consistent vector inverse theory applicable in electromagnetic scattering, however there still exist open avenues utilizing complete polarization information contained in the scattering matrices [S] and [M] defined in (7.36,41), respectively, as well as analyzing in depth the progress made in introducing a unified vector near-field theory [7.234]. For example, the highly developed PO Fourier transform inverse scattering theories can still be improved in order to increase image fidelity and resolution of radar-target images by incoherent superposition of the four images $|S_{ij}|^2$ that can be extracted from the scattering matrix utilizing the invariant

$$p = \sum_{i}^{A} \sum_{j}^{B} |S_{ij}|^2$$

defined in (7.45). Furthermore, we should be able to implement directly the optimal polarization concept into electromagnetic vector inverse theories such as POFFIS.

The concept of introducing inverse boundary conditions which are necessary but not sufficient conditions for recovering size, shape, and material surface properties of a target provides a new dimension to the inverse problem in that a natural discriminator is implied. Namely, the inverse conditions $A^2 = B^2$ and $\underline{A} \cdot \underline{B}^* = 0$ in (7.86) produce two families of discriminant loci S_n, S_p and only at the desired value of the discriminant will the two conditions provide one and only one set of identical values. The properties of these inverse boundary conditions and the measurability of the associated vectors $\underline{A}$ and $\underline{B}$ in (7.85) need to be further analyzed. Closely related to this problem, is that of vector holography which also promises to provide in addition to shape reconstruction, the reconstruction of material target properties.

7.6.3 Limitations and Omissions

In this very brief review it was not possible to assess all contributions useful to developing a unified vector inverse scattering approach. For example, more emphasis could have been given to the concept of *extended boundary conditions*, introduced first by WATERMAN [7.236] and then further extended to the null field method by BATES and his co-workers [7.237,238]. Since some inverse aspects of scalar treatments of their work have been analyzed in other chapters of this volume, an additional treatment, though on its vector nature, has been omitted.

Another very important aspect of electromagnetic inverse scattering utilizing Radon's projection transform theory should have been emphasized more strongly, since treatment of POFFIS shape reconstruction techniques for the limited aspect, noncooperative target case can best be treated in the mixed Radon representation as shown in [7.239].

Finally, we should note that there do exist strong polarization-dependent nonlinear target scattering effects, such as utilized in [7.240] which should be of future interest for developing electromagnetic inverse scattering techniques.

7.6.4 Recommendations

The main result of this study is to emphasize that, in the mm-to-m wavelength region total polarization information needs to be utilized to describe the characteristic properties of a target. Since complete adaptive scattering matrix measurements including the extraction of optimal polarizations are now available, vector inverse scattering theories can now be developed, which utilize this complete information.

Since in practice, only partial, highly limited information can be made available, the inverse scattering problem should be viewed as a learning process and thus we need to incorporate fundamentals of decision and estimation theory, as well as pattern recognition into the development of applicable inverse techniques as was attempted with some initial success by KSIENSKI [7.23] and further improvement utilizing complete polarization information is imminent.

Acknowledgments. The research resulting in this contribution was supported in parts by grants of the Natural Sciences and Engineering Research Council of Canada (A7240), a Nato Research Grant 1405, and a Humboldt Fellowship. In addition, the partial support received from the Department of Information Engineering, University of Illinois at Chicago Circle is acknowledged. I am grateful for the helpful discussions with Mr. John C. Daley, Dr. Peter T. Gough and Dr. Gus Tricoles during the preparation of the manuscript. In addition, I wish to thank some of my students who contributed with their research to these studies; in particular Dr. Sujeet K. Chaudhuri, Dr. Max L. Gassend, Dr. Harinder P.S. Ahluwalia, Mr. Hossein Ghandeharian, Dr. Yogadish Das, Mr. Chuk-Min Ho, Mr. Chung-Yee Chan and Mrs. Panagiota Mastoris. I also wish to acknowledge the typing skills of the secretaries of the Communications Laboratory at UICC, Ms. Barbara Kapusnik and Mrs. Beatriz Tryfonopoulos who were instrumental in correcting the final manuscript. Finally, I wish to thank the editor Dr. H.P. Baltes for inviting me to contribute this chapter to the volume.

References

7.1 L. Colin: *Mathematics of Profile Inversion*, Proc. of a Workshop, NASA Tech. Memo. X-62.150, Ames Research Center, Moffet Field, CA (1972)
7.2 J.B. Keller: Am. Math. Mon. *83*, 107-118 (1976)
7.3 R.L. Parker: Ann. Res. Earth Planet Sci. *5*, 35-64 (1977)
7.4 K. Chadan, P.C. Sabatier: *Inverse Problems in Quantum Scattering Theory*, Texts and Monographs in Physics (Springer, Berlin, Heidelberg, New York 1977)
7.5 W-M. Boerner: "State of the Art Review on Polarization Utilization in Electromagnetic Inverse Scattering"; Rpt. No.78-3, Communications Laboratory, University of Illinois, Chicago (1978)
7.6 C.L. Bennett: "Inverse Scattering; Time-Domain Solutions Via Integral Equations", Nato Advanced Study Institute on Theoretical Methods for Determining the Interaction of Electromagnetic Waves with Structures, ed. by J.K. Skwircynski (Sijphoff & Noordhoff, Amsterdam 1980)
7.7 J.J. Bowman, T.B.A. Senior, P.L.E. Uslenghi: *Electromagnetic and Acoustic Scattering by Simple Shapes* (North-Holland, Amsterdam 1969)
7.8 J.W. Crispin, Jr., K.M. Siegel: *Methods of Radar Cross Section Analysis* (Academic, New York 1968)
7.9 D.S. Jones: *The Theory of Electromagnetism* (McMillan, New York 1964)
7.10 V.H. Weston, W-M. Boerner: Can. J. Phys. *47*, 1177-1184 (1969)
7.11 W-M. Boerner, H.P.S. Ahluwalia: Can. J. Phys. *50*, 3023-3061 (1972)
7.12 R.G. Newton: "Scattering Theory in the Mixed (Radon Space) Representation", in *Mathematical Methods and Applications of Scattering Theory*, ed. by J.A. DeSanto, A.W. Saenz, W.W. Zachari, Lecture Notes in Physics, Vol.130 (Springer, Berlin, Heidelberg, New York 1980)
7.13 V.H. Weston: "Electromagnetic Inverse Scattering", in *Electromagnetic Scattering*, ed. by P.L.E. Uslenghi (Academic, New York 1978) pp.289-313
7.14 V.H. Weston: "Some Results in Inverse Scattering", presented at 16th Ann. Meeting, Soc. Eng. Sci., Northwestern University, Evanston, IL, Sept.5-7, 1979, Session TM-7

7.15 A.K. Jordan, S-Y. Ahn: "Inverse Scattering Theory and Profile Reconstruction"; NRL-Mem. Rpt.3981, Naval Research Laboratory, Washington, D.C. (1979)
7.16 R.M. Lewis: IEEE Trans. AP-*17*, 308-314 (1969)
7.17 E.M. Kennaugh, R.L. Cosgriff: IRE Natl. Conv. Rec., 72-77 (1958)
7.18 E.M. Kennaugh, D.L. Moffatt: IEEE Proc. NAEC-*53*, 893-901 (1965)
7.19 J.L. Mesla, D.L. Cohn: *Decision and Estimation Theory* (McGraw-Hill, New York 1978)
7.20 J.K. von Schlachta: IEEE Conf. Publ. (London) *155*, 135-139 (1977)
7.21 K.V. Mardia: *Statistics of Directional Data* (Academic, New York 1972)
7.22 Y.T. Lin, A.A. Ksienski: Radio & Electron. Eng. *46*, 472-486 (1976)
7.23 A.A. Ksienski: "Inverse Scattering as a Target Identification Problem", in *International Symp. Recent Developments in Classical Wave Scattering – Focus on the T-Matrix Approach*, Ohio State University, Columbus, Ohio, June 25-27, 1979, ed. by V.K. Varadan, V.V. Varadan (Pergamon, New York 1980)
7.24 J.D. Young: "Approximate Image Reconstruction from Transient Signature", in *International Symp. Recent Developments in Classical Wave Scattering – Focus on the T-Matrix Approach*, Ohio State University, Columbus, Ohio, June 25-27, 1979, ed. by V.K. Varadan, V.V. Varadan (Pergamon, New York 1980)
7.25 N.N. Bojarski: "Three-Dimensional Electromagnetic Short Pulse Inverse Scattering"; Special Projects Lab. Rpt. AD845 126 Syracuse Univ. Res. Corp. Syracuse, N.Y. (Feb. 1967). [Also "Electromagnetic Inverse Scattering Theory", Syracuse Univ. Res. Corp. Rpt. SPL-TR-68-70 (Dec. 1968)]
7.26 G.A. Deschamps, G.A. Cabayan: IEEE Trans. AP-*20*, 168-174 (1972)
7.27 W.L. Perry: IEEE Trans. AP-*22*, 826-829 (1974)
7.28 R.D. Mager, N. Bleistein: IEEE Trans. AP-*26*, 695-699 (1978)
7.29 N. Bleistein, J.K. Cohen: "A Survey of Recent Progress in Inverse Scattering"; Tech. Rpt. MS-R-7806, University of Denver (1978)
7.30 J. Radon: Ber. Verh. Sächs. Akad. Wiss. Leipzig Math. Phys. Kl. *69*, 262-267 (1917)
7.31 Y. Das, W-M. Boerner: IEEE Trans. AP-*26*, 274-279 (1978)
7.32 W-M. Boerner, C-M. Ho: "Development of Physical Optics, Far Field Inverse Scattering and Its Limitations", IEEE-APS 1979 Intern. APS Symp. Proc., Vol.I, pp.240-243
7.33 K.T. Smith, D.C. Solomon, S.L. Wagner: Bull. AMS *83*, 1227-1270 (1977)
7.34 D. Ludwig: Commun. Pure Appl. Math. *19*, 49-81 (1966)
7.35 W-M. Boerner: "Development of Physical Optics Inverse Scattering Theory", in *Mathematical Methods and Applications of Scattering Theory*, ed. by J.A. DeSanto, A.W. Saenz, W.W. Zachari, Lecture Notes in Physics, Vol.130 (Springer, Berlin, Heidelberg, New York 1980)
7.36 A. Papoulis: *Probability, Random Variables and Stochastic Processes* (McGraw-Hill, New York 1965)
7.37 G. Backus, I. Gilbert: Philos. Trans. R. Soc. London Ser. *A266*, 123-192 (1970); also see Geophys. J. R. Astron. Soc. *13*, 247-276 (1967)
7.38 V.G. Makhankov: Phys. Rep. *35*, 1-128 (1978)
7.39 A.C. Scott, F.Y.F. Chu, D.W. McLaughlin: IEEE Proc. NAEC-*61*, 1443-1483 (1973)
7.40 J.M. Bennett, H.E. Bennett: "Polarization", in *Handbook of Optics*, ed. by W.G. Criscoll, W. Vaughan (McGraw-Hill, New York 1978) Chap.10
7.41 J.A. Stratton: *Electromagnetic Theory* (McGraw-Hill, New York 1941)
7.42 M.A. Leontovich: "Appendix", in *Diffraction, Reflection and Refraction of Radio Waves*, 13 Papers by V.A. Fock, ed. by N. Logan, P. Blacksmith; Rpt. No. AD-117276 (U.S. Government Printing Office, Washington, D.C. 1957); also see: *Investigation of Propagation of Radio Waves, II* (Sovetskoye Radio Press, Moscow 1948) [in Russian]
7.43 R.S. Berkowitz: *Modern Radar, Analysis, Evaluation, and System Design* (Wiley, New York 1965)
7.44 F.E. Nathanson: *Radar Design Principles, Signal Processing and the Environment* (McGraw-Hill, New York 1969)
7.45 M.I. Skolnik: *Introduction to Radar Systems* (McGraw-Hill, New York 1962); also see: M.I. Skolnik: *Radar Handbook* (McGraw-Hill, New York 1978)

7.46 G.T. Ruck, D.E. Barrick, W.D. Stuart, C.K. Krichbaum: *Radar Cross Section Handbook*, Vols.I,II (Plenum, New York 1970)
7.47 J.A. Stratton, L.J. Chu: Phys. Rev. *56*, 99-107 (1939)
7.48 L.B. Felsen (ed.): *Transient Electromagnetic Fields*, Topics in Applied Physics, Vol.10 (Springer, Berlin, Heidelberg, New York 1976)
7.49 H. Hönl, A.W. Maue, K. Westpfahl: "Theorie der Beugung", in *Crystal Optics. Diffraction*, Encyclopedia of Physics, Vol.XXV/1 (Springer, Berlin Göttingen, Heidelberg 1961)
7.50 N. Wiener: Acta Math. *55*, 117-258 (1930); also see reprint (MIT Press, Cambridge, MA 1964)
7.51 B. Van der Pol, H. Bremmer: *Operational Calculus* (Cambridge University Press, Cambridge 1955)
7.52 M. Kline, I.W. Kay: *Electromagnetic Theory and Geometrical Optics*, Pure and Applied Mathematics, Vol.XII (Wiley-Interscience, New York 1965)
7.53 M. Born, E. Wolf: *Principles of Optics* (Pergamon, New York 1959)
7.54 F.A. Jenkins, H.E. White: *Fundamentals of Optics*, 3rd ed. (McGraw-Hill, New York 1951)
7.55 P. Beckmann, A. Spizzichino: *The Scattering of Electromagnetic Waves from Rough Surfaces* (McMillan, New York 1963)
7.56 P. Beckmann: *The Depolarization of Electromagnetic Waves* (Golem, Boulder, CO 1968)
7.57 M.W. Long: *Radar Reflectivity of Land and Sea* (Lexington Books, Heath & Co., Lexington, MA 1975)
7.58 D.B. Kanareykin, N.F. Pavlov, U.A. Potekhin: *The Polarization of Radar Signals* (Sovetskoye Radio Press, Moscow 1966) [in Russian] [English transl.: *Radar Polarization Effects* (C. Collier & MacMillan, New York 1974) Chaps. 10-12]
7.59 G.A. Deschamps: Proc. IRE *39*, 543-548 (1951)
7.60 J.R. Huynen: "Radar Target Sorting Based upon Polarization Signature Analysis"; Lockheed Missiles and Space Division Rpt.28-82-16 (1960); AD-318597 (1972)
7.61 J.R. Huynen: Proc. IEEE *53*, 936-946 (1965)
7.62 J.R. Huynen: "Phenomenological Theory of Radar Targets"; Ph.D. Thesis, Technical University Delft (1970)
7.63 J.R. Huynen: "Radar Target Phenomenology in Electromagnetic Scattering", in *Electromagnetic Scattering*, Selected Papers of a NH Conf., UICC, Chicago, June 15-18, 1976, ed. by P.L.E. Uslenghi (Academic, New York 1978)
7.64 A.B. Schneider, P.D.L. Williams: Radio Electron. Eng. *47*, 11-29 (1976)
7.65 G. Sinclair: "Modification of the Radar Range Equation for Arbitrary Targets and Arbitrary Polarization"; Rpt.302-19, Antenna Laboratory, Ohio State University, Columbus (1948)
7.66 P.J. Napier, R.H.T. Bates: Proc. IEE *120*, 30-34 (1973); Proc. IREEE (Australia) *32*, 164-165 (1971); Int. J. Eng. Sci. *9*, 1107-1121, 1193-1208 (1971)
7.67 K.E. Maffett: "Radar Polarization Properties", in *Methods of Radar Cross Section Analysis*, ed. by J.W. Crispin, K.M. Siegel (Academic, New York 1968)
7.68 C.D. Graves: Proc. IRE *44*, 248-252 (1956)
7.69 A. Ishimaru: *Wave Propagation and Scattering in Random Media*, Vol.1: *Single Scattering and Transport Theory*; Vol.11: *Multiple Scattering, Turbulence, Rough Surfaces, and Remote Sensing* (Academic, New York 1978)
7.70 J.A. Stiles, J.C. Holtzman (eds.): "Radar Backscatterer from Terrain"; Tech. Rpt. RSLR374-2, Remote Sensing Laboratory, Fort Belvoir, VA (1979)
7.71 E.M. Kennaugh: "Effects of Type of Polarization on Echo Signals"; Tech. Rpt.389-9, Antenna Laboratory, Ohio State University, Columbus (1951)
7.72 E.M. Kennaugh: "Polarization Properties of Target Reflection"; Tech. Rpt. 389-2, Griffis AFB (1952)
7.73 V.H. Rumsey, G.Y. Deschamps, M.L. Kales, J.I. Bohnert: Proc. IRE *39*, 535-553 (1951)
7.74 H. Gent: "Elliptically Polarized Waves and Their Reflection from Radar Targets: a Theoretical Analysis"; Memo 584 Telecommunications Research Establishment (1954)

7.75 J.R. Copeland: Proc. IRE *48*, 1290-1296 (1960)
7.76 G.H. Knittel: IEEE Trans. AP-*15*, 217-221 (1967)
7.77 J.W. Crispin, Jr.: "The Measurement and Use of Scattering Matrices"; Tech. Rpt.2500-3T, Radiation Laboratory, Ann Arbor (1960)
7.78 J.R. Huynen: IRE Conv. Rec. Pt. *5*, Vol.10, 3-11 (1962)
7.79 N.R. Landry: "Measuring the Phase and the Amplitude of Backscattered Radar Energy on a Static Range"; Tech. Rpt. RADC-TDR-65-24, Rome Air Development Center, NY (1964)
7.80 J.A. Webb, W.P. Allan: "Precision Measurements of the Radar Scattering Matrix"; Tech. Rpt. RADC-TDR-64-25, Rome Air Development Center, NY (1964)
7.81 O. Lowenschuss: Proc. IEEE *53*, 988-992 (1965)
7.82 F. Kuhl, R. Covelli: Proc. IEEE *53*, 1110-1115 (1965)
7.83 D.B. Kanareykin, V.A. Potekhin, I. Shishkin: *Maritime Polarimetry* (Sudostroyeniye, Moscow 1968) [in Russian]
7.84 Yu.B. Kobzarev (ed.): *Modern Radar* (Sovetskoye Radio Press, Moscow 1969) [in Russian]
7.85 Ya.D. Shirman (ed.): *Theoretical Fundamentals of Radar* (Sovetskoye Radio Press, Moscow 1970) [in Russian]
7.86 M.M. Gorshkov: *Ellipsometry* (Sovetskoye Radio Press, Moscow 1974) [in Russian]
7.87 L.A. Zhivotovskiy: Radio Eng. *31*, 46-53 (1976)
7.88 See Chap.2
7.89 L.M. Spetner, I. Katz: IEEE Trans. AP-*8* , 242-246 (1960)
7.90 W.S. Ament: IEEE Trans. AP-*8*, 167-174 (1960)
7.91 H.C. Ko: Proc. IRE *50*, 1950-1951 (1962)
7.92 A.K. Fung: Proc. IEEE *54*, 996-998 (1966)
7.93 A.K. Fung: Proc. IEEE *54*, 395-396 (1966)
7.94 A.K. Fung: Planet Space Sci. SA-*15*, 1337-1347 (1967)
7.95 J.W. Strohbehn (ed.): *Laser Beam Propagation in the Atmosphere*, Topics in Applied Physics, Vol.25 (Springer, Berlin, Heidelberg, New York 1978)
7.96 T. Hagfors: Radio Sci. *2*, 445-465 (1967)
7.97 A. Stogryn: Radio Sci. *2*, 415-428 (1967)
7.98 V. Evans: Bull. Va. Polytech. Inst. *4*, 237-259 (1962)
7.99 H.C. Van de Hulst: *Light Scattering by Small Particles* (Wiley, New York 1957)
7.100 G.C. McCormick, A. Hendry: Radio Sci. *11*, 731-740 (1976)
7.101 J.C. Daley, W.T. Davis, N.R. Mills: "Radar Sea Return in High Sea States"; Tech. Rpt. No.7142, National Research Laboratories (1970)
7.102 F.G. Bass, I.M. Fuks: "Scattering of Waves by Statistically Irregular Surfaces", Parts I,II [transl. from Russian] JPRS Rpt. No.66061-1, NTIS (1975)
7.103 G.R. Valenzuela: Boundary-Layer Meteorol. *13*, 61-85 (1978)
7.104 G.R. Valenzuela: "Scattering of Electromagnetic Waves from the Ocean", in *Surveillance of Environmental Pollution and Resources by Electromagnetic Waves*, ed. by T. Lund (Reidel, Dordrecht, Holland 1978) pp.199-226
7.105 G. Kortüm: *Reflectance Spectroscopy* (Springer, Berlin, Heidelberg, New York 1969)
7.106 J.C. Leader: Radio Sci. *13*, 441-457 (1978)
7.107 P.T. Gough, W-M. Boerner: J. Opt. Soc. Am. *69*, 1212-1217 (1979)
7.108 J.C. Daley: "Radar Target Detection Based on Polarization Effects", NAVAIR Review (1978)
7.109 G.A. Ioannidis, D.E. Hammer: IEEE Trans. AP-*27*, 357-363 (1979)
7.110 L.A. Morgan: "Radar Cross Section Scattering Matrix", NAVAIR Review (1978)
7.111 R.O. Harger: *Synthetic Aperture Radar Systems: Theory and Design* (Academic, New York 1970)
7.112 S. Marder: "Synthetic Aperture Radar", in *Atmospheric Effects on Radar Target Identification and Imaging* (Reidel, Dordrecht, Holland 1976)
7.113 F.T. Ulaby: IEEE Trans. AP-*22*, 257-266 (1974)
7.114 F.M. Dickey, R.K. Moore, C. King, J. Holtzman: "Moisture Dependency of Radar Backscattering from Irrigated and Non-Irrigated Fields at 400 MHz and 13 GHz"; CRES Tech. Rpt.177-33, Center of Res., University of Kansas, Lawrence, KA (1972)
7.115 R. Rawson, F. Smith, R. Larson: in *Proceedings of the 1975 IEEE International Radar Conference* (Washington, D.C. 1975) pp.505-509

7.116 R.W. Larson, F. Smith, R. Lawson, M.L. Brian: "Multispectral Microwave Imaging Radar for Remote Sensing Applications", in *Microwave Scattering and Emission from the Earth*, Proc. URSI Special Meeting, ed. by E. Schanda (University of Bern 1974) pp.305-315

7.117 J.R. McCauley: "Surface Configuration as an Explanation for Lithology-Related Cross Polarized Radar Image Anomalies"; Tech. Rpt.177-36, Space Technology Laboratory, Kansas (1973)

7.118 H. Gniss, K. Magura: "mm-Wave Images of Ground-Based Objects", Rpt.1-78-153, FHP-FGAN, Wachtberg-Werthoven, FRG (1978)

7.119 H. Gniss, J. Magura, R. Karg, H. Ermert, H. Brand, W-M. Boerner: "Polarization Dependence of Image Fidelity in Microwave Mapping Systems", in *1978 International IEEE-APS Symposium Proceedings*, pp.38-42

7.120 R.H.T. Bates, M.J. McDonell: Astrophys. J. *208*, 443-452 (1976)

7.121 R.G. Koujoumjian: Proc. IEEE *53*, 864-876 (1965)

7.122 R. Kell: Proc. IEEE *53*, 983-988 (1965)

7.123 J.W. Strutt: *Lord Rayleigh: Collected Scientific Papers* (Dover, New York 1964)

7.124 J.S. Asvestas, R.E. Kleinman: "Low Frequency Electromagnetic Scattering", in *Electromagnetic Scattering*, ed. by P.L.E. Uslenghi (Academic, New York 1978)

7.125 P.J. Wyatt: Appl. Opt. *7*, 1879-1896 (1968)

7.126 N.A. Logan: Proc. IEEE *53*, 773-785 (1965)

7.127 R.F. Harrington: "Characteristic Modes for Antennas and Scatterers", in *Numerical and Asymptotic Techniques in Electromagnetics*, ed. by R. Mittra, Topics in Applied Physics, Vol.3 (Springer, Berlin, Heidelberg, New York 1975) pp.51-87

7.128 C.E. Baum: Proc. IEEE *64*, 1598-1616 (1976)

7.129 F.M. Tesche: IEEE Trans. AP-*21*, 53-62 (1973)

7.130 L. Marin: IEEE Trans. AP-*21*, 266-274 (1973)

7.131 L. Marin: IEEE Trans. AP-*21*, 809-818 (1973)

7.132 A.J. Berni: IEEE Trans. AES-*11*, 147-154 (1975)

7.133 D.L. Moffatt, R.K. Mains: IEEE Trans. AP-*23*, 358-367 (1975)

7.134 C.W. Chuang, D.L. Moffatt: Abstracts of 1975 URSI NRC Meeting, June 3-5, 1975, UIUC, Urbana, IL, pp.67-68

7.135 Electro-Science Lab.: "Radar Target Identification I,II"; Class Notes, Sept.1976, Ohio State University, Columbus, Ohio

7.136 A. Freedman: Radio Electron. Eng. *25*, 51-64 (1963)

7.137 R.M. Bracewell: *The Fourier Transform and Its Applications* (McGraw-Hill, New York 1965)

7.138 J.W. Goodman: *Introduction to Fourier Optics* (McGraw-Hill, New York 1968)

7.139 J.B. Keller: "On the Use of Short-Pulse Broad Band Radar for Target Identification"; Rpt., Feb.17, 1965, RCA Moorestown, NJ

7.140 W-M. Boerner, C-M. Ho: Wave Motion *4* (1980) in press

7.141 D.L. Moffatt, J.D. Young: IEEE Trans. AP-*17*, 337-344 (1975)

7.142 R.G. Newton: Physica *96A*, 271-279 (1979)

7.143 N.N. Bojarski: "Electromagnetic Inverse Scattering", NAVAIR Systems Command, AD-775 235/1-4 (1972-4)

7.144 N. Bleistein, R.A. Handelsmann: *Asymptotic Expansions of Integrals* (Holt, Reinhart, and Winston, New York 1975)

7.145 J.D. Young: "Target Imaging from Multiple-Frequency Radar Returns"; Tech. Rpt.2268-5, AD-728235, Electro-Science Laboratory, Ohio State University, Columbus, Ohio (1971); also see: IEEE Trans. AP-*24*, 276-282 (1976)

7.146 S.K. Chaudhuri, W-M. Boerner: Appl. Phys. *11*, 337-350 (1976); also see: IEEE Trans. AP-*25*, 505-511 (1977)

7.147 A. Majda: Commun. Pure Appl. Math. *30*, 165-194 (1977)

7.148 J. Picht: *Optische Abbildung* (Vieweg, Braunschweig 1931)

7.149 R.K. Luneburg: *Mathematical Theory of Optics* (Brown University Notes, Providence, R.I. 1944)

7.150 I. Kay, J.B. Keller: J. Appl. Phys. *25*, 876-883 (1954)

7.151 J.B. Keller: *The Geometrical Theory of Diffraction*, URSI Symposium on Microwave Optics (McGill University, Montreal 1953)

7.152 P.Ya. Ufimtsev: "Method of Edge Waves in the Physical Theory of Diffraction"; [transl. from Russian] Foreign Technology Service, Sept., 1971, Wright-Patterson AFB
7.153 C.E. Ryan, W.L. Peters, Jr.: "The Relation of Creeping Wave Phenomena to the Shadow Zone Geometry", Proc. GISAT II Symp., Pct.2-7, 1967, Mitre Corp., Bedford, MA, Vol.II, Pt.1, pp.315-349
7.154 R.F. Miller: Radio Sci. *8*, 785-796 (1973)
7.155 V.A. Borovikov, B.Ye. Kinber: Proc. IEEE *62*, 1416-1437 (1974)
7.156 J.B. Keller: IEEE Trans. AP-*7*, 146 (1959)
7.157 J.L. Altman, R.H.T. Bates, E.N. Fowle: "Introductory Notes Relating to Electromagnetic Inverse Scattering"; Tech. Rpt. SR-121, Mitre Corp. (1964)
7.158 J.B. Keller, R.M. Lewis: Commun. Pure Appl. Math. *9*, 207-265 (1956)
7.159 M.R. Weiss: J. Opt. Soc. Am. *58*, 1524-1528 (1968)
7.160 B.Ye. Kinber: Akust. Zh. *1*(3), 221-225 (1955)
7.161 L.A. Blasberg: "Short-Pulse Signature Analysis for Midcourse Discrimination". Internal Rpt., 1963, RCA Moorestown, NJ
7.162 H. Minkowski: Math. Ann. *57*, 447-495 (1903)
7.163 L. Nirenberg: Commun. Pure Appl. Math. *6*, 337-394 (1953)
7.164 J.J. Stoker: Commun. Pure Appl. Math. *3*, 231-257 (1950)
7.165 P.C. Waterman, M.R. Weiss: "Inverse Scattering and the Minkowski Problem", Proc. GISAT II Symp., Oct.2-4, 1967, Mitre Corp., Bedford, MA, Vol.2, Pt.1, pp.371-376
7.166 W.T. Payne: "Determination of Shape and Size of Non-Axisymmetric Conducting Targets by Geometrical Optics", Proc. GISAT II Symp., Oct.2-4, 1967, Mitre Corp., Bedford, MA, Vol.2, Pt.1, pp.303-313
7.167 F.H. Vandenberghe, W-M. Boerner: Radio Sci. *6*, 1163-1171 (1971)
7.168 C.L. Bennett, J.D. Delorenzo, A.M. Auckenthaler: "Integral Equation Approach to Wide-Band Inverse Scattering"; Final Rpt. Contract No. F30602-69-C-0322, Sperry-Rand Res. Centre, Sudberry, MA (1970)
7.169 C.L. Bennett, A.M. Auckenthaler, R.S. Smith, J.D. Delorenzo: "Space-Time Integral Equation Approach to the Large Body Scattering Problem", Rpt. No. SCRCR-Cr-73-1, Sperry-Rand Res. Centre, Sudberry, MA (1973)
7.170 M.E. Bechtel, R.A. Ross: "Radar Scattering Analysis"; Tech. Rpt. EK-IRIS-10 (1966)
7.171 J.A. Hammer: "A Method to Determine the Scattering Centers from the Back Scatterer Pattern of a Body", Proc. GISAT II Symp., Oct.2-4, 1967, Mitre Corp., Bedford, MA, Vol.2, Pt.1, pp.223-235; AD 839-700
7.172 S.H. Bickel: "Polarization Studies and Scattering Matrix Applications", Proc. GISAT II Symp., Oct.2-4, 1967, Mitre Corp., Bedford, MA, Vol.2, Pt.1, pp.119-133
7.173 S.H. Bickel, J.F.A. Ormsby: Proc. IEEE *53*, 1067-1089 (1965)
7.174 N.M. Tomljanovich, H.S. Ostrowsky, J.F.A. Ormsby: "Narrow-Band Interferometer Imaging"; Project 4966, Mitre Corp. (1968); AD 679-208
7.175 J.C. Dainty (ed.): *Laser Speckle and Related Phenomena*, Topics in Applied Physics, Vol.9 (Springer, Berlin, Heidelberg, New York 1975)
7.176 L.J. Porcello, J.L. Allan (eds.): Proc. IEEE *62* (1974) (Special Issue, Modern Radar Technology and Applications)
7.177 R.M. Bracewell: Austr. J. Phys. *9*, 198-217 (1956)
7.178 J.B. DeVelis, G. Reynolds: *Theory and Application of Holography* (Academic, New York 1969)
7.179 H.M. Smith: *Principles of Holography* (Wiley-Interscience, New York 1969)
7.180 W.G. Stroke: *An Introduction to Coherent Optics and Holography* (Academic, New York 1969)
7.181 W.E. Kock: *Radar, Sonar and Holography* (Academic, New York 1973)
7.182 N. George, A. Jain, R.D.S. Melville, Jr.: Appl. Phys. *6*, 65-70 (1975)
7.183 H. Ghandeharian, W-M. Boerner: J. Opt. Soc. Am. *68*, 931-934 (1978)
7.184 R.P. Porter, W.C. Schwab: J. Opt. Soc. Am. *61*, 789-796 (1971)
7.185 D. Gabor: Proc. R. Soc. (London) *A197*, 454-487 (1949); Nature *161*, 777-778 (1948)
7.186 D. Gabor: Proc. Phys. Soc. (London) *B64*, 449-482 (1951)
7.187 A.W. Lohmann: Appl. Opt. *4*, 1667-1668 (1965)
7.188 C.N. Kurtz: Appl. Phys. Lett. *14*, 59-61 (1969)

7.189 G.L. Rogers: J. Opt. Soc. Am. *56*, 831 (1966); Nature *177*, 613 (1956)
7.190 O. Bryngdahl: J. Opt. Soc. Am. *58*, 702 (1968); *57*, 545-546 (1967)
7.191 M.W. Fourney, A.P. Waggoner, K.V. Mate: J. Opt. Soc. Am. *58*, 701-702 (1968)
7.192 W.H. Cater, P.D. Engeling, A.A. Dougal: IEEE Trans. QE-*2*, 44-46 (1966)
7.193 E.N. Leith, J. Upatnieks: J. Opt. Soc. Am. *54*, 1295-1302 (1964)
7.194 S.C. Som, R.A. Lassard: Appl. Phys. Lett. *17*, 381-382 (1970)
7.195 K.S. Pennington, R.J. Collier: Appl. Phys. Lett. *8*, 14-16, 44 (1966)
7.196 W.A. Shurcliff: *Polarized Light* (Harvard University Press, Cambridge, MA 1962)
7.197 M.L.A. Gassend: "Holographic Imaging of Trans-Illuminated Particle Fields Using a Local Optically Processed Reference Wave"; Ph.D. Thesis, University of Manitoba, Winnipeg, Canada (1976)
7.198 N. George, A. Jain, R.D.S. Melville, Jr.: Appl. Phys. *7*, 157-169 (1975)
7.199 M.L. Varshavshuk, V.O. Kobak: Radio Eng. Electron. Phys. *16*, 201-205 (1971)
7.200 H. Ghandeharian, W-M. Boerner: Opt. Acta *24*, 1087-1097 (1977)
7.201 M.L.A. Gassend, W-M. Boerner: Appl. Phys. *13*, 71-79 (1977)
7.202 R.P. Porter: Proc. IEEE *59*, 307-308 (1971)
7.203 R.P. Porter: J. Opt. Soc. Am. *60*, 1051-1059 (1970)
7.204 R.P. Porter: Phys. Lett. *2A*, 193-194 (1969)
7.205 A.F. Methrell (ed.): *Acoustical Holography*, Vol.3 (Plenum, New York 1971)
7.206 G. Tricoles, N.H. Farhat: Proc. IEEE *65*, 108-121 (1977)
7.207 N.H. Farhat: Opt. Eng. *14*, 490-505 (1975)
7.208 K. Iizuka: Proc. IEEE *57*, 812-814 (1969)
7.209 W.E. Kock: Appl. Opt. *14*, 1471-1472 (1975)
7.210 G. Graf: IEEE Trans. AP-*24*, 378-381 (1976)
7.211 K. Magura: "Probleme bei der holographischen Abbildung im Mikrowellenbereich"; Tech. Rpt.6-72-11, HFP-FGAN Wachtberg Werthoven (1972)
7.212 R. Karg: Arch. Elektr. Übertr. *31*, 150-156 (1976)
7.213 J. Detlefsen: Nachrichtentech. Z. *30*, 723-725 (1977)
7.214 K. Iizuka: Proc. IEEE *5*, 812-814 (1967)
7.215 P.M. Woodward: J. IEE *93-111A*, 1554 (1943) [also see: IEE *95-111A*, 363-370 (1948)]
7.216 D.R. Rhodes: Proc. IEEE *53*, 1013-1021 (1965); IEEE Trans. AP-*19*, 162-166 (1971); AP-*20*, 143-145 (1972)
7.217 R.A. Hurd: Proc. IEE *121*, 32-48 (1974)
7.218 G. Tricoles, D.C. Yue: "Feasibility of Microwave Holographic Imaging Systems"; ARPA Tech. Rpt. F33615-72C-1655, General Dynamics, Electr. Div., San Diego, CA (1972)
7.219 D.D. Howard: IEEE Trans. AES-*11*, 749-755 (1975)
7.220 G. Tricoles: Acoust. Hologr. *6*, 469-484 (1975)
7.221 W-M. Boerner, F.H. Vandenberghe: Can. J. Phys. *49*, 1507-1535 (1971)
7.222 W-M. Boerner, O.A. Aboul-Atta: Util. Math. *3*, 163-273 (1973)
7.223 V.H. Weston, W-M. Boerner: "Inverse Scattering Investigations"; Tech. Rpt. 8575, University of Michigan Radiation Lab. (1968)
7.224 V.H. Weston, J.J. Bowman, E. Ar: Arch. Ration. Mech. Anal. *31*, 199-213 (1968)
7.225 H.P.S. Ahluwalia, W-M. Boerner: IEEE Trans. AP-*21*, 672-673 (1973)
7.226 H.P.S. Ahluwalia, W-M. Boerner: IEEE Trans. AP-*22*, 663-682 (1974)
7.227 H.P.S. Ahluwalia: "Application of the Concept of Electromagnetic Inverse Boundary Conditions to Profile Characteristics Inversion of Conducting Shapes"; Ph.D. Thesis, University of Manitoba, Winnipeg, Canada (1973)
7.228 H.M. Smith (ed.): *Holographic Recording Materials*, Topics in Applied Physics, Vol.20 (Springer, Berlin, Heidelberg, New York 1977)
7.229 G.A. Yerokhin, V.G. Kocherzhevsky: Radio Eng. Electron. Phys. *19*, 17-23 (1974)
7.230 D.V. Skobel'tsyn: Sov. Phys. Usp. *16*, 381-401 (1973)
7.231 P.F. Wacker: "Non-Planar Near-Field Measurements: Spherical Scanning"; Tech. Rpt. NSRIR 75-809, National Bureau of Standards, Boulder, CO (1975)
7.232 A.C. Newell, R.C. Baird, P.F. Wacker: IEEE Trans. AP-*21*, 418-431 (1973)
7.233 J. Brown, E.V. Jull: Proc. IEE *108* B, 635-644 (1961)

7.234 P.F. Wacker: "A Qualitative Survey of Near Field Analysis and Measurement"; Tech. Rpt. NBSIR-79-1602, National Bureau of Standards, Boulder, CO (1979)
7.235 Y.T. Lo, W.F. Richards, P. Simon: Proceedings Intern. Con. IEEE APS Symp., June 18-22, 1979, Seattle, WA, pp.11-12
7.236 P.C. Waterman: Proc. IEEE *53*, 805-812 (1965)
7.237 R.H.T. Bates, D.J.N. Wall: Philos. Trans. R. Soc. (London) *A287*, 45-114 (1977)
7.238 R.H.T. Bates: "General Introduction to the Extended Boundary Conditions", in *Recent Developments in Classical Wave Scattering*, Intern. Symposium & Workshop, June 25-27, 1979, Ohio State University (Academic, to be published)
7.239 W-M. Boerner, C-M. Ho: "Use of Radon's Projection Theory in Electromagnetic Inverse Scattering", IEEE Trans. AP-*29* (in preparation)
7.240 R.F. Elsner: "Vehicular Variable Parameter"; METRRA Systems Final Rpt., Rpt. No. ITR-I-E6224, Illinois Institute of Technology, Tech. Research Center (1974); AD-782-214
7.241 A.J. Poelman: IEEE Trans. AES-*11*, 660-662 (1975)
7.242 A.J. Poelman: IEEE Trans. AES-*12*, 674-682 (1976)
7.243 A.J. Poelman: Electron. Lett. *13*, 533-534 (1977)
7.244 A.J. Poelman: Tijdschr. Ned. Electron. Radio-genoot. *44*, 93-106 (1979)
7.245 IEEE Trans. AP-*26* (1978) (Special Issue on EMP)
7.246 IEEE Proc. *62* (1974) (Special Issue on Diffraction)
7.247 S. Weisbrod, L.A. Morgan: "RCS Matrix Studies of Sea Clutter", NAVAIR Systems Command, Rpt. R2-79, Teledyne Micronetics, San Diego, CA (1979)
7.248 R. Rosien, D. Hammer, G. Ioannidis, J. Bell, J. Nemit: "Implementation Techniques for Polarization Control in ECCM"; RADC-TR-79-4, Griffis AFB, Rome, NY (1979)
7.249 W-M. Boerner: "Polarization Microwave Holography: An Extension of Scalar to Vector Holography", Intern. Opt. Computing Cong., SPIE's Techn. Symposium East, April 9, 1980, Washington D.C., Paper No.231-23
7.250 W-M. Boerner: "Polarization Dependence in Electromagnetic Inverse Problems", IEEE Trans. AP-*29* (in preparation)

Additional References with Titles

Chapter 1

De Santo, J.A., Sâenz, A.W., Zachary, W.W. (eds.): *Mathematical Methods and Applications of Scattering Theory*, Lecture Notes in Physics, Vol.130 (Springer, Berlin, Heidelberg, New York 1980)

Herman, G.T.: *Image Reconstruction from Projections: Implementation and Applications*, Topics in Applied Physics, Vol.32 (Springer, Berlin, Heidelberg, New York 1979)

Van Schoonefeld, C. (ed.): *Image Formation from Coherence Function in Astronomy* (Reidel Pub. Comp., Dordrecht 1979)

Mesla, J.L., Cohn, D.L.: *Decision and Estimation Theory* (McGraw Hill, New York 1978)

Bates, R.H.T., Milane, R.P.: Time domain approach to inverse scattering. IEEE-AP, submitted

Nussenzveig, H.M.: Causality and Analyticity in Optics. Proc. ICO Meeting Optics in Four Dimensions, Ensenada, Mexico 1980

Schmidt-Weinmar, H.G.: Superresolution with coherent light. Proc. Intern. Conf. Lasers 1979 (STS Press, McLean, Virginia 1980) in press

Schmidt-Weinmar, H.G., Gunn, D.W.K., Schmidt-Weinmar, M.L.: Sources and the plane-wave representation of the electromagnetic field. Can. J. Phys. (to be published)

Schmidt-Weinmar, H.G., Gunn, D.W.K., Schmidt-Weinmar, M.L.: Complex planar frequencies and causality in the halfspace representation of the electromagnetic field of sources of wavelength dimensions. Project Report, Dept. Electr. Eng., University of Alberta, Edmonton, Canada (1980)

Boivin, R., Boivin, A.: Optimized amplitude filtering for superresolution over a restricted field. I. Achievement of maximum central irradiance under an energy constraint. Opt. Acta *27*, 587-610 (1980)

Roger, A.: Grating profile optimizations by inverse scattering methods. Opt. Commun. *32*, 11-13 (1980)

Huiser, A.M.: On the influence of partial coherent illumination on the solvability of the phase problem in electron microscopy. Optik *55*, 241-252 (1980)

Barakat, R.: Moment estimator approach to the retrieval problem in coherence theory. J. Opt. Soc. Am. *70*, 688-694 (1980)

Hoenders, B.J.: The unique solution of the inverse diffraction problem. Opt. Commun. *30*, 327-328 (1979)

Hoenders, B.J.: On the inversion of an integral equation relating two wavefunctions in planes of an optical system suffering from an arbitrary number of aberrations. Opt. Acta *26*, 711-730 (1979)

Bates, R.H.T.: Fringe visibility intensities may uniquely define brightness distributions. Astron. Astrophys. *70*, L27-L29 (1978)

Bates, R.H.T., Cady, F.M.: Towards true imaging by wideband speckle interferometry. Opt. Commun. (in press)

Collet, E., Wolf, E.: New equivalence theorems for planar sources that generate the same distributions of radiant intensity. J. Opt. Soc. Am. *69*, 942-950 (1979)

Dekkers, N.H.: Object wave reconstruction in STEM. Optik *53*, 131-142 (1979)

Gilbert, A.D., Scott, T.C.: The deduction of surface profiles from the reflection of horizontal light sources. II. Calculation of the surface profile. Opt. Acta *27*, 767-781 (1980)

Deans, S.R.: A unified radon inversion formula. J. Math. Phys. *19*, 2346-2349 (1978)

Weston, V.H.: Non-linear approach to inverse scattering. J. Math. Phys. *20*, 53-59 (1979)

Weston, V.H.: Inverse problem for reduced wave equation with fixed incident field. J. Math. Phys. *21*, 758-764 (1980)
Weston, V.H.: Inverse problem for reduced wave equation with fixed incident field, part II. Submitted to J. Math. Phys.
Hansen, E.W., Goodman, J.W.: Optical reconstruction from projections via circular harmonic expansion. Opt. Commun. *24*, 268-275 (1978)
Hofer, J.: Optical reconstruction from projections via deconvolution. Opt. Commun. *29*, 22-26 (1979)
Jordan, A.K., Ahn, S.: Inverse scattering theory and profile reconstruction. Proc. IEE *126*, 945-950 (1979)
Oldenburg, D.W., Samson, J.C.: Inversion of interferometric data from cylindrically symmetric, refractionless plasmas. J. Opt. Soc. Am. *69*, 927-942 (1979)
Cohen, A., Cooney, J., Raviv, G., Wolfson, N.: Mathematical inversion of angular multiple light scattering data. Appl. Opt. *18*, 2466-2469 (1979)
Shaw, G.E.: Inversion of optical scattering and spectral extinction measurements to recover aerosol size spectra. Appl. Opt. *18*, 988-993 (1979)
Walters, P.T.: Practical Applications of inverting spectral turbidity data to provide aerosol size distributions. Appl. Opt. *19*, 2353-2365 (1980)
Farina, J.D., Narducci, L.M., Collett, E.: Generation of highly directional beams from a globally incoherent source. Opt. Commun. *32*, 203-208 (1980)
Levine, R.D.: An information theoretical approach to inversion problems. J. Phys. A*13*, 91-108 (1980)
Quattropani, A., Schwendimann, P., Baltes, H.P.: Sub-Poissonian statistics of an anharmonic oscillator in thermal equilibrium. Opt. Acta *27*, 135-138 (1980)

Chapter 2

Lakatos, I.: *Mathematics, Science and Epistomology*, ed. by J. Worral, G. Currie (Cambridge University Press 1978)
Grene, M.: *The Knower and the Known* (Faber and Faber, London 1966)
Ronkin, L.I.: An analogy of the canonical product for entire functions of several complex variables. Trans. Moscow Math. Soc. *18* (1968)
Bochner, S., Martin, W.: *Several Complex Variables* (Princeton University Press 1948)
Nussenzveig, H.M.: Phase problem in coherence theory. J. Math. Phys. *8*, 561 (1967)
Hoenders, B.J.: On the solution of the phase retrieval problem. J. Math. Phys. *16*, 1719 (1975)
Wolf, E.: Is a complete determination of the energy spectrum of light possible from measurements of the complex degree of coherence? Proc. Phys. Soc. *80*, 1269 (1962)
Helstrom, C.W.: Resolvable degrees of freedom in observation of a coherent object. J. Opt. Soc. Am. *67*, 833 (1977)
Mitsui, T.: "X-ray diffraction studies of membranes", in *Advances in Biophysics*, ed. by Masao Kotani, Vol.10 (University Park Press 1978) p.115
O'Neill, E.L., Walther, A.: The question of phase in image formation. Opt. Acta *10*, 33 (1963)
Saxton, W.O.: *Computer Techniques for Image Processing in Electron Microscopy* (Academic Press, New York 1978)
de Bruijn, N.G.: The roots of trigonometric integrals. Duke. Math. J. *17*, 197 (1950)
de Brange, L.: *Hilbert Spaces of Entire Functions* (Prentice-Hall, New York 1968)
Bond, F.E., Cahn, C.R.: On sampling the zeros of bandwidth limited signals. I.R.E. Trans. Information Theory, September 110 (1958)
Levinson, N.: Gap and density theorems. Amer. Math. Soc. Colloq. Publ. *26* (1940)
Cartwright, M.L.: On the zeros of certain integral functions I and II. Quart. J. Math. (Oxford Series) (1) *1*, 38 (1930) and *2*, 113 (1931)
Cartwright, M.L.: *Integral Functions*, Cambridge tracts in mathematics and mathematical physics, No.44 (Cambridge University Press 1955)
Stein, R.S., Wilson, P.R.: Scattering of light by polymer films possessing correlated orientation fluctuations. J. Appl. Phys. *33*, 1914 (1962)
Stein, R.S., Hotta, T.: Light scattering from oriented polymer films. J. Appl. Phys. *35*, 2237 (1964)
Stein, R.S., Erhardt, P.F., Clough, S.B., Adams, G.: Scattering of light by films having non-random orientation fluctuations. J. Appl. Phys. *37*, 3980 (1966)

Roe, R.J., Krigbaum, W.R.: Description of crystalline orientation in polycrystalline materials having fibre texture. J. Chem. Phys. *40*, 2608 (1964)
Krigbaum, W.R., Roe, R.J.: Crystalline orientation in materials having fibre texture II. A study of strained samples of cross linked polyethylene. J. Chem. Phys. *41*, 737 (1964)
Roe, R.J.: Description of crystalline orientation in polycrystalline materials III. General solution to pole figure inversion. J. Appl. Phys. *36*, 2024 (1965)
Roe, R.J. Methods of description of orientation in polymers. J. Polymer Sci., Pt. A-2, *8*, 1187 (1970)
Nomura, S., Kawai, H., Kumira, I., Kagiyama, M.: General description of orientation factors in terms of expansion of orientation distribution function in a series of spherical harmonics. J. Polymer Sci. Pt. A-2, *8*, 383 (1970)
Goldstein, M., Michalik, E.R.: Theory of scattering by an inhomogeneous solid possessing fluctuations in density and anisotropy. J. Appl. Phys. *26*, 1450 (1955)
Debye, P., Bueche, A.M.: Scattering by an inhomogeneous solid. J. Appl. Phys. *20*, 518 (1949)
Ward, I.M.: Optical and mechanical anisotropy of crystalline polymers. Proc. Phys. Soc. *80*, 1176 (1962)
Nieto-Vesperinas, M., Ross, G., Fiddy, M.A.: "The theory of light scattering in absorbing media", in: *Optica Hoy y Mañana*, ed. by J. Bescos, A. Hidalgo, L. Plaza, J. Santamaria (Instituto de Optica, Madrid 1978) pp.747-750
Ross, G., Fiddy, M.A., Nieto-Vesperinas, M., Manolitsakis, I.: "The nature of the analytically of scattered fields in the Fresnel and Fraunhofer region", in: *Optica Hoy y Mañana*, ed. by J. Bescos, A. Hidalgo, L. Plaza, J. Santamaria (Instituto de Optica, Madrid 1978) pp.739-742
Fiddy, M.A., Ross, G., Nieto-Vesperinas, M.: "The encoding of structural information, in scattering and image formation, by the zeros of entire functions", in: *Optica Hoy y Mañana*, ed. by J. Bescos, A. Hidalgo, L. Plaza, J. Santamaria (Instituto de Optica, Madrid 1978) pp.743-746
Ross, G., Fiddy, M.A., Moezzi, H., Nieto-Vesperinas, M: "The inverse problem in light scattering", in: *Antennas and Propagation , Vol.I.*, International Symposium Digest (IEEE 1979) pp.232-235
Fiddy, M.A., Ross G.: Analytic Fourier optics: the encoding of information by complex zeros. Opt. Acta *26*, 1139-1146 (1979)
Ross, G., Jarvis, D.A.: The characterization of orientation in anisotropic media by means of elastic light scattering. I. The inhomogeneity of anisotropic media and its description. Opt. Acta *27*, 359-370 (1980)
Ross, G., Jarvis, D.A.: The characterization of orientation in anisotropic media by means of elastic light scattering. II. The elastic scattering of light by an inhomogeneous medium. Opt. Acta *27*, 371-384 (1980)
Ross, G., Fiddy, M.A., Moezzi, H.: The solution to the inverse scattering problem, based on zero location from two measurements. Opt. Acta (in press)
Ross, G., Nieto-Vesperinas, M.: Theory of light scattering by amorphous absorbing media. Opt. Acta (in press)
Fienup, J.R.: Reconstruction of an object from the modulus of its Fourier transform. Opt. Lett. *3*, 27-29 (1978)
Bruck, Yu. M., Sodin, L.G.: On the ambiguity of the image reconstruction problem. Opt. Commun. *30*, 304-308 (1979)
Boucher, R.H.: "Convergence of algorithms for phase retrieval from two intensity distributions", in: *Proc. I.O.C.C. Meeting, Washington April 8, 1980*, to be published by S.P.I.E.
Misell, D.L.: "The phase problem in electron microscopy", in: *Advances in Optical and Electron Microscopy*, Vol.7, ed. by V.E. Cosslett, R. Barer (Academic Press, London 1978) pp.185-219

Chapter 3

Pike, E.R.: The Malvern correlator. Case study in development. Phys. Technol. *10*, 104-109 (1979)
Parry, G., Walker, J.G., Scaddan, R.J.: On the statistics of stellar speckle patterns and pupil plane scintillation. Opt. Acta *26*, 563-574 (1979)
Oliver, C.J.: Spectral analysis with short data batches. J. Phys. A*12*, 591-617 (1979)

Fujii, H.: Contrast variation of non-gaussian speckle. Opt. Acta *27*, 409-418 (1980)
Jakeman, E.: Statistics of integrated gamma-lorentzian intensity fluctuations. Opt. Acta *27*, 735-741 (1980)

Chapter 4

Davies, E.B.: *Quantum Theory of Open Systems* (Academic Press, London 1976) Chap.5, pp.69-84
Srinivas, M.D., Davies, E.B.: Photon counting probabilities in quantum optics. To be published
Aoki, T., Sakurai, K.: Photon statistics of partially polarized Gaussian light. Phys. Rev. A*20*, 1593-1598 (1979)
Barakat, R.: Intensity correlations in the presence of additive background noise. Opt. Commun. *28*, 4-6 (1979)
Mandel, L.: Inversion problem in photon counting with dead time. J. Opt. Soc. Am. *70*, 873-874 (1980)
McGuire, D.: Coherent detection of partially coherent sources. Opt. Letts. *5*, 73-75 (1980)
Selloni, A., Quattropani, A., Schwendimann, P., Baltes, H.P.: Temperature effects in photodetection: additional results. Phys. Rev. A*22*, 315-317 (1980)

Chapter 5

Anger, G. (ed.): *Inverse and Improperly Posed Problems in Differential Equations*, Proc. of the Conference on Mathematical and Numerical Methods held in Halle/Saale (GDR) (Akademie-Verlag, Berlin 1979)
Barakat, R., Blackman, E.: Application of the Tichonov regularization algorithm to object restoration. Opt. Commun. *9*, 252-256 (1973)
Bertero, M., De Mol, C.: Stability problems in inverse diffraction, Preprint 1980
Ciulli, S., Pomponiu, C., Sabba-Stefanescu, I.: Analytic extrapolation techniques and stability problems in dispersion relation theory. Phys. Rep. *17*, 133-224 (1975)
Colton, D.: The inverse scattering problem for a cylinder. Proc. Roy. Soc. of Edinburgh *84A*, 135-143 (1979)
Majda, A.: High frequency asymptotics for the scattering matrix and the inverse problem of acoustic scattering. Comm. Pure Appl. Math. *30*, 165-194 (1977)
Nashed, M.Z. (ed.): *Ill-Posed Problems: Theory and Practice*, Proc. of the International Symposium held at the University of Delaware, October 2-6, 1979 (in preparation)
Ramm, A.G.: On resolution ability of optical systems. Optics and Spectroscopy *29*, 422-424 (1970)
Sabatier, P.C.: Positivity constraints in linear inverse problems I and II. Geophys. J. R. Astr. Soc. *48*, 415-442 and 443-469 (1977)
Krishnaprasad, P.S., Barakat, R.: A descent approach to a class of inverse problems. J. Computational Phys. *24*, 339-347 (1977)

Chapter 6

Kanal, M., Moses, H.E.: Direct-inverse problems in transport theory, the inverse albedo problem for a finite medium. J. Math. Phys. *19*, 2641-2645 (1978)
Gupta, R.K.: Infrared remote temperature measurements: its physics with reference to complexities, approximations and limitations involved. II - Temperature profile retrieval. J. Indian Institute of Science *60*, 253-284 (1978)
Santoro, R.J., Emmerman, P.J., Goulard, R., Semerjian, H.G., Shabahang, R.: Multiangular absorption measurements in a methane diffusion jet. To be published in Proc. of the Symposium on "Laser Probes for Combustion Chemistry", ed. by D.R. Crosley

Chapter 7

De Santo, J.A. (ed.): *Ocean Acoustics*, Topics in Applied Physics, Vol.32 (Springer, Berlin, Heidelberg, New York 1979)
Silvia, M.T., Robinson, E.A.: "Deconvolution of Geophysical Time Series in the Exploration for Oil and Natural Gas", Developments in Petroleum Science 10 (Elsevier 1979)

Gal'perin, E.I.: "Vertical Seismic Profiling", translated by A.J. Hermont, ed. by J.E. White (Society of Exploration Geophysicists, Special Publication No.12 1974)

Preston, K., Jr., Taylor, K.J.W., Johnson, S.A., Ayers, W.R. (eds.): *Medical Imaging Techniques - A Comparison* (Plenum Press, New York 1979)

Larsen, L.E., Jacobi, J.H.: Microwave interrogation of dielectric targets. Part I: By scattering parameters; Part II: By microwave time delay spectroscopy. Med. Phys. *5* (6), 500-513 (1979)

Larsen, L.E., Jacobi, J.H.: Microwave scattering parameter imagery of an isolated carine kidney. Med. Phys. *6* (5), 394-403 (1979)

Van der Pol, B., Bremmer, H.: *Operational Calculus* (Based on the Two-Sided Laplace Transform) (Cambridge University Press 1955)

Båth, M.: *Mathematical Aspects of Seismology* (Elsevier, Amsterdam 1968)

Båth, M.: *Introduction to Seismology* (Birkhäuser, Basel and Stuttgart 1973)

Båth, M.: *Spectral Analysis in Geophysics* (Elsevier, Amsterdam 1974)

Dobrin, M.B.: *Introduction to Geophysical Prospecting*, 3rd edn. (McGraw-Hill, New York 1976)

Jenkins, G.M., Watts, D.G.: *Spectral Analysis and Its Applications* (Holden-Day, San Francisco, Calif. 1968)

Grant, F.S., West, G.F.: *Interpretation Theory in Applied Geophysics* (McGraw-Hill, New York 1965)

Dike, G.A., Burt, E.C., Wallenberg, R.F.: "An Application of GTD and PO to Object Identification Through Inverse Scattering", Intern. Symp. IEEE-APS/URSI-B, Québec, Special Session III on Inverse Scattering 1980

Cherry, S.M., Goddard, J.W.F., Hall, M.P.M.: "Examination of rain dron sizes using a dual-polarization radar", 19th Conf. on Radar Meteorology, 15-18 April 1980, Miami Beach, Fl.

Babič, V.M., Kirpičnikova, N.Y.: *The Boundary-Layer Method in Diffraction Problems*, Springer Series in Electrophysics, Vol.3 (Springer, Berlin, Heidelberg, New York 1979)

Babič, V.M.: *Mathematical Questions in the Theory of Wave Diffraction and Propagation*, STEKLO/114-LC. 74-2363, ISBN 0-8218-3015-5, AMS Catalogue 1979

Subject Index